MECHANICS
OF
ELASTIC STRUCTURES

MECHANICS
OF
ELASTIC STRUCTURES

*Classical and Finite
Element Methods*

JOE G. EISLEY

University of Michigan

Prentice Hall, Englewood Cliffs, New Jersey 07632

Library of Congress Cataloging-in-Publication Data

EISLEY, JOE.
   Mechanics of elastic structures: classical and finite element
methods / by Joe G. Eisley.
       p.   cm.
   Bibliography: p.
   Includes index.
   ISBN 0-13-567470-0
   1. Elastic analysis (Theory of structures) 2. Finite element
method. I. Title.
TA653.E37 1987   88-19677
624.1'71—dc19   CIP

Editorial/production supervision and
  interior design: Patrice Fraccio
Cover design: Ben Santora
Manufacturing buyer: Mary Noonan

© 1989 by Prentice-Hall, Inc.
A Division of Simon & Schuster
Englewood Cliffs, New Jersey 07632

Printed in the United States of America
10 9 8 7 6 5 4 3 2 1

ISBN 0-13-567470-0

PRENTICE-HALL INTERNATIONAL (UK) LIMITED, *London*
PRENTICE-HALL OF AUSTRALIA PTY. LIMITED, *Sydney*
PRENTICE-HALL CANADA INC., *Toronto*
PRENTICE-HALL HISPANOAMERICANA, S.A., *Mexico*
PRENTICE-HALL OF INDIA PRIVATE LIMITED, *New Delhi*
PRENTICE-HALL OF JAPAN, INC., *Tokyo*
SIMON & SCHUSTER ASIA PTE. LTD., *Singapore*
EDITORA PRENTICE-HALL DO BRASIL, LTDA., *Rio de Janeiro*

To Marilyn, Paul, and Susan

CONTENTS

APPENDIXES

REFERENCES

INDEX

PREFACE

This book is designed primarily as a text for use at the upper undergraduate and beginning graduate level, although many practicing structural analysts may find it useful as a reference. To use this text effectively, the reader must have studied and mastered to a degree certain prerequisite material that is common to all mechanics-based curricula, such as is found in mechanical, aerospace, civil, and related fields of engineering. An understanding of calculus through a first course in differential equations is a must, and familiarity with elementary matrix theory is highly desirable. While the subject matter on the behavior of solid bodies under applied load is reasonably self-contained in the text, it is assumed, nevertheless, that the reader has taken a first course in the mechanics of materials.

Several dozen books on finite element analysis are currently available. There are also a number of books on classical methods of structural analysis. What justification is there for another text? It is the belief that far too many engineers attempt the study of the finite element method without a sufficiently strong background in classical methods. The usual first course in the mechanics of materials is generally not sufficient. This text bridges that gap to show the strong foundation of finite element analysis in classical theory and methods.

In the study of the mechanics of structures, certain material about basic definitions and equations must be presented first, but it soon becomes a matter of choice on how to proceed. Perhaps a few remarks on the several ways structures and the analysis of structures can be classified will help in understanding the organization of this text. The structures themselves have two main types of classification: one according to geometry and the other according to material properties. From geometry, we have slender bars, also called rods, shafts, beams, and columns, which are characterized by having one dimension, the length, much larger that the other two, the thickness and width. Then there are plates and shells, which have one dimension, the thickness, much smaller than the other two. Finally, we have general three-dimensional bodies, which are none of the above. From the material properties, we have elastic bodies, which may be either linearly or nonlinearly elastic, and inelastic bodies, which have numerous subclassifications. The material may be homogeneous or nonhomogeneous, and it may be isotropic, orthotropic, or anistropic.

In each of these classifications, the methods of analysis may differ, and the analysis can be classified according to the mathematical methods used in finding solutions. One principal formulation, usually based upon Newton's laws of motion, is in the form of differential equations that often may be solved by exact, approximate analytical, or approximate numerical means. Another major formulation, often based upon the principle of virtual work, is in the form of integral equations that also may be solved by exact, approximate analytical, or approximate numerical means.

There are still two more classifications of structures to each part of which each of the above classifications apply. In this classification, structures are categorized according to whether they are stable or unstable, and according to whether they are static or dynamic. It is easy enough to identify the static from the dynamic. Time is simply an independent variable in the one case and not the other. It is harder to explain instability from stability without resorting to examples, so we shall save that for the appropriate chapter.

From these various categories, we can explain the choices made for the organization of this particular text. First, we have chosen to concentrate on linearly elastic, homogeneous, and isotropic bodies undergoing small deformations, except in a very few instances. This was not done because excluded subject areas are less important; certainly to the manufacturing engineer dealing in metal forming, for example, the inelastic region and large deformations are of more importance. This was done because the operational performance of vehicles and other structures is the primary motivating influence on the choice of material. Aircraft, automobiles, machinery, and so on, tend to be made (or the components are made) of homogeneous and isotropic materials that operate in the linearly elastic and small deflection range. This is changing with the development of composite materials but is still largely true. Second, we have chosen to emphasize the analysis of slender bars in various configurations, pay some attention to plates and shells, but treat three dimension solid bodies only in a limited way. These choices are based on two considerations: the great practical utility of the structural forms selected, and the value of the lessons learned in studying them. Finally, nearly equal weight is given to classical differential equation methods that have analytical solutions, and modern approximate methods, principally the finite elements methods, that have numerical solutions.

This choice of methods needs further explanation. In an earlier time much more emphasis would have been given to approximate analytical methods because so few exact analytical solutions had been found and numerical methods were often laborious and expensive. The problems would have been refined to a great degree by analytical approaches, such as the Rayleigh-Ritz method, before the numerical analysis was attemped. With the advent of the electronic digital computer, times have changed. Of course, once one inserts numbers even the most analytical approaches become numerical, in a sense, but the digital computer permits analytical formulations that are relatively easy to obtain, but that require relatively large amounts of computation in exhange. The computation is now so fast and so inexpensive that the trade-off is quite favorable. These are the modern numerical methods, and foremost among them is the finite element method (FEM), which has truly revolutionized structural mechanics.

All the material here is well known in the general technical literature. The references cited are all general references, and no attempt has been made to identify original sources. The original sources are often identified in the references. I cannot attempt to acknowledge who among my colleagues and students have influenced my approach and choice of materials, except to note that William J. Anderson at the University of Michigan has contributed significantly to the way I approach the finite element method.

Joe G. Eisley
Ann Arbor, Michigan

MECHANICS
OF
ELASTIC STRUCTURES

CLASSICAL EQUATIONS OF ELASTICITY

1

1.1 INTRODUCTION

Structural mechanics is concerned with the behavior of solid bodies or assemblies of solid bodies under the action of applied loads. That behavior is usually described by mathematical models from which the internal forces and displacements are found in terms of the spatial coordinates and, in the case of dynamics of structures, as a function of time. In this chapter, we shall develop sets of equations that are the mathematical model for a general three-dimensional solid body at rest with certain restrictions on the nature of the deformation and material properties of the body. We shall then examine the possibility of finding exact analytical solutions for the behavior of the body and, after finding few, we shall look for more easily solved sets of equations made possible by simplifications in the geometry, applied loads, and constraints.

The material in this chapter may be found in many other books. For example, References 1, 2, and 3 on elementary mechanics of materials or References 4 and 5, more advanced books on structural mechanics, provide excellent coverage. Then there are several books on the theory of elasticity that have been useful references for many years, such as References 6, 7, and 8.

1.2 GENERAL THEORY OF ELASTICITY

Fundamental to the mathematical analysis of solid bodies is an understanding of the concepts of *stress* and *strain*. In the next section, we define stress and express the relationships among the stress components, body forces, and applied loads, which are known as the *equations of equilibrium*. This is followed by the definition of strain in terms of the displacements or the *strain-displacement equations*. At this point, we establish the physical law that relates stress to strain, the *stress-strain equations*, which for a linearly elastic solid is known as *Hooke's law*.

When given appropriate *boundary conditions* in terms of *applied loads* and *geometric constraints* on the body, the three sets of equations, equilibrium, strain-displacement, and stress-strain, completely define the behavior of the body. Symbolic representations of the quantities and sets of equations identified by the italicized words are shown in Figure 1.2.1. The braces { } identify column matrices and brackets [ ] identify rectangular matrices. A summary of matrix notation and operations is given in Appendix A.

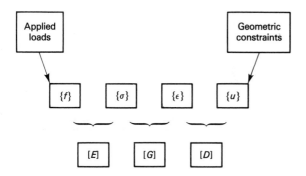

Figure 1.2.1 Symbolic Representation of the Equations of Solid Mechanics.

In Figure 1.2.1, the column matrices represent the following quantities:

$\{f\}$ body forces
$\{\sigma\}$ stress components
$\{\epsilon\}$ strain components
$\{u\}$ displacements

These quantities are joined together to form the matrix equations:

$$[E]\{\sigma\} + \{f\} = 0, \qquad \text{equilibrium equations}$$

$$\{\sigma\} = [G]\{\epsilon\}, \qquad \text{stress-strain equations} \qquad (1.2.1)$$

$$\{\epsilon\} = [D]\{u\}, \qquad \text{strain-displacement equations}$$

These equations must be solved consistent with the boundary conditions expressed in terms of applied loads and the geometric constraints. We shall not give symbolic representation to loads and constraints just yet, because they are better understood when given for specific examples in the solution of practical problems where the boundary edges and surfaces are clearly defined.

1.2.1 Force and Stress. Equilibrium

For solid bodies at rest, Newton's laws require that the sum of all forces and moments be zero. This can be expressed in rectangular cartesian coordinates by the sum of the three force components and the three moment components, as follows:

$$F_x = 0, \qquad F_y = 0, \qquad F_z = 0$$
$$M_x = 0, \qquad M_y = 0, \qquad M_z = 0 \tag{1.2.2}$$

These equations apply to the whole body or to any part of it. We distinguish between two types of forces: external, or *applied loads*, and internal. Some of the applied loads may be on the surface of the body with units of force per unit area, while others may be volume dependent, called *body forces*, with units of force per unit volume. Generally, the external forces are known in the statement of the problem, although there are some exceptions, as we shall see.

   This brings us to the important concept of internal surface forces, that is, forces that occur on internal, and generally unseen, surfaces. If we imagine the body to be cut in two, we can expose these surfaces, at least in our minds, and consider the forces acting on them. One of the best ways of expressing these forces is to consider the force per unit area on the internal surfaces. We call these *stresses*. When broken down into components normal and parallel to the surface in question, they are called *stress components*, although we often simply use the term stresses to mean stress components as well. Generally, the stresses are unknowns in the statement of the problem.

   The stress and the stress components on an internal surface are depicted in Figure 1.2.2. The force acting on the element of area ΔA is denoted by ΔF. Furthermore, if we resolve ΔF into components normal to and tangential to the surface, ΔF_n and ΔF_t, respectively, and allow ΔA to shrink to a point, we can define a *normal* component of stress and a *tangential* or *shearing* stress component, as shown in Equation 1.2.3.

$$\sigma = \lim_{\Delta A \to 0} \frac{\Delta F_n}{\Delta A}, \qquad \tau = \lim_{\Delta A \to 0} \frac{\Delta F_t}{\Delta A} \tag{1.2.3}$$

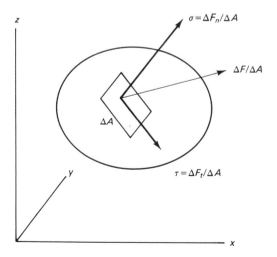

Figure 1.2.2 Stress and Stress Components on an Internal Surface.

To depict the complete state of stress at a point in rectangular cartesian coordinates, consider a rectangular element oriented with respect to the coordinate axes as shown in Figure 1.2.3. While a rectangular element, seemingly of finite size, is shown in order to show the surfaces on which the stresses act, think of it as shrinking to a point to depict the state of stress at a point.

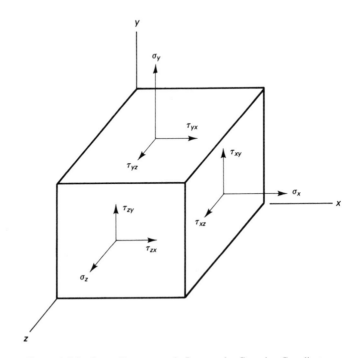

Figure 1.2.3   Stress Components in Rectangular Cartesian Coordinates.

Since the value of a stress component depends both on its direction and the surface on which it acts, we use a double-subscript notation to differentiate the several stress components. The first subscript identifies the direction of the normal to the surface on which the stress component acts and the second subscript, the direction of the stress component. Since all normal stress components have repeated subscripts, one is considered redundant and is usually dropped. The shearing stress components always have two different subscripts, as shown. It is common practice to refer to the stress components simply as normal and shearing stresses. At this time it would appear that the complete state of stress at a point in rectangular cartesian coordinates is given by three normal stresses, σ_x, σ_y, σ_z, and six shearing stresses, τ_{xy}, τ_{yz}, τ_{zx}, τ_{yx}, τ_{zy}, τ_{xz}.

The sign convention adopted for stresses needs some explanation. First, we call a surface whose outwardly directed normal acts in a positive coordinate direction a *positive surface*, and when it acts in a negative coordinate direction, a *negative surface*. Then a stress is defined as positive if its direction and the direction of the surface on which it acts are both positive or both negative and is defined as negative if its direction and the direction of the surface are of opposite sign. Only positive stresses are depicted in Figure 1.2.3. A further convention is that positive normal stresses are called *tensile stresses*, and negative normal stresses are called *compressive stresses*. No separate names are given to positive and negative shearing stresses.

The stresses just defined are related to each other by the equations of equilibrium, as given symbolically in Equation 1.2.1. This condition applies to any portion of the body, as well as to the whole body, so let us consider a differential element as depicted in Figure 1.2.4. On the surface of the element are the stresses shown and, in addition, there may be body forces present. Before summing forces, we must consider the nature of the stress distribution. First, we assume that the stresses are continuous functions of the coordinates. It follows, in general, that the stresses on one face will differ from those on an opposite face.

We note that dx, dy, and dz are differentials, and we can write a Taylor's series expansion presenting stresses on one face in terms of those on the opposite face.

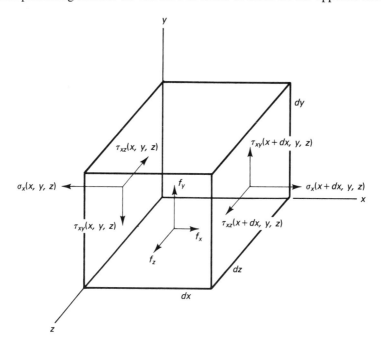

Figure 1.2.4 Forces on an Element.

Thus

$$\sigma_x(x + dx, y, z) = \sigma_x(x, y, z) + \frac{\partial \sigma_x}{\partial x} dx$$

$$\tau_{xy}(x + dx, y, z) = \tau_{xy}(x, y, z) + \frac{\partial \tau_{xy}}{\partial x} dx \qquad (1.2.4)$$

$$\tau_{xz}(x + dx, y, z) = \tau_{xz}(x, y, z) + \frac{\partial \tau_{xz}}{\partial x} dx$$

when higher-order terms are neglected. Similar results are obtained for the other two pairs of faces; however, for clarity, they are not shown on the figure.

First, let us sum the forces in the x-direction:

$$\left(\sigma_x + \frac{\partial \sigma_x}{\partial x}\,dx\right)dy\,dz - \sigma_x\,dy\,dz + \left(\tau_{xy} + \frac{\partial \tau_{xy}}{\partial y}\,dy\right)dx\,dz$$
$$- \tau_{xz}\,dx\,dz + \left(\tau_{xz} + \frac{\partial \tau_{xz}}{\partial z}\,dz\right)dx\,dy - \tau_{xz}\,dx\,dy + f_x\,dx\,dy\,dz = 0 \tag{1.2.5}$$

This simplifies to

$$\frac{\partial \sigma_x}{\partial x} + \frac{\partial \tau_{xy}}{\partial y} + \frac{\partial \tau_{xz}}{\partial z} + f_x = 0 \tag{1.2.6a}$$

Two additional equations are obtained by summation of forces in the y- and z-directions:

$$\frac{\partial \tau_{yx}}{\partial x} + \frac{\partial \sigma_y}{\partial y} + \frac{\partial \tau_{yz}}{\partial z} + f_y = 0 \tag{1.2.6b}$$

$$\frac{\partial \tau_{zx}}{\partial x} + \frac{\partial \tau_{zy}}{\partial y} + \frac{\partial \sigma_z}{\partial z} + f_z = 0 \tag{1.2.6c}$$

Moment equilibrium can be established by taking moments about each coordinate axis in turn. Note that no body moments were assumed because these rarely occur in nature. For moments about the z-axis,

$$\left(\tau_{xy} + \frac{\partial \tau_{xy}}{\partial x}\,dx\right)dy\,dz\,\frac{dx}{2} + \tau_{xy}\,dy\,dz\,\frac{dx}{2}$$
$$- \left(\tau_{yx} + \frac{\partial \tau_{yx}}{\partial y}\,dy\right)dz\,dx\,\frac{dy}{2} - \tau_{yx}\,dz\,dx\,\frac{dy}{2} = 0 \tag{1.2.7}$$

which simplifies to

$$\tau_{xy} + \frac{\partial \tau_{xy}}{\partial x}\,\frac{dx}{2} - \tau_{yx} - \frac{\partial \tau_{yx}}{\partial y}\,\frac{dy}{2} = 0 \tag{1.2.8}$$

and which in the limit as $dx \to 0$ and $dy \to 0$ becomes

$$\tau_{xy} = \tau_{yx} \tag{1.2.9a}$$

Two additional equations are determined by moments about the x and y axes. They are

$$\tau_{yz} = \tau_{zy}, \qquad \tau_{zx} = \tau_{xz} \tag{1.2.9b,c}$$

The six equations in Equations 1.2.6 and 1.2.9 couple the nine stresses. It is common practice to use Equations 1.2.9 to eliminate three of the stresses from Equation 1.2.6. In this form, the three Equations 1.2.6 are known as the differential *equations of equilibrium*. Since there are now only three equations and six unknown stresses, the equations cannot yet be solved.

In keeping with the matrix notation in Figure 1.2.1 and Equation 1.2.1, we represent the stresses and the body forces as column matrices $\{\sigma\}$ and $\{f\}$, respectively, and $[E]$ as a rectangular matrix. The equations of equilibrium in this form are

$$[E]\{\sigma\} = -\{f\}$$

or

$$\begin{bmatrix} \dfrac{\partial}{\partial x} & 0 & 0 & \dfrac{\partial}{\partial y} & 0 & \dfrac{\partial}{\partial z} \\[2ex] 0 & \dfrac{\partial}{\partial y} & 0 & \dfrac{\partial}{\partial x} & \dfrac{\partial}{\partial z} & 0 \\[2ex] 0 & 0 & \dfrac{\partial}{\partial z} & 0 & \dfrac{\partial}{\partial y} & \dfrac{\partial}{\partial x} \end{bmatrix} \begin{bmatrix} \sigma_x \\ \sigma_y \\ \sigma_z \\ \tau_{xy} \\ \tau_{yz} \\ \tau_{zx} \end{bmatrix} = - \begin{bmatrix} f_x \\ f_y \\ f_z \end{bmatrix} \qquad (1.2.10)$$

In this derivation, we appear to assume that each component of stress is uniform across any given face of the element. Actually, even if we permit such variation across a face, the result is the same. For a discussion of this, see Reference 7, pages 2–7.

Finally, we wish to point out that the state of stress at a point with respect to one orientation of the coordinate system will uniquely determine the state of stress at the same point with respect to any other rotational orientation of the same coordinate system. The rules for the transformation of stress and the concept of principal stresses are well documented in books on elementary mechanics of materials (for example, see References 1, 2, and 3). These rules have been summarized in Appendix B for reference.

1.2.2 Displacement and Strain

The *displacement components*, or simply *displacements*, of a point in a solid body are measured with respect to a fixed coordinate system by the distance the point moves in the x-, y-, and z-directions when the body is loaded, relative to the position of the point when the body is not loaded. The displacements are denoted by u, v, and w, respectively. The body is said to have deformed when there is relative motion between two or more points in the body. Note that if all points on the body move the same amount the body has displaced but has not deformed. This is referred to as rigid body displacement.

Consider the points A and B in the element of the undeformed body shown in Figure 1.2.5. We shall examine the relative motion of these two points when the body has deformed. The position of point A may be expressed in terms of the coordinates x, y, and z of the point with respect to the fixed rectangular cartesian coordinate system. Point B is located at $x + dx$, $y + dy$, and $z + dz$.

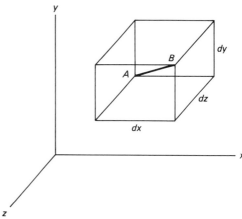

Figure 1.2.5 An Element before Straining.

The two points and a line connecting them are shown also in Figure 1.2.6. The distance between the two points may be seen to be given by

$$ds^2 = dx^2 + dy^2 + dz^2 \qquad (1.2.11)$$

In the deformed position, the points are labeled A' and B' and similarly the distance between them is given by

$$ds^2 = dx^2 + dy^2 + dz^2 \qquad (1.2.12)$$

The displacement of point A is given by

$$u = x - x, \qquad v = y - y, \qquad w = z - z \qquad (1.2.13)$$

and, in general, u, v and w are differentiable functions of x, y, and z. If we solve the equations for x, y, and z and substitute the results into Equation 1.2.12,

$$ds^2 = dx^2 + dy^2 + dz^2 + 2\,du\,dx + 2\,dv\,dy + 2\,dw\,dz$$
$$+ du^2 + dv^2 + dw^2 \qquad (1.2.14)$$

and it follows that

$$ds^2 - ds^2 = 2(du\,dx + dv\,dy + dw\,dz) + du^2 + dv^2 + dw^2 \qquad (1.2.15)$$

From the rules of partial differentiation

$$du = \frac{\partial u}{\partial x}dx + \frac{\partial u}{\partial y}dy + \frac{\partial u}{\partial z}dz$$

$$dv = \frac{\partial v}{\partial x}dx + \frac{\partial v}{\partial y}dy + \frac{\partial v}{\partial z}dz \qquad (1.2.16)$$

$$dw = \frac{\partial w}{\partial x}dx + \frac{\partial w}{\partial y}dy + \frac{\partial w}{\partial z}dz$$

If we substitute Equations 1.2.16 into 1.2.15 and simplify, we obtain

$$ds^2 - ds^2 = 2\epsilon_x\,dx^2 + 2\epsilon_y\,dy^2 + 2\epsilon_z\,dz^2$$
$$+ 2\gamma_{xy}\,dx\,dy + 2\gamma_{yz}\,dy\,dz + 2\gamma_{zx}\,dz\,dx \qquad (1.2.17)$$

where

$$\epsilon_x = \frac{\partial u}{\partial x} + \frac{1}{2}\left[\left(\frac{\partial u}{\partial x}\right)^2 + \left(\frac{\partial v}{\partial x}\right)^2 + \left(\frac{\partial w}{\partial x}\right)^2\right]$$

$$\epsilon_y = \frac{\partial v}{\partial y} + \frac{1}{2}\left[\left(\frac{\partial u}{\partial y}\right)^2 + \left(\frac{\partial v}{\partial y}\right)^2 + \left(\frac{\partial w}{\partial y}\right)^2\right]$$

$$\epsilon_z = \frac{\partial w}{\partial z} + \frac{1}{2}\left[\left(\frac{\partial u}{\partial z}\right)^2 + \left(\frac{\partial v}{\partial z}\right)^2 + \left(\frac{\partial w}{\partial z}\right)^2\right]$$

$$(1.2.18)$$

$$\gamma_{xy} = \frac{\partial v}{\partial x} + \frac{\partial u}{\partial y} + \frac{\partial u}{\partial x}\frac{\partial u}{\partial y} + \frac{\partial v}{\partial x}\frac{\partial v}{\partial y} + \frac{\partial w}{\partial x}\frac{\partial w}{\partial y}$$

$$\gamma_{yz} = \frac{\partial w}{\partial y} + \frac{\partial v}{\partial z} + \frac{\partial u}{\partial y}\frac{\partial u}{\partial z} + \frac{\partial v}{\partial y}\frac{\partial v}{\partial z} + \frac{\partial w}{\partial y}\frac{\partial w}{\partial z}$$

$$\gamma_{zx} = \frac{\partial u}{\partial z} + \frac{\partial w}{\partial x} + \frac{\partial u}{\partial z}\frac{\partial u}{\partial x} + \frac{\partial v}{\partial z}\frac{\partial v}{\partial x} + \frac{\partial w}{\partial z}\frac{\partial w}{\partial x}$$

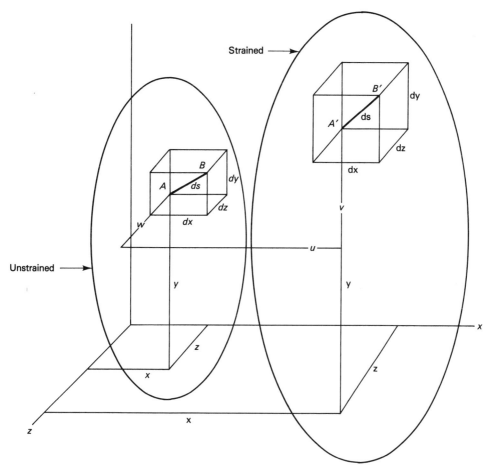

Figure 1.2.6 Deformation of Points A and B.

When the deflections u, v, and w are small compared to the characteristic dimensions of the body, the squares and products of the first partial derivatives of the displacements are small compared to the partial derivatives themselves. We can greatly simplify the analysis that follows by taking advantage of small displacements and dropping squares and products of the derivatives of the displacements. By dropping the higher-order terms, we obtain

$$\epsilon_x = \frac{\partial u}{\partial x}, \qquad \epsilon_y = \frac{\partial v}{\partial y}, \qquad \epsilon_z = \frac{\partial w}{\partial z}$$

$$\gamma_{xy} = \frac{\partial v}{\partial x} + \frac{\partial u}{\partial y}, \qquad \gamma_{yz} = \frac{\partial w}{\partial y} + \frac{\partial v}{\partial z}, \qquad \gamma_{zx} = \frac{\partial u}{\partial z} + \frac{\partial w}{\partial x} \qquad (1.2.19)$$

The quantities ϵ and γ are called the *strain components*, or more simply the *strains*. The Equations 1.2.19 are called the *linear strain-displacement equations*.

In matrix form (see Figure 1.2.1 and Equation 1.2.1), these equations are

$$\{\epsilon\} = [D]\{u\}$$

or

$$
\begin{bmatrix} \epsilon_x \\ \epsilon_y \\ \epsilon_z \\ \gamma_{xy} \\ \gamma_{yz} \\ \gamma_{zx} \end{bmatrix} = \begin{bmatrix} \dfrac{\partial}{\partial x} & 0 & 0 \\[6pt] 0 & \dfrac{\partial}{\partial y} & 0 \\[6pt] 0 & 0 & \dfrac{\partial}{\partial z} \\[6pt] \dfrac{\partial}{\partial y} & \dfrac{\partial}{\partial x} & 0 \\[6pt] 0 & \dfrac{\partial}{\partial z} & \dfrac{\partial}{\partial y} \\[6pt] \dfrac{\partial}{\partial z} & 0 & \dfrac{\partial}{\partial x} \end{bmatrix} \begin{bmatrix} u \\ v \\ w \end{bmatrix} \qquad (1.2.20)
$$

When the displacements and their derivatives are not so small, the higher-order terms may have to be retained. There are many important situations in structural mechanics when this must be done; however, in this text primary use is made of the simplified or *linearized* strain-displacement equations shown in Equations 1.2.19 and 1.2.20.

These newly defined quantities, which we call strains, are useful in relating the displacements to the stresses, as we shall see in the next section. When written in the linearized form, as above, they have simple physical interpretations. The component ϵ_x is the rate of change of displacement u with respect to x and similarly ϵ_y and ϵ_z are rates of change with respect to y and z, respectively. These three strain components are called the *normal* strains.

The other three components are called the *shearing* strains. We shall see that they have a physical interpretation as changes in angular orientation of points in the body. For convenience, let us look at a two-dimensional body before and after strain, as shown in Figure 1.2.7.

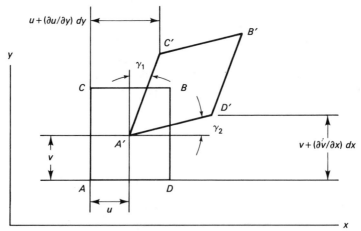

Figure 1.2.7 Shearing Strains in an Element.

After deformation the line elements AC and AD have rotated through the angles γ_1 and γ_2 to the positions $A'C'$ and $A'D'$. Since the derivatives of the displacements are small, these angles are approximately equal to their sines, and it may be seen that the decrease in the right angle formed by AC and AD when the body deforms to the position $A'C'$ and $A'D'$ is equal to the shearing strain γ_{xy}.

$$\gamma_1 + \gamma_2 = \frac{\partial u}{\partial y} + \frac{\partial v}{\partial x} = \gamma_{xy} \qquad (1.2.21)$$

In a three-dimensional body, the lines move through angles in the y-z and z-x planes as well. For small displacements, however, the angular displacement of the projections of the lines on the x-y plane will provide the same interpretation. Similar interpretations apply to γ_{yz} and γ_{zx}.

The transformation of strain at a point and the concept of *principal strains* can be established in much the same manner as for stress. This is presented in Appendix C for reference.

Some interesting relationships called the *equations of compatibility* are found by eliminating the displacements from the strain-displacement equations by substitution. For example, we can take the strain-displacement equation for the γ_{xy} component of the shearing strain and perform the operations indicated next to obtain Equation 1.2.22.

$$\frac{\partial^2 \gamma_{xy}}{\partial x \partial y} = \frac{\partial^2}{\partial x \, \partial y} \frac{\partial u}{\partial y} + \frac{\partial^2}{\partial x \, \partial y} \frac{\partial v}{\partial x} = \frac{\partial^2}{\partial y^2} \frac{\partial u}{\partial x} + \frac{\partial^2}{\partial x^2} \frac{\partial v}{\partial y} = \frac{\partial^2 \epsilon_x}{\partial y^2} + \frac{\partial^2 \epsilon_y}{\partial x^2} \quad (1.2.22)$$

By similar operations, we can obtain five more equations, as shown in Equation 1.2.23. Together these are known as the *equations of compatiblity*.

$$
\begin{aligned}
\frac{\partial^2 \gamma_{yz}}{\partial y \, \partial z} &= \frac{\partial^2 \epsilon_y}{\partial z^2} + \frac{\partial^2 \epsilon_z}{\partial y^2}, \qquad \frac{\partial^2 \gamma_{zx}}{\partial z \, \partial x} = \frac{\partial^2 \epsilon_z}{\partial x^2} + \frac{\partial^2 \epsilon_x}{\partial z^2} \\[2mm]
2\frac{\partial^2 \epsilon_x}{\partial y \, \partial z} &= \frac{\partial}{\partial x}\left(-\frac{\partial \gamma_{yz}}{\partial x} + \frac{\partial \gamma_{zx}}{\partial y} + \frac{\partial \gamma_{xy}}{\partial z} \right) \\[2mm]
2\frac{\partial^2 \epsilon_y}{\partial z \, \partial x} &= \frac{\partial}{\partial y}\left(+\frac{\partial \gamma_{yz}}{\partial x} - \frac{\partial \gamma_{zx}}{\partial y} + \frac{\partial \gamma_{xy}}{\partial z} \right) \\[2mm]
2\frac{\partial^2 \epsilon_z}{\partial x \, \partial y} &= \frac{\partial}{\partial z}\left(-\frac{\partial \gamma_{yz}}{\partial x} + \frac{\partial \gamma_{zx}}{\partial y} - \frac{\partial \gamma_{xy}}{\partial z} \right)
\end{aligned}
\qquad (1.2.23)
$$

We shall have some use for these equations later.

1.2.3 Material Properties. Hooke's Law

Thus far, all the equations we have developed are independent of the material of the body. They are applicable to any continuous solid body undergoing small displacements. The material properties now enter in the form of a particular relationship between stress and strain, which we shall call, not surprisingly, the *stress-strain equations*. There are many materials with widely differing properties and, consequently, many different stress-strain relationships are possible. We shall restrict ourselves, for the most part, to materials that are *homogeneous* and *isotropic*; that is, their properties are independent of position and of direction, respectively, in the body. These form the bulk of structural materials in use today or are, at least, close approximations to many materials, but because of the growing importance of materials, such as *composites*, which may be neither homogeneous nor isotropic, a review of stress-strain equations for nonhomogeneous and nonisotropic materials is included in Appendix D.

We shall also reserve most of the following discussion to materials that are *elastic*, in fact, *linearly elastic*. By elastic, we mean the stress-strain relation is the same for both loading and unloading. In addition, the restriction to linear elasticity requires that the stress be proportional to the strain. Fortunately, many materials of interest in structural applications possess the preceding properties, at least under certain conditions. Further discussion of the properties of nonlinear elastic and inelastic materials and of the justification of the announced restrictions is given in Appendix D. Finally, we shall not take into account temperature variations; that is, the body is at constant temperature and the material properties are given for that temperature. The effect of temperature changes is also mentioned in Appendix D.

The general form of the linearly elastic stress-strain equations that express that stress is proportional to strain, as given in Equations 1.2.1, is

$$\{\sigma\} = [G]\{\epsilon\} \qquad (1.2.24)$$

or, in inverse form,

$$\{\epsilon\} = [C]\{\sigma\} \qquad (1.2.25)$$

where $[C]$ is the inverse of $[G]$. Materials that have properties that follow this format are said to obey *Hooke's law*.

To find the $[G]$ and $[C]$ matrices, consider a rectangular element of a body made from a material that is homogeneous, isotropic, and linearly elastic and subjected only to a uniform σ_x stress. Both its unloaded and undeformed and its loaded and deformed shapes are shown in Figure 1.2.8. By careful observation in an actual experiment, we would see that the element elongates in the x-direction and contracts in the y- and z-directions. The amount of elongation can be expressed in terms of the normal strain times the original length in each direction, as shown in the figure. The normal strain in each direction is proportional to the stress, σ_x. This may be expressed symbolically as

$$\epsilon_x = \frac{\sigma_x}{E}, \qquad \epsilon_y = -\frac{\upsilon\sigma_x}{E}, \qquad \epsilon_z = -\frac{\upsilon\sigma_x}{E} \qquad (1.2.26)$$

in which E is a constant of proportionality known as *Young's modulus* and υ is called *Poisson's ratio*.

The values of υ and E for any given material are found by test. These particular coefficients were chosen to express the proportionality between stress and strain because of their convenient physical interpretations. Young's modulus is the slope of the familiar stress-strain curve, and Poisson's ratio is the ratio of lateral to longitudinal strain for the element in Figure 1.2.8. This is explained in detail in any elementary mechanics of materials book and is briefly reviewed in Appendix D.

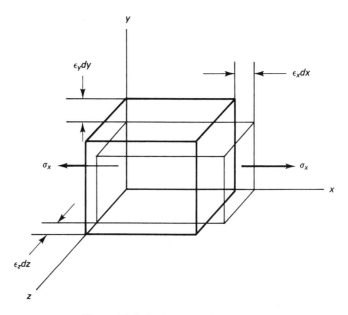

Figure 1.2.8 Deformation of an Element.

If σ_x is removed and σ_y and σ_z are applied in turn, similar results are obtained. If all three stresses are applied simultaneously, the results for each individual stress may be superimposed, and we have

$$\epsilon_x = \frac{1}{E}[\sigma_x - v(\sigma_y + \sigma_z)], \qquad \epsilon_y = \frac{1}{E}[\sigma_y - v(\sigma_z + \sigma_x)]$$
$$\epsilon_z = \frac{1}{E}[\sigma_z - v(\sigma_x + \sigma_y)] \tag{1.2.27}$$

It may be noted that there is no angular distortion of the element and, therefore, normal stresses do not produce shearing strains. If, however, we apply purely shearing stresses, we observe angular distortion. This may be expressed symbolically as

$$\gamma_{xy} = \frac{\tau_{xy}}{G}, \qquad \gamma_{yz} = \frac{\tau_{yz}}{G}, \qquad \gamma_{zx} = \frac{\tau_{zx}}{G} \tag{1.2.28}$$

where G is called the *shear modulus*. It can be shown (see Appendix D) that

$$G = \frac{E}{2(1 + v)} \tag{1.2.29}$$

Hooke's law as given in Equations 1.2.27 and 1.2.28 may also be expressed in the matrix form of Equation 1.2.25 as

$$\{\epsilon\} = [C]\{\sigma\}$$

or

$$
\begin{bmatrix} \epsilon_x \\ \epsilon_y \\ \epsilon_z \\ \gamma_{xy} \\ \gamma_{yz} \\ \gamma_{zx} \end{bmatrix} = \begin{bmatrix} \dfrac{1}{E} & -\dfrac{v}{E} & -\dfrac{v}{E} & 0 & 0 & 0 \\ -\dfrac{v}{E} & \dfrac{1}{E} & -\dfrac{v}{E} & 0 & 0 & 0 \\ -\dfrac{v}{E} & -\dfrac{v}{E} & \dfrac{1}{E} & 0 & 0 & 0 \\ 0 & 0 & 0 & \dfrac{1}{G} & 0 & 0 \\ 0 & 0 & 0 & 0 & \dfrac{1}{G} & 0 \\ 0 & 0 & 0 & 0 & 0 & \dfrac{1}{G} \end{bmatrix} \begin{bmatrix} \sigma_x \\ \sigma_y \\ \sigma_z \\ \tau_{xy} \\ \tau_{yz} \\ \tau_{zx} \end{bmatrix} \qquad (1.2.30)
$$

The inverse form, as given in Equation 1.2.24, which is repeated here, is

$$\{\sigma\} = [G]\,\{\epsilon\}$$

or

$$
\begin{bmatrix} \sigma_x \\ \sigma_y \\ \sigma_z \\ \tau_{xy} \\ \tau_{yz} \\ \tau_{zx} \end{bmatrix} = \begin{bmatrix} \lambda + 2G & \lambda & \lambda & 0 & 0 & 0 \\ \lambda & \lambda + 2G & \lambda & 0 & 0 & 0 \\ \lambda & \lambda & \lambda + 2G & 0 & 0 & 0 \\ 0 & 0 & 0 & G & 0 & 0 \\ 0 & 0 & 0 & 0 & G & 0 \\ 0 & 0 & 0 & 0 & 0 & G \end{bmatrix} \begin{bmatrix} \epsilon_x \\ \epsilon_y \\ \epsilon_z \\ \gamma_{xy} \\ \gamma_{yz} \\ \gamma_{zx} \end{bmatrix} \qquad (1.2.31)
$$

where

$$\lambda = \frac{vE}{(1 + v)(1 - 2v)} \qquad (1.2.32)$$

1.2.4 General Equations. Applied Loads. Constraints

The three equilibrium equations, the six strain-displacement equations, and the six equations of Hooke's law, Equations 1.2.10, 1.2.20, and 1.2.31, respectively, together with the appropriate boundary conditions for both applied loads and geometric constraints (see the following), form a complete mathematical model for the behavior of a continuous, homogeneous, isotropic, linearly elastic structure for small deflections. That behavior is expressed in terms of the fifteen dependent variables composed of the six components of stress, the six components of strain, and the three components of displacement. These are known as the three-dimensional *equations of elasticity*.

The boundary conditions are a statement of applied load or displacement at every point on the surface of the body. An applied load boundary condition requires that the internal stresses in a body at the boundary be equal to the surface forces per unit area applied to the boundary. Similarly, a displacement boundary condition requires that the internal displacement at the boundary of the body be equal to the prescribed boundary displacement. Note that at every point on the boundary *either* a surface load *or* a displacement may be prescribed, but not both.

In specifying applied loads, we may note that Figure 1.2.2 is applicable in representing the applied loads if ΔA is an element of boundary surface and ΔF is the force on that surface. Let us designate the force per unit area on the surface of a body, F_s, by its components in the three coordinate directions as follows:

$$\{F_s\} = \begin{bmatrix} X_s \\ Y_s \\ Z_s \end{bmatrix} \tag{1.2.33}$$

These are sometimes called *surface tractions*. By relating the surface tractions and the internal stresses to their respective areas at any point on the surface and then invoking equilibrium, it is possible to find the following relationship:

$$\begin{bmatrix} X_s \\ Y_s \\ Z_s \end{bmatrix} = \begin{bmatrix} \sigma_x & \tau_{xy} & \tau_{zx} \\ \tau_{xy} & \sigma_y & \tau_{yz} \\ \tau_{zx} & \tau_{yz} & \sigma_z \end{bmatrix} \begin{bmatrix} 1 \\ m \\ n \end{bmatrix} \tag{1.2.34}$$

where l, m, and n are the direction cosines of the normal to the surfaces on which the forces act. It should be noted that while X_s, Y_s, and Z_s have units of force per unit area they are not necessarily normal or tangential to that area, while normal and shearing stress, by definition, are normal and tangential to the areas with respect to which they are defined. In many cases, the external surfaces are oriented with respect to rectangular cartesian coordinates; for example, for rectangular bodies, all surfaces have normals that point in one of the coordinate directions. Then, on any given surface, one of the direction cosines is 1 and the others are 0, and the preceding equations simplify accordingly.

At some point we must note that in this formulation when the body is unloaded and unconstrained it is totally stress free. In actual practice, material forming and treating processes can provide us with a structural member that is not stress free in the unloaded and unconstrained configuration. These stresses are called *residual* stresses. In all that we do here, we assume that the residual stresses are zero. The addition of residual stresses is given in several of the references.

1.2.5 Solutions of the Equations of Elasticity

In a typical practical problem, the geometry of the body, the material properties, the constraints on displacement on portions of the boundary of the body, the applied loads on the rest of the boundary, and the body forces are known. The stress, strains, and displacements are the unknowns. Unfortunately, these equations do not yield the values of the stresses, strains, and displacements easily and cannot be solved in general. Only a few exact analytical solutions to the full three-dimensional equations of elasticity have been found and then only when severe restrictions have been imposed on the kind and form of the loading, the geometry of the body, and the geometric constraints.

To gain insight into what these equations can tell us, we shall give as the first example an inverse problem; that is, we shall take a known solution and use the equations we have derived to find all we can about the problem that it solves. Then, as our second example, we shall do a forward solution.

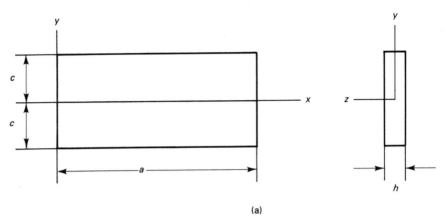

(a)

Example 1.2.1

Given the solid body shown in figure (a) and the following values for the displacements and the constraints, find out what you can about the strains, stresses, and applied loads.

The displacements are

$$u(x, y, z) = \frac{k}{E}\left(x - \frac{a}{2}\right)y, \qquad v(x, y, z) = \frac{k}{2E}(ax - x^2 - vy^2)$$

$$w(x, y, z) = -\frac{vkyz}{E}$$

(a)

The constraints are

$$u(0, 0, 0) = v(0, 0, 0) = v(a, 0, 0) = w(x, y, 0) = 0 \qquad \text{(b)}$$

The first three of these constraints serve to attach the structure to the coordinate axes. The last one is essentially a symmetry condition.

Solution: First check that the displacements satisfy the constraints. It may be seen that they do. Next, let us find the strains from the displacements using the strain-displacement equations, Equations 1.2.19 or 1.2.20:

$$\epsilon_x = \frac{\partial u}{\partial x} = \frac{ky}{E}, \qquad \epsilon_y = \frac{\partial v}{\partial y} = -\frac{vky}{E}, \qquad \epsilon_z = \frac{\partial w}{\partial z} = -\frac{vky}{E}$$

$$\gamma_{xy} = \frac{\partial v}{\partial x} + \frac{\partial u}{\partial y} = 0, \qquad \gamma_{yz} = \frac{\partial w}{\partial y} + \frac{\partial v}{\partial z} = -\frac{vkz}{E}$$

(c)

$$\gamma_{zx} = \frac{\partial u}{\partial z} + \frac{\partial w}{\partial x} = 0$$

The stresses are then found from these strains by Hooke's law, Equation 1.2.30:

$$
\begin{bmatrix}
\lambda + 2G & \lambda & \lambda & 0 & 0 & 0 \\
\lambda & \lambda + 2G & \lambda & 0 & 0 & 0 \\
\lambda & \lambda & \lambda + 2G & 0 & 0 & 0 \\
0 & 0 & 0 & G & 0 & 0 \\
0 & 0 & 0 & 0 & G & 0 \\
0 & 0 & 0 & 0 & 0 & G
\end{bmatrix}
\begin{bmatrix}
\dfrac{ky}{E} \\
-\dfrac{vky}{E} \\
-\dfrac{vky}{E} \\
0 \\
-\dfrac{vkz}{E} \\
0
\end{bmatrix}
=
\begin{bmatrix}
\sigma_x \\
\sigma_y \\
\sigma_z \\
\tau_{xy} \\
\tau_{yz} \\
\tau_{zx}
\end{bmatrix}
$$

(d)

This reduces to

$$\sigma_x = ky, \qquad \sigma_y = 0, \qquad \sigma_z = 0$$

$$\tau_{xy} = 0, \qquad \tau_{yz} = -\frac{\upsilon kz}{2(1 + \upsilon)}, \qquad \tau_{zx} = 0 \tag{e}$$

Let us now check to see if the equilibrium equations are satisfied. Substituting these stresses into Equations 1.2.10, we get

$$\begin{bmatrix} \dfrac{\partial}{\partial x} & 0 & 0 & \dfrac{\partial}{\partial y} & 0 & \dfrac{\partial}{\partial z} \\[2mm] 0 & \dfrac{\partial}{\partial y} & 0 & \dfrac{\partial}{\partial x} & \dfrac{\partial}{\partial z} & 0 \\[2mm] 0 & 0 & \dfrac{\partial}{\partial z} & 0 & \dfrac{\partial}{\partial y} & \dfrac{\partial}{\partial x} \end{bmatrix} \begin{bmatrix} ky \\ 0 \\ 0 \\ 0 \\ -\dfrac{\upsilon kz}{2(1 + \upsilon)} \\ 0 \end{bmatrix} = - \begin{bmatrix} f_x \\ f_y \\ f_z \end{bmatrix} \tag{f}$$

This reduces to

$$f_x = f_z = 0, \qquad f_y = \frac{\upsilon k}{2(1 + \upsilon)} \tag{g}$$

Thus, equilibrium is satisfied if the body forces have the values shown.

Loads are prescribed on all portions of the surface where constraints are not. These loads may be found by evaluating the stresses at the surfaces. First, for the left and right ends, $x = 0$ and a, we have the following loading:

$$\sigma_x(0, y, z) = \sigma_x(a, y, z) = ky$$

$$\tau_{xy}(0, y, z) = \tau_{xy}(a, y, z) = \tau_{zx}(0, y, z) = \tau_{zx}(a, y, z) = 0 \tag{h}$$

On the top and bottom surfaces, $y = \pm c$:

$$\sigma_y(x, c, z) = \sigma_y(x, -c, z) = 0, \qquad \tau_{xy}(x, c, z) = \tau_{xy}(x, -c, z) = 0$$

$$\tau_{yz}(x, c, z) = \tau_{yz}(x, -c, z) = -\frac{\upsilon kz}{2(1 + \upsilon)} \tag{i}$$

On the front and back faces, $z = \pm h/2$:

$$\sigma_z\left(x, y, \frac{h}{2}\right) = \sigma_z\left(x, y, -\frac{h}{2}\right) = 0, \qquad \tau_{zx}\left(x, y, \frac{h}{2}\right) = \tau_{zx}\left(x, y, -\frac{h}{2}\right) = 0$$

$$\tau_{yz}\left(x, y, \frac{h}{2}\right) = -\frac{\upsilon kh}{4(1 + \upsilon)}, \qquad \tau_{yz}\left(x, y, -\frac{h}{2}\right) = \frac{\upsilon kh}{4(1 + \upsilon)} \tag{j}$$

This example has illustrated an inverse solution; that is, given the answer the problem is found. Its value is in showing how the various equations (strain-displacement, stress-strain, and equilibrium) may be used to gain information about a structure. It also gives some insight into how boundary conditions are stated.

In the next example, a very elementary three-dimensional elasticity problem is posed, which can be solved. This is a forward solution in that the problem is stated in conventional form with the geometry, constraints, and applied loads given. It is then solved for the unknown displacements, strains, and stresses. While this example will provide additional experience in using the three-dimensional equations of elasticity, it will not, unfortunately, give much insight as to how more complex elasticity problems are solved, since, in general, the internal stress distribution cannot be guessed from the form of the applied loads, as is done in this case. Nevertheless, it can be a valuable exercise for learning more about the equations of elasticity.

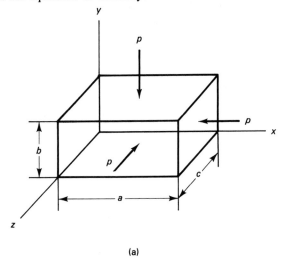

(a)

Example 1.2.2

A rectangular three-dimensional solid body has the dimensions shown in figure (a). It is acted on by a uniform hydrostatic pressure, p, on the three positive faces. The three negative faces are attached to the fixed coordinate system. Find the stresses, strains, displacements, and support reactions of the body if there are no body forces present.

Solution: Equations 1.2.34 for the boundary conditions for applied loads are especially simple. For example, for the face at $x = a$ with a normal in the x-direction,

$$\begin{bmatrix} X_s \\ Y_s \\ Z_s \end{bmatrix} = \begin{bmatrix} -p \\ 0 \\ 0 \end{bmatrix} = \begin{bmatrix} \sigma_x & \tau_{xy} & \tau_{zx} \\ \tau_{xy} & \sigma_y & \tau_{yz} \\ \tau_{zx} & \tau_{yz} & \sigma_z \end{bmatrix} \begin{bmatrix} 1 \\ 0 \\ 0 \end{bmatrix} \qquad \text{(a)}$$

from which the boundary conditions for applied loads on that face are seen to be

$$\sigma_x(a, y, z) = -p, \qquad \tau_{xy}(a, y, z) = \tau_{zx}(a, y, z) = 0$$

$$0 < y < b; \quad 0 < z < c \qquad \text{(b)}$$

On the face at $y = b$ with a normal in the y-direction, we have

$$\begin{bmatrix} X_s \\ Y_s \\ Z_s \end{bmatrix} = \begin{bmatrix} 0 \\ -p \\ 0 \end{bmatrix} = \begin{bmatrix} \sigma_x & \tau_{xy} & \tau_{zx} \\ \tau_{xy} & \sigma_y & \tau_{yz} \\ \tau_{zx} & \tau_{yz} & \sigma_z \end{bmatrix} \begin{bmatrix} 0 \\ 1 \\ 0 \end{bmatrix} \qquad \text{(c)}$$

or

$$\sigma_y(x, b, z) = -p; \qquad \tau_{yz}(x, b, z) = \tau_{xy}(x, b, z) = 0$$

$$0 < x < a; \quad 0 < z < c \qquad \text{(d)}$$

Finally, on the face at $z = c$ with a normal in the z-direction, we have

$$\begin{bmatrix} X_s \\ Y_s \\ Z_s \end{bmatrix} = \begin{bmatrix} 0 \\ 0 \\ -p \end{bmatrix} = \begin{bmatrix} \sigma_x & \tau_{xy} & \tau_{zx} \\ \tau_{xy} & \sigma_y & \tau_{yz} \\ \tau_{zx} & \tau_{yz} & \sigma_z \end{bmatrix} \begin{bmatrix} 0 \\ 0 \\ 1 \end{bmatrix} \tag{e}$$

or

$$\sigma_z(x, y, c) = -p, \qquad \tau_{zx}(x, y, c) = \tau_{yz}(x, y, c) = 0,$$
$$0 < x < a; \quad 0 < y < b \tag{f}$$

In addition, there are the boundary conditions based on geometric constraints:

$$u(0, y, z) = 0, \qquad 0 < y < b, \quad 0 < z < c$$
$$v(x, 0, z) = 0, \qquad 0 < x < a, \quad 0 < z < c \tag{g}$$
$$w(x, y, 0) = 0, \qquad 0 < x < a, \quad 0 < y < b$$

The structural behavior is modeled by the 15 equations of elasticity. By inspection (that is, someone thought to try the following), trial values for the stresses are suggested to be

$$\sigma_x(x, y, z) = \sigma_y(x, y, z) = \sigma_z(x, y, z) = -p$$
$$\tau_{xy}(x, y, z) = \tau_{yz}(x, y, z) = \tau_{zx}(x, y, z) = 0 \tag{h}$$

When these are substituted into the equations of equilibrium and it is noted that

$$f_x = f_y = f_z = 0 \tag{i}$$

we find that equilibrium is satisfied identically.

When the stresses are substituted into Hooke's law, we find the normal and shearing strains based on these stresses are

$$\epsilon_x(x, y, z) = \epsilon_y(x, y, z) = \epsilon_z(x, y, z) = -(1 - 2v)\frac{p}{E} = e$$
$$\gamma_{xy}(x, y, z) = \gamma_{yz}(x, y, z) = \gamma_{zx}(x, y, z) = 0 \tag{j}$$

If we integrate the three normal strain-displacement equations, we get

$$\epsilon_x = \frac{\partial u}{\partial x} = e \longrightarrow u(x, y, z) = ex + f(y, z)$$

$$\epsilon_y = \frac{\partial v}{\partial y} = e \longrightarrow v(x, y, z) = ey + g(z, x) \tag{k}$$

$$\epsilon_z = \frac{\partial w}{\partial z} = e \longrightarrow w(x, y, z) = ez + h(x, y)$$

When these displacements are substituted into the geometric boundary conditions, we get

$$u(0, y, z) = f(y, z) = 0$$
$$v(x, 0, z) = g(z, x) = 0 \tag{l}$$
$$w(x, y, 0) = h(x, y) = 0$$

Therefore, the displacements are

$$u(x, y, z) = ex, \qquad v(x, y, z) = ey, \qquad w(x, y, z) = ez \tag{m}$$

These displacements are valid only if they also satisfy identically the shear strain–displacement equations (which, fortunately, they do), that is, if

$$\frac{\partial u}{\partial y} + \frac{\partial v}{\partial x} = 0, \qquad \frac{\partial v}{\partial z} + \frac{\partial w}{\partial y} = 0, \qquad \frac{\partial w}{\partial x} + \frac{\partial u}{\partial z} = 0 \qquad \text{(n)}$$

In summary, the stresses given in equation (h), the strains given in equation (j), and the displacements given in equation (m) satisfy the 15 equations of elasticity: Equations 1.2.10, 1.2.20, and 1.2.31 and the boundary conditions on applied loads given in equations (b), (d), and (f) and on geometric constraints given in equation (g). This is a complete and exact analytical solution to the stated problem.

From these results, we can find the forces on the constrained surfaces, that is, the forces on the surfaces where displacements are prescribed as given in equations (g). Evaluating the stresses on these surfaces, we get

$$\sigma_x(0, y, z) = -p, \qquad \tau_{xy}(0, y, z) = \tau_{zx}(0, y, z) = 0$$

$$\sigma_y(x, 0, z) = -p, \qquad \tau_{yz}(x, 0, z) = \tau_{xy}(x, 0, z) = 0 \qquad \text{(o)}$$

$$\sigma_z(x, y, 0) = -p, \qquad \tau_{zx}(x, y, 0) = \tau_{yz}(x, y, 0) = 0$$

The geometry, applied loads, and constraints were carefully chosen in Example 1.2.2 so that an exact analytical solution could be found. While there are a few published exact solutions to three-dimensional problems in elasticity, for example, in References 6, 7, and 8, they do not begin to cover the range of structures of practical interest. Even when useful solutions can be found for a given solid body, minor changes in loading, shape, or constraint will usually preclude finding an exact analytical solution to the slightly altered problem. The method of inspection we have used, in short, taking an educated guess at the solution, does not carry us very far. The example was offered to show all the components of an exact analytical solution, but not as a guide to the solution of practical problems.

There are several ways we can turn at this point, and eventually we are going to emphasize approximations to these equations. Before doing so, we are going to examine certain classes of solid bodies with simplified geometry for which a few exact analytical solutions are possible.

We should note that all the derivations, so far, have been referred to rectangular cartesian coordinates. Depending on the geometry of the body being studied, development of the equations in other coordinate systems has been fruitful. Cylindrical and polar coordinate systems have been particularly useful, especially for bodies of revolution that are frequently used as structural components. A lack of space in this text dictates that we refrain from any extensive coverage of topics in these coordinates. A brief review of the equations in cylindrical and polar coordinates is given in Appendix E, and the reader is referred to other sources for solutions.

1.3 SIMPLIFIED EQUATIONS

Since exact analytical solutions to the full equations of elasticity are generally not possible, we look to simplifying assumptions that reduce the number of dependent variables and hence the number of equations to be solved. In many cases, the combination of the shape of the body and the loads applied to it will suggest that several of the unknown quantities are, indeed, zero. We shall now examine two such classes of problems that have wide applicability.

1.3.1 Plane Problems in Elasticity

Prominent among the simplifying assumptions are those that reduce the problem to two dimensions, that is, eliminate the need to carry one of the three independent variables. These are called plane problems, and there are two standard forms, *plane stress* and *plane strain*.

Suppose we arbitrarily restrict the nonzero stresses, applied loads, and body forces to those lying in the x-y plane and assume the nonzero quantities are not functions of z. Thus, we have

$$\sigma_z = \tau_{zx} = \tau_{yz} = f_z = 0 \qquad (1.3.1)$$

and also that σ_x, σ_y, τ_{xy}, u, v, f_x, and f_y are functions only of x and y. Such a state of stress is known as *plane stress*. It can apply to the type of body shown in Figure 1.3.1, where the body is cylindrical with generators parallel to the z-axis and with two plane boundaries at $z = \pm h/2$, where h is the thickness. The equilibrium equations become

$$
\begin{bmatrix}
\dfrac{\partial}{\partial x} & 0 & 0 & \dfrac{\partial}{\partial y} & 0 & \dfrac{\partial}{\partial z} \\[2mm]
0 & \dfrac{\partial}{\partial y} & 0 & \dfrac{\partial}{\partial x} & \dfrac{\partial}{\partial z} & 0 \\[2mm]
0 & 0 & \dfrac{\partial}{\partial z} & 0 & \dfrac{\partial}{\partial y} & \dfrac{\partial}{\partial x}
\end{bmatrix}
\begin{bmatrix}
\sigma_x \\ \sigma_y \\ 0 \\ \tau_{xy} \\ 0 \\ 0
\end{bmatrix}
= -
\begin{bmatrix}
f_x \\ f_y \\ 0
\end{bmatrix}
\qquad (1.3.2)
$$

which reduces to

$$\frac{\partial \sigma_x}{\partial x} + \frac{\partial \tau_{xy}}{\partial y} + f_x = 0, \qquad \frac{\partial \tau_{xy}}{\partial x} + \frac{\partial \sigma_y}{\partial y} + f_y = 0 \qquad (1.3.3)$$

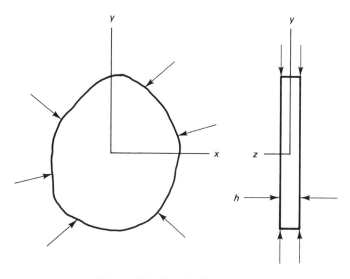

Figure 1.3.1 Body in Plane Stress.

Since the stresses are functions of only x and y, it follows from Hooke's law that the strains are also functions of x and y only. Hooke's law for plane stress becomes

$$
\begin{bmatrix}
\dfrac{1}{E} & -\dfrac{v}{E} & -\dfrac{v}{E} & 0 & 0 & 0 \\[2mm]
-\dfrac{v}{E} & \dfrac{1}{E} & -\dfrac{v}{E} & 0 & 0 & 0 \\[2mm]
-\dfrac{v}{E} & -\dfrac{v}{E} & \dfrac{1}{E} & 0 & 0 & 0 \\[2mm]
0 & 0 & 0 & \dfrac{1}{G} & 0 & 0 \\[2mm]
0 & 0 & 0 & 0 & \dfrac{1}{G} & 0 \\[2mm]
0 & 0 & 0 & 0 & 0 & \dfrac{1}{G}
\end{bmatrix}
\begin{bmatrix}
\sigma_x \\[2mm] \sigma_y \\[2mm] 0 \\[2mm] \tau_{xy} \\[2mm] 0 \\[2mm] 0
\end{bmatrix}
=
\begin{bmatrix}
\epsilon \\[2mm] \epsilon_y \\[2mm] \epsilon_z \\[2mm] \gamma_{xy} \\[2mm] \gamma_{yz} \\[2mm] \gamma_{zx}
\end{bmatrix}
\tag{1.3.4}
$$

or

$$
\epsilon_x = \frac{1}{E}(\sigma_x - v\sigma_y), \qquad \epsilon_y = \frac{1}{E}(\sigma_y - v\sigma_x)
$$
$$
\epsilon_z = -\frac{v}{E}(\sigma_x + \sigma_y), \qquad \gamma_{xy} = \frac{\tau_{xy}}{G}, \qquad \gamma_{yz} = \gamma_{zx} = 0
\tag{1.3.5}
$$

The inverse form of Hooke's law for plane stress is

$$
\sigma_x = \frac{E}{1-v^2}(\epsilon_x + v\epsilon_y), \qquad \sigma_y = \frac{E}{1-v^2}(\epsilon_y + v\epsilon_x)
$$
$$
\tau_{xy} = G\sigma_{xy}, \qquad \sigma_z = \tau_{yz} = \tau_{zx} = 0
\tag{1.3.6}
$$

The strain–displacement equations are now

$$
\epsilon_x = \frac{\partial u}{\partial x}, \qquad \epsilon_y = \frac{\partial v}{\partial y}, \qquad \epsilon_z = \frac{\partial w}{\partial z}
$$
$$
\gamma_{xy} = \frac{\partial u}{\partial y} + \frac{\partial v}{\partial x}, \qquad \gamma_{yz} = \gamma_{zx} = 0
\tag{1.3.7}
$$

It may be noted that, while all the stresses, strains, and the in-plane displacements are assumed to be functions only of x and y, the displacement w is not so restricted. The displacements must satisfy the preceding equations and the geometric constraints.

In summary, of the fifteen stress, strain, and displacement components of the equations of elasticity, five are zero:

$$
\sigma_z = \tau_{zx} = \tau_{yz} = \gamma_{zx} = \gamma_{yz} = 0
\tag{1.3.8}
$$

Of the remaining equations, eight are coupled in terms of eight dependent variables, σ_x, σ_y, τ_{xy}, ϵ_x, ϵ_y, γ_{xy}, u, v, all of which are functions only of x and y, and the other two equations, dealing with the dependent variables ϵ_z and w, are uncoupled from the other eight. In matrix form, we have for equilibrium

$$
[E]\{\sigma\} = -\{f\}
$$

or

$$\begin{bmatrix} \dfrac{\partial}{\partial x} & 0 & \dfrac{\partial}{\partial y} \\[2mm] 0 & \dfrac{\partial}{\partial y} & \dfrac{\partial}{\partial x} \end{bmatrix} \begin{bmatrix} \sigma_x \\ \sigma_y \\ \tau_{xy} \end{bmatrix} = -\begin{bmatrix} f_x \\ f_y \end{bmatrix} \tag{1.3.9}$$

For Hooke's law,

$$\{\sigma\} = [G]\,\{\epsilon\}$$

or

$$\begin{bmatrix} \sigma_x \\[2mm] \sigma_y \\[2mm] \tau_{xy} \end{bmatrix} = \begin{bmatrix} \dfrac{E}{1-v^2} & \dfrac{vE}{1-v^2} & 0 \\[2mm] \dfrac{vE}{1-v^2} & \dfrac{E}{1-v^2} & 0 \\[2mm] 0 & 0 & G \end{bmatrix} \begin{bmatrix} \epsilon_x \\[2mm] \epsilon_y \\[2mm] \gamma_{xy} \end{bmatrix} \tag{1.3.10}$$

For strain–displacement,

$$\{\epsilon\} = [D]\,\{u\}$$

or

$$\begin{bmatrix} \epsilon_x \\ \epsilon_y \\ \gamma_{xy} \end{bmatrix} = \begin{bmatrix} \dfrac{\partial}{\partial x} & 0 \\[2mm] 0 & \dfrac{\partial}{\partial y} \\[2mm] \dfrac{\partial}{\partial y} & \dfrac{\partial}{\partial x} \end{bmatrix} \begin{bmatrix} u \\ v \end{bmatrix} \tag{1.3.11}$$

Once these eight equations are solved consistent with the boundary conditions, an exact analytical solution for the plane stress case has been found. Note that once the eight equations are solved we can find ϵ_z from Equations 1.3.5 and use the strain–displacement relations to find the deflection from the following:

$$\epsilon_z = -\frac{v}{E}\,(\sigma_x + \sigma_y) = \frac{\partial w}{\partial z} \tag{1.3.12}$$

These are the two uncoupled equations that round out the set. The boundary conditions complete the statement of the problem. Geometric constraints are imposed directly on the displacements. For applied loads, the boundary conditions are given by a reduced form of Equations 1.2.34, or

$$\begin{bmatrix} X_s \\ Y_s \end{bmatrix} = \begin{bmatrix} \sigma_x & \tau_{xy} \\ \tau_{xy} & \sigma_y \end{bmatrix} \begin{bmatrix} 1 \\ m \end{bmatrix} \tag{1.3.13}$$

since all z-components of stress are zero.

Unfortunately, as we have already noted, we cannot, generally, specify all the geometry, applied loads, and geometric constraints in advance and hope to find an exact solution. Some success has been achieved with what is called the inverse method. In the inverse method, a solution to the preceding equations is proposed, usually in simple polynomial form, and the problem for which it is a solution is determined; that is, the geometry, applied loads, and geometric constraints compatible with the solution are determined. Surprisingly, some useful solutions have been found that way.

The next two examples illustrate two of a relatively small number of known exact solutions to plane stress problems in rectangular cartesian coordinates. They will be used later to help justify further simplifying assumptions. For now, their purpose is to illustrate all the steps in an exact solution.

Example 1.3.1

A thin rectangular plate with uniform thickness, h, and other dimensions as shown in figure (a) has a uniform distributed load applied to two opposite edges as shown in figure (b) and is constrained so that no rigid body motions are permitted.

The boundary normal stresses have the value A where A is a known constant. Find the stresses, strains, and deflections throughout the plate according to the plane stress assumptions if no body forces are present. Check to see if the solution found is also a solution to the full three-dimensional equations of elasticity.

Solution: Treating this as a plane stress problem, we immediately set

$$\sigma_z = \tau_{zx} = \tau_{yz} = f_z = 0 \qquad \text{(a)}$$

and since there are no body forces, we also have

$$f_y = f_x = 0 \qquad \text{(b)}$$

The equations to be satisfied are

$$[E]\,\{\sigma\} + \{f\} = 0$$
$$\{\sigma\} = [G]\,\{\epsilon\} \qquad \text{(c)}$$
$$\{\epsilon\} = [D]\,\{u\}$$

where $[E]$, $[G]$, and $[D]$ are given in Equations 1.3.9, 1.3.10, and 1.3.11, respectively.

The applied load or stress boundary conditions for this plate are

$$
\begin{aligned}
\sigma_x(0, y) &= \sigma_x(a, y) = A, & -c < y < c \\
\tau_{xy}(0, y) &= \tau_{xy}(a, y) = 0, & -c < y < c \\
\sigma_y(x, -c) &= \sigma_y(x, +c) = 0, & 0 < x < a \\
\tau_{xy}(x, -c) &= \tau_{xy}(x, +c) = 0, & 0 < x < a
\end{aligned}
\qquad \text{(d)}
$$

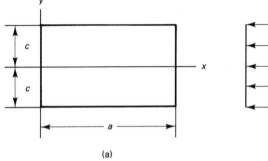

(a)

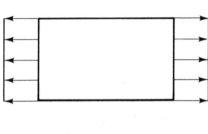

(b)

Let us wait until we need to specify the geometric constraints before we do so, since we are somewhat restricted in what we can specify and still get an exact analytical answer to the problem.

Let us try a solution of the form

$$\sigma_x(x, y) = A, \qquad \sigma_y(x, y) = 0, \qquad \tau_{xy}(x, y) = 0 \tag{e}$$

Of course, we have no assurance that this state of stress will be true until we can confirm that it satisfies all the equations and all the boundary conditions, but our experience with Example 1.2.1 suggests it may be good. Certainly, this state of stress will satisfy all the boundary conditions on applied loading, which we can readily check. Now let us see if it satisfies the equilibrium equations. The equation

$$[E] \{\sigma\} = -\{f\} \tag{f}$$

becomes

$$\begin{bmatrix} \dfrac{\partial}{\partial x} & 0 & \dfrac{\partial}{\partial y} \\[2mm] 0 & \dfrac{\partial}{\partial y} & \dfrac{\partial}{\partial x} \end{bmatrix} \begin{bmatrix} A \\ 0 \\ 0 \end{bmatrix} = \begin{bmatrix} 0 \\ 0 \end{bmatrix} \tag{g}$$

which also is seen to be satisfied.

Since we now have the stresses and want to find the strains, we need the form of Hooke's law for plane stress, as follows:

$$\{\epsilon\} = [C] \{\sigma\} \tag{h}$$

which is

$$\begin{bmatrix} \epsilon_x \\[2mm] \epsilon_y \\[2mm] \gamma_{xy} \end{bmatrix} = \begin{bmatrix} \dfrac{1}{E} & -\dfrac{v}{E} & 0 \\[2mm] -\dfrac{v}{E} & \dfrac{1}{E} & 0 \\[2mm] 0 & 0 & \dfrac{1}{G} \end{bmatrix} \begin{bmatrix} A \\ 0 \\ 0 \end{bmatrix} \tag{i}$$

From this the strains are seen to be

$$\frac{\partial u}{\partial x} = \epsilon_x(x, y) = \frac{A}{E}, \qquad \frac{\partial v}{\partial y} = \epsilon_y(x, y) = -v\frac{A}{E},$$

$$\frac{\partial u}{\partial y} + \frac{\partial v}{\partial x} = \gamma_{xy} = 0 \tag{j}$$

Note that

$$\epsilon_z = -\frac{vA}{E} \tag{k}$$

The strain–displacement equations, also included in equation (j), may then be integrated and the following obtained:

$$u(x, y) = \frac{Ax}{E} + f(y), \qquad v(x, y) = -\frac{vAy}{E} + g(x) \tag{l}$$

What follows may not be so obvious, but bear with it and see what happens. Let us substitute the results shown in equation (l) into the third of equations (j):

$$\frac{\partial u}{\partial y} + \frac{\partial v}{\partial x} = \frac{\partial f}{\partial y} + \frac{\partial g}{\partial x} = 0 \qquad \text{(m)}$$

It follows that

$$\frac{\partial g}{\partial x} = -\frac{\partial f}{\partial y} = C_1 \qquad \text{(n)}$$

from the argument that everything to the left of the first equal sign is a function of x only and that to the right is a function of y only; therefore, both must be equal to a constant. If we change to total derivatives and integrate to find $f(y)$ and $g(x)$, we get

$$-\frac{df}{dy} = C_1 \longrightarrow f(y) = -C_1 y + C_2$$

$$\frac{dg}{dx} = C_1 \longrightarrow g(x) = C_1 x + C_3 \qquad \text{(o)}$$

from which

$$u(x, y) = \frac{Ax}{E} - C_1 y + C_2, \qquad v(x, y) = -\frac{vAy}{E} + C_1 x + C_3 \qquad \text{(p)}$$

Now we are ready for geometric boundary conditions. Note that the preceding integration process produced three constants. We can specify three discrete geometric conditions to obtain values for these three constants. Let us specify zero displacement in both the x- and y-directions at the origin of the coordinates; thus

$$u(0, 0) = C_2 = 0, \qquad v(0, 0) = C_3 = 0 \qquad \text{(q)}$$

Note that these two geometric conditions prevent rigid body translations. A third condition is needed to prevent rigid body rotation. For example, let the y-deflection on the x-axis at the opposite edge of the plate be zero, as follows:

$$v(a, 0) = C_1 a = 0 \longrightarrow C_1 = 0 \qquad \text{(r)}$$

So, finally,

$$u(x, y) = \frac{Ax}{E}, \qquad v(x, y) = -\frac{vAy}{E} \qquad \text{(s)}$$

We could just as well have chosen the third geometric constraint to be at some other point in the body. For example, let the deflection in the x-direction at the upper-left corner of the body be zero, as follows:

$$u(0, c) = C_1 c = 0 \longrightarrow C_1 = 0 \qquad \text{(t)}$$

In this case it has the same result and $u(x, y)$ and $v(x, y)$ are exactly the same.

A plot of the deflection (exaggerated in magnitude for illustration) presented in figure (c) shows a simple uniform elongation in the x-direction and a uniform contraction in the y-direction, as might be expected. This provides an opportunity to point out another feature of the linear theory of elasticity. No distinction is made between the coordinates of a point on the body in the deflection and undeflected condition. For example, the deflection of a point on the right edge of the plate is given by

$$u(a, y) = \frac{Aa}{e} \qquad \text{(u)}$$

Note that the value of $x = a$ for the undeflected position is entered for the x-coordinate even though the point has moved to $a + Aa/E$ in the deflected position.

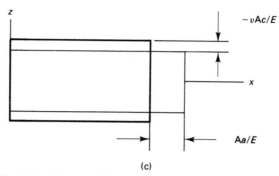

(c)

Having solved the eight coupled equations for plane stress, we can now solve the remaining two uncoupled equations, Equations 1.3.12.

$$\epsilon_z(x, y, z) = -\frac{v}{E}(\sigma_x + \sigma_y) = -\frac{vA}{E} \tag{v}$$

$$\frac{\partial w}{\partial z} = -\frac{vA}{E} \longrightarrow w(x, y, z) = -\frac{vA}{E}z + h(x, y)$$

From symmetry arguments, an appropriate constraint condition is

$$w(x, y, 0) = 0 + h(x, y) = 0 \tag{w}$$

Therefore,

$$w(x, y, z) = -\frac{vAz}{E} \tag{x}$$

Finally, let us check to see if the answers obtained satisfy the full three-dimensional equations of elasticity. Let us use the displacements found from the plane stress assumptions as if they are the displacements for a true three-dimensional body and see if the full equations of elasticity are satisfied. Thus, we let

$$u(x, y, z) = \frac{Ax}{E}, \qquad v(x, y, z) = -\frac{vAy}{E}, \qquad w(x, y, z) = -\frac{vAz}{E} \tag{y}$$

be the displacements of the full three-dimensional body. To check, first substitute the displacements in the full strain-displacement equations. We obtain

$$\epsilon_x(x, y, z) = \frac{A}{E}, \qquad \epsilon_y(x, y, z) = -\frac{vA}{E}, \qquad \epsilon_z(x, y, z) = -\frac{vA}{E}$$

$$\gamma_{xy}(x, y, z) = \gamma_{yz}(x, y, z) = \gamma_{zx}(x, y, z) = 0 \tag{z}$$

This leads to the stress values of

$$\sigma_x(x, y, z) = \frac{(\lambda + 2G - 2\lambda v)A}{E} = A, \qquad \sigma_y = \sigma_z = 0$$

$$\tau_{xy} = \tau_{yz} = \tau_{zx} = 0 \tag{aa}$$

Such a state of stress obviously satisfies the equations of equilibrium, and since Hooke's law and strain–displacement are also satisfied, all 15 equations of elasticity are satisfied. Since the displacements satisfy all constraints and the stresses at the surfaces match all applied loads, we do, indeed, have a complete and exact solution to the three-dimensional problem.

Sec. 1.3   Simplified Equations

27

Example 1.3.2

A thin rectangular plate with uniform thickness, h, and other dimensions as shown in figure (a) has linearly varying tractions applied to two opposite edges as shown in figure (b) and is constrained so that no rigid body motions are permitted.

The boundary normal stresses shown have the value ky, where k is a known constant. What are the stresses, strains, and deflections throughout the plate if no body forces are present?

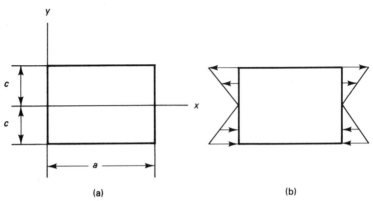

(a)   (b)

Solution: This solution proceeds exactly as in Example 1.3.1. Treating this as a plane stress problem, we immediately set

$$\sigma_z = \tau_{yz} = \tau_{zx} = f_z = 0 \tag{a}$$

and, since there are no body forces, we also have

$$f_x = f_y = 0 \tag{b}$$

The equations to be satisfied are

$$[E]\{\sigma\} + \{f\} = 0$$
$$\{\sigma\} = [G]\{\epsilon\} \tag{c}$$
$$\{\epsilon\} = [D]\{u\}$$

where $[E]$, $[G]$, and $[D]$ are given in Equations 1.3.9, 1.3.10, and 1.3.11, respectively.

The stress boundary conditions for this plate are

$$
\begin{aligned}
\sigma_x(0, y) &= \sigma_x(a, y) = ky, & -c < y < c \\
\tau_{xy}(0, y) &= \tau_{xy}(a, y) = 0, & -c < y < c \\
\sigma_y(x, -c) &= \sigma_y(x, +c) = 0, & 0 < x < a \\
\tau_{yx}(x, -c) &= \tau_{yx}(x, +c) = 0, & 0 < x < a
\end{aligned} \tag{d}
$$

Again let us wait until we need to specify the geometric constraints before we do so, since we are somewhat restricted in what we can specify and still get an exact analytical answer to the problem.

Let us try a solution of the form

$$\sigma_x(x, y) = ky \tag{e}$$

with all other stresses equal to zero and see if it satisfies the equilibrium equations.

$$[E]\{\sigma\} + \{f\} = 0 \tag{f}$$

becomes

$$\begin{bmatrix} \dfrac{\partial}{\partial x} & 0 & \dfrac{\partial}{\partial y} \\[2mm] 0 & \dfrac{\partial}{\partial y} & \dfrac{\partial}{\partial x} \end{bmatrix} \begin{bmatrix} ky \\ 0 \\ 0 \end{bmatrix} = \begin{bmatrix} 0 \\ 0 \end{bmatrix} \tag{g}$$

which is seen to be satisfied.

Since we now have the stresses and want to find the strains, we need the form of Hooke's Law for plane stress, as follows:

$$\{\epsilon\} = [C]\,\{\sigma\} \tag{h}$$

which is

$$\begin{bmatrix} \epsilon_x \\[2mm] \epsilon_y \\[2mm] \gamma_{xy} \end{bmatrix} = \begin{bmatrix} \dfrac{1}{E} & -\dfrac{v}{E} & 0 \\[2mm] -\dfrac{v}{E} & \dfrac{1}{E} & 0 \\[2mm] 0 & 0 & \dfrac{1}{G} \end{bmatrix} \begin{bmatrix} ky \\ 0 \\ 0 \end{bmatrix} \tag{i}$$

From this the strains are seen to be

$$\frac{\partial u}{\partial x} = \epsilon_x(x, y) = \frac{ky}{E}, \qquad \frac{\partial v}{\partial y} = \epsilon_y(x, y) = -v\frac{ky}{E}$$

$$\frac{\partial u}{\partial y} + \frac{\partial v}{\partial x} = \gamma_{xy} = 0 \tag{j}$$

We pause to note that the z-component of strain is

$$\epsilon_z = -\frac{vky}{E} \tag{k}$$

The strain–displacement equations, also included in equation (j), may then be integrated and the following obtained:

$$u(x, y) = \frac{kxy}{E} + f(y), \qquad v(x, y) = -\frac{vky^2}{2E} + g(x) \tag{l}$$

Let us substitute the results shown in equation (l) into the third of equations (j):

$$\frac{\partial u}{\partial y} + \frac{\partial v}{\partial x} = \frac{kx}{E} + \frac{\partial f}{\partial y} + \frac{\partial g}{\partial x} = 0 \tag{m}$$

It follows that

$$\frac{\partial g}{\partial x} + \frac{kx}{E} = -\frac{\partial f}{\partial y} = C_1 \tag{n}$$

from the argument that everything to the left of the first equal sign is a function of x only and the next term is a function of y only; therefore, both must be equal to a constant. If we change to total derivatives and integrate to find $f(y)$ and $g(x)$ we get

$$-\frac{df}{dy} = C_1 \longrightarrow f(y) = -C_1 y + C_2$$

$$\frac{dg}{dx} = C_1 - \frac{kx}{E} \longrightarrow g(x) = C_1 x - \frac{kx^2}{2E} + C_3 \tag{o}$$

from which

$$u(x, y) = \frac{kxy}{E} - C_1 y + C_2$$

$$v(x, y) = -\frac{\upsilon k y^2}{2E} + C_1 x - \frac{kx^2}{2E} + C_3 \tag{p}$$

We can specify three discrete geometric conditions to obtain values for these three constants. Let us specify zero displacement in both x- and y-directions at the origin of the coordinates; thus,

$$u(0, 0) = C_2 = 0, \qquad v(0, 0) = C_3 = 0 \tag{q}$$

A third condition is needed to prevent rigid body rotation. For example, let the y-deflection on the x-axis at the opposite edge of the plate be zero, as follows:

$$v(a, 0) = C_1 a - \frac{ka^2}{2E} = 0 \longrightarrow C_1 a = \frac{ka}{2E} \tag{r}$$

So, finally,

$$u(x, y) = \frac{kxy}{E} - \frac{kay}{2E} = \frac{ky}{E}\left(x - \frac{a}{2}\right)$$

$$v(x, y) = -\frac{\upsilon k y^2}{2E} + \frac{kax}{2E} - \frac{kx^2}{2E} = \frac{k}{2E}(ax - x^2 - \upsilon y^2) \tag{s}$$

As before, we could just as well have chosen the third geometric constraint to be at some other point in the body. For example, let the deflection in the x-direction at the upper-left corner of the body be zero, as follows:

$$u(0, c) = -C_1 c = 0 \longrightarrow C_1 = 0 \tag{t}$$

and the deflections become

$$u(x, y) = \frac{kxy}{E}$$

$$v(x, y) = -\frac{\upsilon k y^2}{2E} - \frac{kx^2}{2E} = -\frac{k}{2E}(x^2 + \upsilon y^2) \tag{u}$$

In this case, it appears that the answers are quite different for $u(x, y)$ and $v(x, y)$. Actually, the two answers provide exactly the same shape for the two cases, but, in effect, because of the different orientation of the coordinate axes, one is rotated with respect to the other. This may be seen when the two results are plotted as shown in figures (c) and (d).

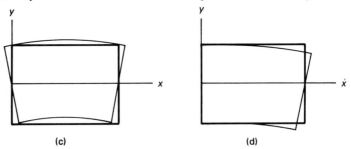

(c)　　　　　　　　　　(d)

Once again, note that the deflections are specified in terms of the coordinates of points on the undeflected plate.

A question remains whether this solution for plane stress is also a solution to the full three–dimensional equations of elasticity. First, we must complete the plane stress solution by solving for the z-component of deflection using the uncoupled equation for the displacement w. From equation (k) and the strain–displacement equations, we have

$$\epsilon_z = \frac{\partial w}{\partial z} = -\frac{\upsilon k y}{E} \longrightarrow w(x, y, z) = -\frac{\upsilon k y}{E} z + h(x, y) \tag{v}$$

From symmetry arguments, an appropriate constraint condition is

$$w(x, y, 0) = 0 + h(x, y) = 0 \tag{w}$$

Therefore,

$$w(x, y, z) = -\frac{\upsilon k y}{E} z \tag{x}$$

Choosing the first of the constraint conditions, we now have

$$u(x, y, z) = \frac{k y}{E}\left(x - \frac{a}{2}\right), \qquad v(x, y, z) = \frac{k}{2E}(ax - x^2 - \upsilon y^2) \tag{y}$$

and equation (x) as possible solutions to the full three–dimensional equations. Actually, we have already checked this out. If you look back to Example 1.2.1, you will see that this is precisely the solution we used in the inverse problem of finding what problem this solution fitted. The conclusion in Example 1.2.1 is that

$$\tau_{xy} = -\frac{\upsilon k z}{2(1 + \upsilon)} \tag{z}$$

which is in violation of the assumptions of plane stress. Furthermore, there must be a body force

$$f_y = -\frac{\upsilon k}{2(1 + \upsilon)} \tag{aa}$$

which is a violation of the stated conditions of the problem. This stress distribution implies that there are applied loads on the top and bottom edges and on the front and back surfaces, as follows:

$$\tau_{yz}(x, c, z) = \tau_{yz}(x, -c, z) = -\frac{\upsilon k z}{2(1 + \upsilon)}$$

$$\tau_{yz}\left(x, y, \frac{h}{2}\right) = -\frac{\upsilon k h}{4(1 + \upsilon)}, \qquad \tau_{yz}\left(x, y, -\frac{h}{2}\right) = \frac{\upsilon k h}{4(1 + \upsilon)} \tag{bb}$$

Clearly, the full three-dimensional equations of elasticity do not have the same solution as the plane stress solution for this case; however, it may be argued that the differences are small for a thin plate. The thickness, h, is very much smaller than a and c. Thus, the applied loads at the surfaces in equation (bb) and the internal shearing stress in equation (z), where z is always less than $h/2$, are small compared to other surfaces loads and internal stresses and may be neglected in many problems. And if we neglect the shearing stress, the body force, f_z, is zero from the equilibrium equations. Thus, the plane stress solution is a good approximation when the plate thickness is sufficiently small.

Except for a few simple cases, the exact plane stress solution is not also the exact solution for the full three–dimensional equations of elasticity, but as the plate becomes thinner the plane stress solution becomes a more accurate approximation. This approximation is so good, in fact, that plane stress solution is generally accepted without challenge for thin plates. This approximation is discussed at greater length in References 7 and 8, among others.

This will satisfy our immediate need to understand plane stress; however, for those interested, more solutions may be found in References 7 and 8 and in other publications, which may be sought in the library.

Before leaving plane problems, we shall briefly discuss another plane problem known as *plane strain*. Suppose it is possible to say

$$w = \epsilon_z = \gamma_{yz} = \gamma_{zx} = f_z = 0 \tag{1.3.14}$$

and also that u, v, ϵ_x, ϵ_y, γ_{xy}, σ_x, σ_y, τ_{xy}, f_x, and f_y are functions only of x and y. Such a state of strain is known as *plane strain*. The restriction on the deformation to the x-y plane is again arbitrary. It would be just as convenient to let $u = 0$ and require that v and w be functions of y and z only.

The type of body for which the plane strain assumptions are valid may be the same as that shown in Figure 1.3.1; however, it is noted that the boundary conditions on the faces $z = \pm h/2$ must necessarily differ. To explain this, let us look at the stresses for plane strain. From Hooke's law,

$$
\begin{bmatrix} \sigma_x \\ \sigma_y \\ \sigma_z \\ \tau_{xy} \\ \tau_{yz} \\ \tau_{zx} \end{bmatrix} =
\begin{bmatrix}
\lambda + 2G & \lambda & \lambda & 0 & 0 & 0 \\
\lambda & \lambda + 2G & \lambda & 0 & 0 & 0 \\
\lambda & \lambda & \lambda + 2G & 0 & 0 & 0 \\
0 & 0 & 0 & G & 0 & 0 \\
0 & 0 & 0 & 0 & G & 0 \\
0 & 0 & 0 & 0 & 0 & G
\end{bmatrix}
\begin{bmatrix} \epsilon_x \\ \epsilon_y \\ 0 \\ \gamma_{xy} \\ 0 \\ 0 \end{bmatrix}
\tag{1.3.15}
$$

from which we conclude

$$\sigma_x = (\lambda + 2G)\epsilon_x + \lambda\epsilon_y, \qquad \sigma_y = \lambda\epsilon_x + (\lambda + 2G)\epsilon_y$$

$$\sigma_z = \lambda(\epsilon_x + \epsilon_y), \qquad \tau_{xy} = G\gamma_{xy}, \qquad \tau_{yz} = \tau_{zx} = 0 \tag{1.3.16}$$

It is clear that $\sigma_z \neq 0$ in plane strain, and stresses must occur on the faces at $z = \pm h/2$ for the condition $w = 0$ to be maintained. It is appropriate to think of the plate in Figure 1.3.1 as a slice of a long cylinder with its axis along the z-axis. If the ends of this cylinder are prevented from moving in the z-direction, then w is zero everywhere.

The strain-displacement equations and the equilibrium equations are the same as for plane stress; therefore, in summary, of the fifteen stress, strain, and displacement components of the equations of elasticity, six are zero:

$$w = \epsilon_z = \tau_{zx} = \tau_{yz} = \gamma_{zx} = \gamma_{yz} = 0 \tag{1.3.17}$$

Of the remaining equations, eight are coupled in terms of eight dependent variables (σ_x, σ_y, τ_{xy}, ϵ_x, ϵ_y, γ_{xy}, u, v), all of which are functions only of x and y, and the other equation, dealing with the dependent variable σ_z, is uncoupled from the other eight. In matrix form, we have for equilibrium

$$[E]\{\sigma\} = -\{f\}$$

or

$$\begin{bmatrix} \dfrac{\partial}{\partial x} & 0 & \dfrac{\partial}{\partial y} \\ 0 & \dfrac{\partial}{\partial y} & \dfrac{\partial}{\partial x} \end{bmatrix} \begin{bmatrix} \sigma_x \\ \sigma_y \\ \tau_{xy} \end{bmatrix} = -\begin{bmatrix} f_x \\ f_y \end{bmatrix} \tag{1.3.18}$$

and for Hooke's law

$$\{\sigma\} = [G]\{\epsilon\}$$

or

$$\begin{bmatrix} \sigma_x \\ \sigma_y \\ \tau_{xy} \end{bmatrix} = \begin{bmatrix} \lambda + 2G & \lambda & 0 \\ \lambda & \lambda + 2G & 0 \\ 0 & 0 & G \end{bmatrix} \begin{bmatrix} \epsilon_x \\ \epsilon_y \\ \gamma_{xy} \end{bmatrix} \tag{1.3.19}$$

and for strain–displacement

$$\{\epsilon\} = [D]\{u\}$$

or

$$\begin{bmatrix} \epsilon_x \\ \epsilon_y \\ \gamma_{xy} \end{bmatrix} = \begin{bmatrix} \dfrac{\partial}{\partial x} & 0 \\ 0 & \dfrac{\partial}{\partial y} \\ \dfrac{\partial}{\partial y} & \dfrac{\partial}{\partial x} \end{bmatrix} \begin{bmatrix} u \\ v \end{bmatrix} \tag{1.3.20}$$

The boundary conditions for the coupled equations are the same as in Equations 1.3.13; however, after these equations are solved the normal stress in the z-direction and the necessary boundary conditions on the z-faces may be found from

$$\sigma_z = \lambda(\epsilon_x + \epsilon_y) = Z_s \tag{1.3.21}$$

Solutions for plane strain can be found in exactly the same way as for plane stress; however, because Hooke's law differs, the solutions will differ. We shall illustrate this difference in the next example.

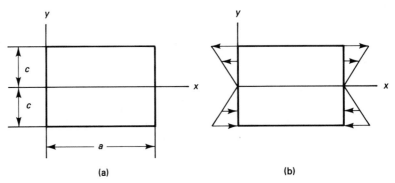

(a)   (b)

Example 1.3.3

A long rectangular prism with its ends restrained to zero displacement has a cross section as shown in figure (a). This body has linearly varying tractions applied to two opposite edges as shown in figure (b) and is constrained so that no rigid body motions are permitted. The boundary applied loads shown have the value ky, where k is a known constant. What are the stresses, strains, and deflections throughout the body if no body forces are present?

Solution: This solution proceeds exactly as in Example 1.3.2, but with changes necessary to treat this as a plane strain problem. First we set

$$w = \epsilon_z = \gamma_{xz} = \gamma_{yz} = f_z = 0 \tag{a}$$

and, since there are no body forces, we also have

$$f_x = f_y = 0 \tag{b}$$

The equations to be satisfied are

$$[E]\{\sigma\} + \{f\} = 0$$
$$\{\sigma\} = [G]\{\epsilon\} \tag{c}$$
$$\{\epsilon\} = [D]\{u\}$$

where $[E]$, $[G]$, and $[D]$ are given in Equations 1.3.18, 1.3.19, and 1.3.20, respectively. The applied load boundary conditions in terms of stresses at the surfaces for this body are

$$\sigma_x(0, y) = \sigma_x(a, y) = ky \qquad -c < y < c$$
$$\tau_{xy}(0, y) = \tau_{xy}(a, y) = 0 \qquad -c < y < c$$
$$\sigma_y(x, -c) = \sigma_y(x, +c) = 0 \qquad 0 < x < a \tag{d}$$
$$\gamma_{xy}(x, -c) = \gamma_{xy}(x, +c) = 0, \qquad 0 < x < a$$

The same geometrical constraints will be applied as in Example 1.3.2 in addition to the plain strain requirement that $w = 0$; that is,

$$u(0, 0) = v(0, 0) = v(a, 0) = 0 \tag{e}$$

Let us try a solution of the form

$$\sigma_x(x, y) = ky, \qquad \sigma_y = \tau_{xy} = 0 \tag{f}$$

and see if it satisfies the equilibrium equations.

$$[E]\{\sigma\} + \{f\} = \{0\} \tag{g}$$

becomes

$$\begin{bmatrix} \dfrac{\partial}{\partial x} & 0 & \dfrac{\partial}{\partial y} \\[2mm] 0 & \dfrac{\partial}{\partial y} & \dfrac{\partial}{\partial x} \end{bmatrix} \begin{bmatrix} ky \\ 0 \\ 0 \end{bmatrix} = \begin{bmatrix} 0 \\ 0 \end{bmatrix} \tag{h}$$

which is seen to be satisfied.

Since we now have the stresses and want to find the strains, we need the form of Hooke's law for plane strain, as follows:

$$\{\epsilon\} = [C]\{\sigma\} \tag{i}$$

which is

$$\begin{bmatrix} \epsilon_x \\[2mm] \epsilon_y \\[2mm] \gamma_{xy} \end{bmatrix} = \begin{bmatrix} \dfrac{1 - v^2}{E} & -\dfrac{v(1 + v)}{E} & 0 \\[3mm] -\dfrac{v(1 + v)}{E} & \dfrac{1 - v^2}{E} & 0 \\[3mm] 0 & 0 & \dfrac{1}{G} \end{bmatrix} \begin{bmatrix} ky \\ 0 \\ 0 \end{bmatrix} \tag{j}$$

from which the strains are seen to be

$$\frac{\partial u}{\partial x} = \epsilon_x(x, y) = \frac{(1 - v^2)ky}{E}$$

$$\frac{\partial v}{\partial y} = -\epsilon_y(x, y) = -\frac{v(1 + v)ky}{E} \tag{k}$$

$$\frac{\partial u}{\partial y} + \frac{\partial v}{\partial x} = \gamma_{xy} = 0$$

The strain–displacement equations, also included in equations (j), may then be integrated and the following obtained:

$$u(x, y) = \frac{(1 - v^2)kxy}{E} + f(y), \qquad v(x, y) = -\frac{v(1 + v)ky^2}{2E} + g(x) \tag{l}$$

Let us substitute the results shown in equation (l) into the third of equations (j); thus

$$\frac{\partial u}{\partial y} + \frac{\partial v}{\partial x} = \frac{(1 - v^2)kx}{E} + \frac{\partial f}{\partial y} + \frac{\partial g}{\partial x} = 0 \tag{m}$$

It follows that

$$\frac{\partial g}{\partial x} + \frac{(1 - v^2)kx}{E} = -\frac{\partial f}{\partial y} = C_1 \tag{n}$$

from the argument that everything to the left of the first equal sign is a function of x only and the next term is a function of y only; therefore, both must be equal to a constant. If we change to total derivatives and integrate to find f(y) and g(x), we get

$$-\frac{df}{dy} = C_1 \qquad \longrightarrow f(y) = -C_1 y + C_2$$

$$\frac{dg}{dx} = C_1 - \frac{(1 - v^2)kx}{E} \longrightarrow g(x) = C_1 x - \frac{(1 - v^2)kx^2}{2E} + C_3 \tag{o}$$

from which

$$u(x, y) = \frac{(1 - v^2)kxy}{E} - C_1 y + C_2$$

$$v(x, y) = -\frac{v(1 + v)ky^2}{2E} + C_1 x - \frac{(1 - v^2)kx^2}{2E} + C_3 \tag{p}$$

We have specified three discrete geometric conditions in equations (e) to obtain values for these three constants; thus,

$$u(0, 0) = C_2 = 0, \qquad v(0, 0) = C_3 = 0 \tag{q}$$

and

$$v(a, 0) = C_1 a - \frac{(1 - v^2)ka^2}{2E} = 0 \longrightarrow C_1 = \frac{(1 - v^2)ka}{2E} \tag{r}$$

So, finally,

$$u(x, y) = \frac{(1 - v^2)kxy}{E} - \frac{(1 - v^2)kay}{2E} = \frac{(1 - v^2)ky}{E}\left(x - \frac{a}{2}\right)$$

$$v(x, y) = -\frac{v(1 + v)ky^2}{2E} + \frac{(1 - v^2)kax}{2E} - \frac{(1 - v^2)kx^2}{2E} \tag{s}$$

$$= -\frac{v(1 + v)ky^2}{2E} + \frac{(1 - v^2)k}{2E}(ax - x^2)$$

This solution for the deflections $u(x, y)$ and $v(x, y)$ differs from that for plane stress only by some Poisson's ratio effects. We shall note some of the consequences of that in the next chapter. We should note that the z-direction stress, which was zero in plane stress, is, as a consequence of the restrained ends, now

$$\sigma_z = v(\sigma_x + \sigma_y) = vky \tag{t}$$

1.3.2 Stress Resultants. St. Venant's Principle

While plane problems are usually easier to solve than full three-dimensional problems, they can still prove difficult. Some additional simplifications are sometimes possible based on the effect of applied loads on solid bodies with certain geometrical characteristics. We shall now examine one class of such problems. This section provides a link between the exact solutions from the theory of elasticity given previously and the general theory of slender bars treated in detail in Chapter 2.

Consider a rectangular strip that is long and narrow and has loads applied only to the opposite ends, as shown in Figure 1.3.2. For the moment, let us assume that these loads are symmetrical about the x-axis.

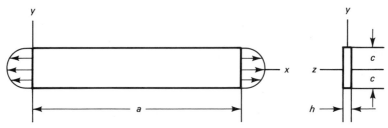

Figure 1.3.2 Long Narrow Strip with Axial Load.

If, for example, the load distribution is a uniform load, precisely the same as in Example 1.3.1, the solution in that example applies exactly to this problem. All we have done for this case is to indicate that a is much larger than $2c$, but that does not affect the solution previously obtained. That solution for stress is, for the uniform end stress equal to A,

$$\sigma_x(x, y) = A \qquad (1.3.22)$$

All other stresses are zero in the structure.

Unfortunately, we cannot extend this approach to other end loads simply because exact analytical solutions have not been found for most other end load distributions. Empirical evidence and the results from numerical methods of solution, which we shall examine in due course in Chapter 4, do provide cause for a very interesting observation. For any other symmetrical load distribution, including the parabolic distribution shown in Figure 1.3.2, for distances well removed from the end of the member, the value of the stress is given approximately, but highly accurately, by Equation 1.3.22. This leads us to define the distributed end load in terms of an *equivalent concentrated force* and the internal distributed stress by a *stress resultant*. Both equivalent concentrated forces and stress resultants are forces or, as we shall see shortly, moments obtained by integrating a distributed applied load or a stress over a prescribed area. The free-body diagram in Figure 1.3.3 of a portion of the structure shown in Figure 1.3.2 illustrates each.

Figure 1.3.3 Equivalent Concentrated Force and Stress Resultant.

The distributed *applied load* on the left end at $x = a$ is replaced by a concentrated force $F_x(a)$ defined as

$$F_x(a) = \int \sigma_x(a, y) \, dA \qquad (1.3.23)$$

where in the notation we take advantage of the fact that the stress at the boundary is equal to the distributed applied load on that boundary. The distributed *internal stress* is replaced by the stress resultant S_x defined as

$$S_x(x) = \int \sigma_x(x, y) \, dA \qquad (1.3.24)$$

Note that when we integrate over the area we are left with S_x being a function of x. From what we have noted previously,

$$\sigma_x(x, y) = \frac{S_x(x)}{A} \qquad (1.3.25)$$

That is, the stress is uniform over the cross section for cross sections sufficiently removed from the ends of the strip. This distribution of internal stress is shown in Figure 1.3.3.

The basis for generalizing the conclusion reached here is enunciated in the *principle of St. Venant*. While this principle cannot be stated in precise and concise terms, it is essentially the following:

If a system of forces acting on a small portion of the surface of an elastic body is replaced by a statically equivalent system, the stresses are changed only in the immediate vicinity of the force application and are virtually unchanged in other parts of the body.

We shall call on this principle many times on subsequent pages as we seek simplified equations for structural analysis.

Example 1.3.4

Find the stress in a slender member, such as is shown in Figure 1.3.2, which has a parabolic end load distribution given by

$$\sigma_x(0, z) = \sigma_x(a, z) = A\left(1 - \frac{y^2}{c^2}\right) \tag{a}$$

Solution: First, we find the stress resultant to be

$$S_x = \int_{-c}^{c} A\left[1 - \left(\frac{y}{c}\right)^2\right] h \, dy = \frac{4Ahc}{3} \tag{b}$$

and the stress at some distance from the ends is

$$\sigma_x = \frac{S_x}{A} = \frac{4Ahc}{3} \div 2ch = \frac{2A}{3} \tag{c}$$

The stress in the vicintity of the ends, that is, in the region of transition from parabolic to uniform, cannot be found by methods introduced so far.

The question remains as to just how quickly the transition takes place from the end load stress distribution to the uniform stress distribution in the interior of the member. As a rough rule of thumb, we shall say that in a distance in from the end that is approximately equal to the height of the member the effect of the true end load distribution will have largely died out. This provides us with a quick measure of what we mean by slender. If $a/(2c) > 10$, for example, the member will be affected over only a small percentage of its total length, and if $a/(2c) < 5$, the uniform stress approximation will be in error over a substantial portion of the member.

Now let us look at the deflection of the member. From Example 1.3.1, we have for the deflection

$$u(x, y) = \frac{Ax}{E}, \qquad v(x, y) = -\frac{vAy}{E} \tag{1.3.26}$$

Because the member is slender, the deflection in the y-direction is never very large, since y can never be larger in magnitude than c. This fact permits an important simplifying assumption, which we shall introduce in Chapter 2.

A study of this deflection provides an opportunity to discuss further the consequences of making no distinction between the coordinates of a point on the body in its undeformed shape and its deformed shape, as was mentioned in Example 1.3.1. In calculating the stress, it was assumed that the area of the cross section was $2ch$, which is true for the undeformed body. In fact, the true area after deformation is decreased due to the Poisson's ratio effect in the y- and z-directions. The true height of the bar is

$$b = 2c - \frac{2vAc}{E} = 2c\left(1 - \frac{vA}{E}\right) \tag{1.3.27}$$

and the true thickness is

$$h = h - \frac{vAh}{E} = h\left(1 - \frac{vA}{E}\right) \qquad (1.3.28)$$

Therefore, the true area after deformation is

$$A = bh = 2ch\left(1 - \frac{vA}{E}\right)^2 \qquad (1.3.29)$$

and the true stress is

$$\sigma_x = \frac{S_x}{2ch[1 - (vA/E)]^2} \qquad (1.3.30)$$

In the linear theory of elasticity, where small deflections are always assumed, this effect is always small and, therefore, this behavior of the structure is ignored. All quantities used in the equations are calculated based on the undeformed coordinates of the body. We shall not refer to this again in this text until we get to Chapter 5, where the coordinates of the deformed body are once again considered.

Let us consider another rectangular strip that is long and narrow, but has loads applied to its opposite ends, as shown in Figure 1.3.4. Let this load be so distributed that we can define two concentrated forces, one an axial force in the x-direction and the other a moment about the z-axis. Note that we are using the word force in a generalized sense; that is, both kind of loads, forces and moments, are called forces. This is quite common practice.

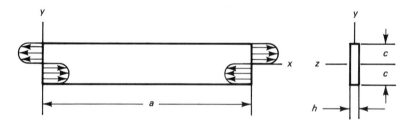

Figure 1.3.4 Long, Narrow Strip with Applied Moment.

For now, let the new value of the equivalent concentrated axial load at each end be zero; thus,

$$F_x(0) = \int \sigma_x(0, z)\, dA = 0, \qquad F_x(a) = \int \sigma_x(a, z)\, dA = 0 \qquad (1.3.31)$$

There will be an equivalent concentrated moment applied about the z-axis at each end, defined by

$$T_z(0) = \int \sigma_x(0, y)\, y\, dA, \qquad T_z(a) = -\int \sigma_x(a, y)\, y\, dA \qquad (1.3.32)$$

The purpose of the minus sign preceding the integral, and some of the stress resultant moments in the following, is a convention that will be made clear in Chapter 2. For now, accept the fact that we are free to define the direction of the moment as we wish and this is how we wish to do it.

If, for example, the end load varies linearly in y, as in Example 1.3.2, the stress distribution in the interior of the body found in that example is

$$\sigma_x(x, y) = ky \qquad (1.3.33)$$

and the stress resultants would be

$$S_x(x) = \int kyh\ dy = 0$$
$$M_z(x) = -\int \sigma_x(x, y)yh\ dy = -k \int y^2\ dA = -kI_{zz} \qquad (1.3.34)$$

where I_{zz} is the area moment of inertia of the cross section of the plate in the y-z plane about the z-axis. Thus,

$$k = -\frac{M_z}{I_{zz}} \qquad (1.3.35)$$

and

$$\sigma_x(x, y) = -\frac{M_z y}{I_{zz}} \qquad (1.3.36)$$

Once again, exact analytical solutions have not been found for other end load distributions, but, once again, St. Venant's principle comes to the rescue. Any end load that produces equivalent concentrated loads as shown in Equations 1.3.31 and 1.3.32 will result in a stress distribution in the interior of the body, as given in Equation 1.3.36. This situation is illustrated in Figure 1.3.5.

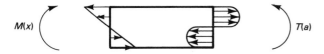

$M(x)$ $T(a)$ **Figure 1.3.5** Equivalent Concentrated Moment and Stress Resultant.

And, once again, the transition takes place within a distance in from the end about equal to the height of the body.

In general, we shall use the notation F_x, F_y, and F_z to represent equivalent concentrated forces and T_x, T_y, and T_z to designate equivalent concentrated moments about the x, y, and z axes, respectively. Similarly, we shall use the notation S_x, S_y, and S_z to represent stress resultant forces and M_x, M_y, and M_z to designate stress resultant moments about the x, y, and z axes, respectively. The location of the forces and moments will be given in terms of the x-coordinate in functional form.

Example 1.3.5

Find the stress in a slender bar that has a double parabolic end load distribution given by

$$\sigma_x(0, y) = \sigma_x(a, y) = k\left(1 - \frac{y}{c}\right)\left(\frac{y}{c}\right), \qquad 0 < y < c$$
$$\sigma_x(0, y) = \sigma_x(a, y) = k\left(1 + \frac{y}{c}\right)\left(\frac{y}{c}\right) \qquad 0 > y > -c \qquad (a)$$

Solution: It can be seen that

$$F_x(0) = \int_{-c}^{c} \sigma_x(0, y)h\ dy = 0, \qquad F_x(a) = \int_{-c}^{c} \sigma_x(a, y)h\ dy = 0 \qquad (b)$$

from the antisymmetry of the preceding functions. To find the equivalent concentrated moments, integrate over the upper half and double it:

$$T_z(0) = 2 \int_0^c k\left(1 - \frac{y}{c}\right)\left(\frac{y}{c}\right) yh \, dy = \frac{khc^2}{6} = -T_z(a) \tag{c}$$

and the stress is

$$\sigma_x(x, y) = \frac{khc^2 y}{6I_{zz}} \tag{d}$$

at points removed from the ends. Once again, we do not have the means as yet to determine the value of the stress in the transition region between the parabolic end loading and linear interior stress.

The deflection found in Example 1.3.2 also applies for any end load distribution that satisfies Equation 1.3.31, except in the immediate vicinity of the ends. And similar statements may be made about the Poisson's ratio effects on the deformed dimensions of the body.

St. Venant's principle is often used in solid mechanics to simplify the solution of a problem by replacing a local set of loads with a statically equivalent set of loads that is easier to deal with. What we have seen here is just one of its many uses. We shall call on it again.

1.3.3 Torsion

For our final example of geometrical and loading simplification in this chapter, we turn to a long cylinder, similar geometrically to the cylinder in plane strain, but with quite different applied loads and geometric constraints. One of the few exact analytical solutions to the three-dimensional equations of elasticity of major interest is the twisting of a cylindrical member with certain cross-sectional shapes and certain restricted conditions of loading. This is one approach to the problem of *torsion*.

Consider a cylindrical body as shown in Figure 1.3.6.

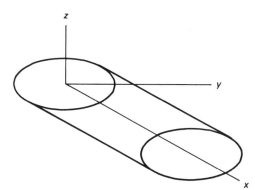

Figure 1.3.6 Cylindrical Body.

While the body is cylindrical at the outset and the end faces are at right angles to the *x*-axis, we make no specification of the cross-sectional shape as yet. Neither do we completely specify the boundary conditions, except to note that there are no stresses on the cylindrical surface, and an applied torque T about the x-axis is twisting the body. This torque is transmitted to the body by shear loads on the two end faces; however, these shear loads will produce no net forces in the y- and z-directions.

The solution is obtained by what is called the semiinverse method, which, as the name implies, is a roundabout way to the answer. It will not necessarily be obvious to the reader why the following steps are taken. Suffice it to say that someone has found it to be a useful approach, and we shall study what they have done.

We start by postulating a simple deformation pattern in the body; that is, a radial line drawn on the cross section remains straight and unextended during deformation and merely rotates about the x-axis, as shown in Figure 1.3.7. From this beginning, we proceed to find out what problem can be solved consistent with this assumption. We shall acquire shapes for the cross section and a set of boundary conditions along the way. Our inspiration for doing this is from the observation of a long, right circular cylinder, or shaft, undergoing twisting; it appears to behave this way.

If the deformation is small, it is reasonable to assume that the angle of twist, β, is proportional to the distance along the x-axis. Thus, if the rotation is restrained at $x = 0$, an implied boundary condition, then

$$\beta = \theta x \tag{1.3.37}$$

where β is the angle of twist or rotation and θ is the twist per unit length, which is assumed to be a constant.

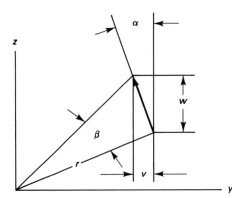

Figure 1.3.7 Deformation in Torsion.

From Figure 1.3.7, we can obtain the following geometrical relations:

$$v = -r\beta \sin \alpha, \qquad w = r\beta \cos \alpha \tag{1.3.38}$$

Since $\sin \alpha = z/r$ and $\cos \alpha = y/r$, we have

$$v = -z\beta = -\theta z x, \qquad w = y\beta = \theta y x \tag{1.3.39}$$

Thus, we see that if we can find θ we know v and w. At this time we will make no demands on the displacement u except to assume that

$$u = u(y, z) \tag{1.3.40}$$

where $u(y, z)$ is called the *warping function*, since displacements normal to the plane of the cross section are referred to as warping of the cross section.

If we insert the preceding displacements into the strain–displacement equations, we find that several of them are identically zero:

$$\epsilon_x = \frac{\partial u}{\partial x} = 0, \qquad \epsilon_y = \frac{\partial v}{\partial y} = 0, \qquad \epsilon_z = \frac{\partial w}{\partial z} = 0$$

$$\gamma_{xy} = \frac{\partial v}{\partial z} + \frac{\partial w}{\partial y} = -\theta x + \theta x = 0$$

(1.3.41)

But two of them are not:

$$\gamma_{xy} = \frac{\partial u}{\partial y} + \frac{\partial v}{\partial x} = \frac{\partial u}{\partial y} - \theta z, \qquad \gamma_{zx} = \frac{\partial u}{\partial z} + \frac{\partial w}{\partial x} = \frac{\partial u}{\partial z} + \theta y \qquad (1.3.42)$$

If we insert Equations 1.3.42 into Hooke's law, we get several stresses that are identically zero,

$$\sigma_x = \sigma_y = \sigma_z = \tau_{yz} = 0 \qquad (1.3.43)$$

and two that are not:

$$\tau_{xy} = G\left(\frac{\partial u}{\partial y} - \theta z\right), \qquad \tau_{zx} = G\left(\frac{\partial u}{\partial z} + \theta y\right) \qquad (1.3.44)$$

Next, we insert the stresses from Equations 1.3.44 into the equilibrium equations and observe that the only equilibrium equation not identically zero in the absence of body forces is

$$\frac{\partial \tau_{xy}}{\partial y} + \frac{\partial \tau_{zx}}{\partial z} = 0 \qquad (1.3.45)$$

and that equation, when expresssed in terms of the displacements, reduces to

$$\frac{\partial^2 u}{\partial y^2} + \frac{\partial^2 u}{\partial z^2} = 0 \qquad (1.3.46)$$

In looking over these various equations, we can see that there are three equations, Equations 1.3.44 and 1.3.46, in terms of four unknown quantities, u, τ_{xy}, τ_{zx}, and θ. We obviously need another equation. It has been found helpful to express the applied torque, which is presumed known, first in terms of the stresses and then, the displacements as follows:

$$T = \iint (\tau_{zx} y - \tau_{xy} z)\, dy\, dz$$

$$= G\iint \left[\frac{\partial u}{\partial z} y - \frac{\partial u}{\partial y} z + \theta(y^2 + z^2)\right] dy\, dx$$

(1.3.47)

This is the fourth equation, which, along with the other three and the boundary conditions, will enable us to solve for the four unknowns and then work backward with Equations 1.3.42 to find the unknown strains and with Equations 1.3.39 to find the remaining displacements.

Now for a word about the boundary conditions. The assumption on displacement in Equation 1.3.37 implies the geometric constraint of no rotation at the one end, or

$$v(0, y, z) = w(0, y, z) = 0 \qquad (1.3.48)$$

We can also claim that

$$u(0, 0, 0) = 0 \qquad (1.3.49)$$

which, in effect, fixes the origin of the coordinates to one end of the body. The condition of no stress on the cylindrical surface may be stated by Equation 1.3.13. This implies that at the boundary there be no normal component of the shear stresses. This may be seen from Figure 1.3.8.

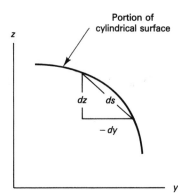

Figure 1.3.8 Boundary Coordinates.

The condition that the shear stresses have no normal component may be stated as

$$\tau_{xy}\frac{dz}{ds} - \tau_{zx}\frac{dy}{ds} = \left(\frac{\partial u}{\partial y} - \theta z\right)\frac{dz}{ds} - \left(\frac{\partial u}{\partial z} + \theta y\right)\frac{dy}{ds} = 0 \qquad (1.3.50)$$

In summary, the four equations, Equations 1.3.44, 1.3.46, and 1.3.47, are solved consistently with the boundary condition on stress given in Equations 1.3.50 and the geometric constraints in Equations 1.3.48 and 1.3.49.

Now let us consider what can be done with these equations in a special case. A known successful procedure for some cross sections of interest is to assume a form of the warping function and find what problem this solves. This is tried in the next example.

Example 1.3.6

A long cylinder of length a is restrained at one end and is acted on by a pure torque at the other. Recognizing this as the classical statement of the torsion problem, what specific problem in torsion is solved when the warping function is a constant?

$$u(y, z) = C \qquad (a)$$

Solution: By substitution, we see immediately that Equation 1.3.46 is solved identically and that the two nonzero stresses are

$$\tau_{xy} = -G\theta z, \qquad \tau_{zx} = G\theta y \qquad (b)$$

From the boundary conditions, we have, first from Equation 1.3.49,

$$u(0, 0, 0) = C = 0 \qquad (c)$$

and then from Equation 1.3.50

$$-z\frac{dz}{ds} - y\frac{dy}{ds} = 0 \qquad (d)$$

from which, by integration, we have

$$\frac{d}{ds}\left(\frac{y^2 + z^2}{2}\right) = 0 \tag{e}$$

or

$$y^2 + z^2 = \text{constant} \tag{f}$$

The interpretation of this result may not be obvious, but think about it for a moment. Since y and z are coordinates of points on the boundary of the cross section and equation (f) is the equation of a circle, it follows that the boundary must be a circle.

Finally, we have

$$T = G\theta \iint (y^2 + z^2)\, dy\, dz = G\theta J \tag{g}$$

where

$$J = \iint (y^2 + z^2)\, dy\, dz \tag{h}$$

is the *torsional constant* of the circular cross section. From this we complete our solution by solving for θ; thus,

$$\theta = \frac{T}{GJ} \tag{i}$$

Note that in this case the torsional constant is also the polar moment of inertia of the cross section. Do not be misled into thinking that this will always be the case for other cross sections, for it will not.

In interpreting our results it is helpful to make the following observation. We can define a stress, τ, in terms of cylindrical coordinates for the circular cross section.

$$\tau = \sqrt{(\tau_{xy})^2 + (\tau_{zx})^2} = \frac{T}{J}\sqrt{y^2 + z^2} = \frac{Tr}{J} \tag{j}$$

where r is any radial point on the cross section. We can also note that for a circular cross section

$$J = \iint (y^2 + z^2)\, dy\, dz = \int r^2 dA = \frac{\pi R^4}{2} \tag{k}$$

where R is the boundary radius of the circular cross section.

Thus, the stress on the cross section is a shearing stress increasing linearly with radial position and, by virtue of the requirement of no component normal to the cylindrical surface, perpendicular to any radial line, as shown in figure (a).

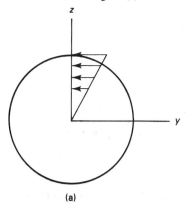

(a)

We can also recognize that the shearing strain in cylindrical coordinates is given by

$$\gamma = \frac{\tau}{G} = \frac{Tr}{GJ} = \theta r \tag{1}$$

If you look back at the derivation of torsion in books on the elementary mechanics of materials for solid circular cylinders, you will see that the assumption of shearing strain in the form given in equation (1) is the starting point for the analysis. This assumption is based on empirical evidence.

In summary, we have solved the problem of a right circular cylinder restrained at one end and loaded with a linearly varying axisymmetric shear stress on the other. The complete displacement, strain, and stress distributions are given in an exact analytical form.

Example 1.3.7

A hollow circular cylinder is loaded with a pure torque as in Example 1.3.6. What are the stresses and displacements?

Solution: In this case, there is no stress on either the outer or the inner cylindrical surface. Note that the solution given in Example 1.3.6 satisfies the condition for the shear stress to have no component normal to the surface by the requirement that the coordinates of any point on the boundary satisfy the condition

$$y^2 + z^2 = \text{constant} \tag{a}$$

or that the boundary be a circle. If both the inner and outer boundaries are circles, this condition is satisfied. Therefore, the solution obtained previously is valid for this case. Thus,

$$\theta = \frac{T}{GJ} \tag{b}$$

where

$$J = \iint (y^2 + z^2)\, dy\, dz \tag{c}$$

and the integration is carried out over the cross section between the inner and outer boundaries. The stresses are given by

$$\tau_{xy} = -G\theta z = \frac{M_x z}{J}, \qquad \tau_{zx} = G\theta y = \frac{M_x y}{J} \tag{d}$$

from which we can obtain

$$\tau = \sqrt{(\tau_{xy})^2 + (\tau_{zx})^2} = \frac{T}{J}\sqrt{y^2 + z^2} = \frac{Tr}{J} \tag{e}$$

where r is any radial point on the cross section between the inner and outer boundaries. A plot of the stress distribution is given in figure (a).

It may be noted that in this case

$$J = \iint (y^2 + z^2)\, dy\, dz = \int r^2 dA = \frac{\pi}{2}(R_o{}^4 - R_i{}^4) \tag{f}$$

where R_o is the outer radius and R_i is the inner radius.

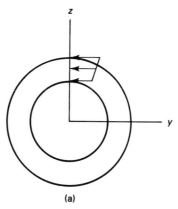

(a)

Other solutions are possible by this method by going to more complicated warping functions. For example, a warping function of the form

$$u(y, z) = Cyz \qquad (1.3.51)$$

will lead to the solution of an elliptical cross section, and

$$u(y, z) = C(z^3 - 3y^2z) \qquad (1.3.52)$$

will lead to the solution of a cross section shaped as an equilateral triangle. These and other solutions may be found in References 4 to 8.

Another approach to the solution of torsional stress distribution has been found to be useful and deserves some attention. Let us introduce a function $\Theta(y, z)$ that has the property that

$$\tau_{xy} = \frac{\partial \Theta}{\partial z}, \qquad \tau_{zx} = -\frac{\partial \Theta}{\partial y} \qquad (1.3.53)$$

This is called a *stress function*, in particular, the *Prandtl torsion function*. If we introduce this into the equilibrium equation, Equation 1.3.45, we see that equilibrium is identically satisfied.

$$\frac{\partial \tau_{xy}}{\partial y} + \frac{\partial \tau_{zx}}{\partial z} = \frac{\partial^2 \Theta}{\partial y \, \partial z} - \frac{\partial^2 \Theta}{\partial y \, \partial z} = 0 \qquad (1.3.54)$$

Combining the expressions for the stress in terms of the deflection given in Equations 1.3.44 with Equations 1.3.53, we have

$$\frac{\partial \Theta}{\partial z} = G\left(\frac{\partial u}{\partial y} - \theta z\right), \qquad \frac{\partial \Theta}{\partial y} = -G\left(\frac{\partial u}{\partial z} + \theta y\right) \qquad (1.3.55)$$

Taking the second derivatives of each of these

$$\frac{\partial^2 \Theta}{\partial z^2} = G\left(\frac{\partial^2 u}{\partial y \, \partial z} - \theta\right), \qquad \frac{\partial^2 \Theta}{\partial y^2} = -G\left(\frac{\partial^2 u}{\partial z \, \partial y} + \theta\right) \qquad (1.3.56)$$

From this, we obtain

$$\nabla^2 \Theta = \frac{\partial^2 \Theta}{\partial z^2} + \frac{\partial^2 \Theta}{\partial y^2} = -2G\theta \qquad (1.3.57)$$

A condition for Θ on the boundary is found by substituting Equation 1.3.53 into Equation 1.3.50. Thus,

$$\frac{\partial \Theta}{\partial y}\, dy + \frac{\partial \Theta}{\partial z}\, dz = d\Theta = 0 \qquad (1.3.58)$$

Since the total differential on the boundary is zero, then

$$\Theta = \text{constant} \qquad (1.3.59)$$

on the boundary. Usually, this constant is taken as zero, as subsequent events will verify.

In solving problems, Equations 1.3.53, 1.3.57, and 1.3.59 are equivalent to and may be used to replace Equations 1.3.44, 1.3.46, and 1.3.50. To complete this formulation, we must also find a replacement for Equation 1.3.47, as follows:

$$
\begin{aligned}
T &= \iint (\tau_{zx}y - \tau_{zy}z)\, dy\, dz = \iint \left(-\frac{\partial \Theta}{\partial y}\, y - \frac{\partial \Theta}{\partial z}\, z \right) dy\, dz \\
&= -\iint \left(\frac{\partial(\Theta y)}{\partial y} + \frac{\partial(\Theta z)}{\partial z} \right) dy\, dz + \iint \Theta\, dy\, dz
\end{aligned}
\qquad (1.3.60)
$$

The latter is merely the rearrangement of terms, where it is noted that

$$\frac{\partial \Theta}{\partial y}\, y = \frac{\partial(\Theta y)}{\partial y} - \Theta \qquad (1.3.61)$$

and so on. If Green's theorem is applied to Equation 1.3.60, we can get

$$T = \int (\Theta z\, dy - \Theta y\, dz)\, ds + 2 \iint \Theta dy\, dz \qquad (1.3.62)$$

The first integral is a line integral around the boundary of the cross section, and since we can take $\Theta = 0$ on the boundary, this integral is zero. What remains is

$$T = 2 \iint \Theta dy\, dz \qquad (1.3.63)$$

If we think of $\Theta(y, z)$ as a surface, an interpretation that will have some physical meaning later, then the applied moment is equal to twice the volume under the surface.

A procedure for solving a problem using the stress function is given in the next example. The solution for an elliptical cross section is demonstrated. In general, we can find solutions for a very limited number of cross sections using either the warping function or the stress function with about equal ease. Thus, Examples 1.3.6 and 1.3.7 could be solved using the stress function (and are in References 6 to 8), and the following example could be solved using the warping function. It is primarily for some further considerations in solving torsion problems, which will be presented following this example, that the stress function has been introduced.

Example 1.3.8

A long cylinder of length a is restrained at the origin and has a pure torque applied to its free end. The cross section is elliptical as shown in figure (a). Find the stress and the rate of twist.

Solution: We must find a $\Theta(y, z)$ that satisfies (1) $\Theta = 0$ on the boundary, and (2) the equation

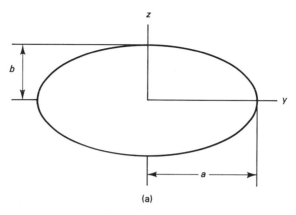

(a)

$$\frac{\partial^2 \Theta}{\partial z^2} + \frac{\partial^2 \Theta}{\partial y^2} = -2G\theta \qquad \text{(a)}$$

Since the boundary of an ellipse is described by

$$\frac{y^2}{a^2} + \frac{z^2}{b^2} = 1 \qquad \text{(b)}$$

it may be observed that a Θ of the form

$$\Theta(y, z) = C\left(\frac{y^2}{a^2} + \frac{z^2}{b^2} - 1\right) \qquad \text{(c)}$$

will satisfy the condition that $\Theta = 0$ on the boundary. Let us substitute this into equation (a) to see what will happen. We find

$$\frac{\partial^2 \Theta}{\partial z^2} + \frac{\partial^2 \Theta}{\partial y^2} = C\left(\frac{2}{a^2} + \frac{2}{b^2}\right) = -2G\theta \qquad \text{(d)}$$

or

$$C = -\frac{a^2 b^2 G\theta}{a^2 + b^2} \qquad \text{(e)}$$

Therefore, Θ for the elliptical bar is

$$\Theta(y, z) = -\frac{a^2 b^2 G\theta}{a^2 + b^2}\left(\frac{y^2}{a^2} + \frac{z^2}{b^2} - 1\right) \qquad \text{(f)}$$

Now let us evaluate the torsional moment in the bar in terms of Θ. Equation 1.3.63 becomes

$$T = 2 \iint \Theta dy\, dz = -2\frac{a^2 b^2 G\theta}{a^2 + b^2} \iint \left(\frac{y^2}{a^2} + \frac{z^2}{b^2} - 1\right) dy\, dz \qquad \text{(g)}$$

which eventually reduces to

$$T = \frac{a^3 b^3 G\theta \pi}{a^2 + b^2} = GJ\theta \qquad \text{(h)}$$

where the torsional constant J is

$$J = \frac{\pi a^3 b^3}{a^2 + b^2} \qquad \text{(i)}$$

Note that the torsional constant is *not* the polar moment of inertia of the cross section.
Let us now evaluate the torsional stresses. From Equations 1.3.53, we have

$$\tau_{xy} = \frac{\partial \Theta}{\partial z} = \frac{2Cz}{b^2} = -2\frac{Tz}{\pi ab^3}$$

$$\tau_{zx} = -\frac{\partial \Theta}{\partial y} = \frac{2Cy}{a^2} = -2\frac{Ty}{\pi a^3 b} \tag{j}$$

If $b < a$, the maximum stress is at $z = \pm b$, or at the outer boundary on the minor axis.
Thus

$$\tau_{max} = \frac{2T}{\pi ab^2} \tag{k}$$

Also, we may note that the resultant stress at any point in the cross section is

$$\tau = \sqrt{(\tau_{xy})^2 + (\tau_{zx})^2} = \frac{2T}{\pi ab}\sqrt{\left(\frac{y^2}{a^4} + \frac{z^2}{b^4}\right)} \tag{l}$$

Finally, if $a = b = R$, we have a circular cross section, and then

$$J = \frac{\pi a^3 b^3}{a^2 + b^2} = \frac{\pi R^4}{2} \tag{m}$$

which is exactly the torsional constant of a circular bar (see Example 1.3.6).

In all the preceding examples, it is possible to find the rate of twist θ from the formula

$$\theta = \frac{d\beta}{dx} = \frac{T}{GJ} \tag{1.3.64}$$

there J is the torsional constant and GJ is called the *torsional stiffness*. The torsional constant for some other cross sections will be found in Chapter 2. When needed, this equation can be integrated to find the angle of twist.

We should note at this point that if the cylinder subjected to pure torque is long and slender, then, from St. Venant's principle, the solutions found for the stress and deflection in the interior of the cylinder are dependent only on the stress resultant for the torque. The actual stress distribution over the ends of the cylinder may be arbitrary as long as it produces only a pure torque. This greatly extends the usefulness of the solutions obtained.

We soon run out of known exact analytical solutions for problems of practical interest and have to introduce additional simplifications or turn to numerical methods for our answers. These will be taken up in subsequent chapters. Before we go on, however, there is a physical analog to the stress function that offers insight into the nature of solutions for torsion in bars with cross sections for which exact analytical solutions have never been found. It is called the *membrane analogy* and results from the fact that the equation for a thin elastic membrane stretched over a region and subjected to a uniform pressure is

$$\frac{\partial u^2}{\partial y^2} + \frac{\partial^2 u}{\partial z^2} = -\frac{p}{T} \tag{1.3.65}$$

where u is the displacement in the x-direction, p is the uniform pressure, and T is the tension in the membrane. If we compare this equation with that for the stress function, we see the similarity where u is analogous to Θ and p/T is analogous to $2G\theta$. If the shape of the region over which the membrane is stretched is the same as the shape of the cross section in torsion, we see that zero deflection of the membrane on the boundary is analogous to the condition $\Theta = 0$.

We may use this similarity to determine the shape of the stress function. It is relatively easy to visualize the shape of a membrane, for example, a soap film or a very thin, balloonlike rubber membrane, when stretched over a hole and pressurized. If we draw contour lines on the deflected membrane we can draw some important conclusions about the stress distribution in torsion. The deflection of the membrane at any point on a contour line is a constant; thus

$$\frac{\partial u}{\partial s} = 0 \tag{1.3.66}$$

along any path s representing a contour line. The corresponding statement for the stress function is

$$\frac{\partial \Theta}{\partial s} = \frac{\partial \Theta}{\partial z}\frac{\partial z}{\partial s} + \frac{\partial \Theta}{\partial y}\frac{\partial y}{\partial s} = \tau_{xy}\frac{\partial z}{\partial s} - \tau_{zx}\frac{\partial y}{\partial s} = 0 \tag{1.3.67}$$

but this is the same statement as Equation 1.3.50; thus, there is no normal component of stress to the contour line. Thus, the tangent to the contour line expresses the direction of the total stress at any point on the contour line. For example, for the elliptical section in Example 1.3.8, the lines of stress direction would be a series of ellipses as shown in Figure 1.3.9.

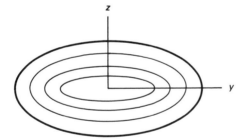

Figure 1.3.9 Membrane Contour Lines and Direction of Total Stress.

Further study will show that the slope of the membrane normal to a contour line is proportional to the magnitude of the total stress at that point, and the volume under the membrane is proportional to the applied moment. From these observations of the membrane, it is possible to get a rather clear picture of the stress distribution in torsion on a cross section with a rather complex boundary. In fact, the membrane analogy has been used to determine experimentally the stress distribution in torsion for odd-shaped cross sections by measuring the slopes, contour line directions, and volumes on an actual membrane.

The membrane analogy may also suggest forms of an analytical solution. Consider a torsional member with a cross section that is a narrow rectangle, as shown in Figure 1.3.10(a). We can make the intuitive argument that since the shear stress cannot have a normal component to the edge and since the two long edges are close together, the shear stress will be nearly parallel to the edges throughout most of the cross section. This suggests a linear distribution of stress, as shown in Figure 1.3.10(b). Such a distribution is confirmed by the membrane analogy. The effect of the short sides of the rectangle would be only local for a membrane stretched over a narrow, rectangular hole and pressurized. By analogy, the stress function would not vary much in the long or y-direction, and we may assume the following approximation:

$$\frac{\partial \Theta}{\partial y} = 0 \tag{1.3.68}$$

which from Equation 1.3.53 is equivalent to neglecting the τ_{zx} component of shear stress. This implies that Θ is a function of z only, and the stress function equation, Equation 1.3.57, reduces to

$$\frac{\partial^2 \Theta}{\partial z^2} = -2G\theta \tag{1.3.69}$$

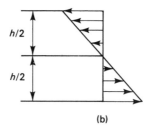

Figure 1.3.10   Torsional Stress for Narrow, Rectangular Cross Section.

Integrating, we obtain

$$\Theta(z) = -G\theta z^2 + C_1 z + C_2 \tag{1.3.70}$$

Applying the boundary conditions

$$\Theta\left(\pm \frac{h}{2}\right) = 0 \tag{1.3.71}$$

we have

$$\Theta = -G\theta\left(z^2 - \frac{h^2}{4}\right) \tag{1.3.72}$$

from which

$$\tau_{xy} = \frac{\partial \Theta}{\partial z} = -2G\theta z, \qquad \tau_{zx} = -\frac{\partial \Theta}{\partial y} = 0 \qquad (1.3.73)$$

The torsional constant is, from Equation 1.3.63,

$$J = \frac{T}{G\theta} = \frac{2}{G\theta} \iint \Theta \, dy \, dz = \frac{2ch^3}{3} \qquad (1.3.74)$$

What we have just found for a narrow, rectangular cross section can be extended to other kinds of open, thin-walled cross sections. For example, if the thin, rectangular section is bent into a circular or rectangular channel, such as those shown in Figure 1.3.11, we can still assume that the shear stresses are primarily parallel to the long edges and linear across the thickness. We can see that the membrane analogy supports this assumption.

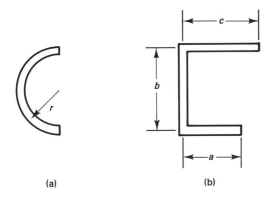

(a)   (b)   **Figure 1.3.11**   Thin-walled Open Sections.

The twist is still given by

$$\theta = \frac{T}{GJ} \qquad (1.3.75)$$

where J is

$$J = \frac{sh^3}{3} \qquad (1.3.76)$$

and s is the length of the path measured along the center line between the long edges of the section. For example, in Figure 1.3.11(a), $s = \pi r$, and in Figure 1.3.11(b), $s = a + b + c$. Furthermore, the stress is still given by Equation 1.3.73 if we interpret the coordinates as s along the center line parallel to the long edges and n normal to s. Then

$$\tau_{xs} = \frac{\partial \Theta}{\partial n} = -2G\theta n, \qquad \tau_{nx} = -\frac{\partial \Theta}{\partial s} = 0 \qquad (1.3.77)$$

1.4 CONCLUSIONS AND SUMMARY

By limiting ourselves to the small deflection of solid bodies at rest that are made of homogeneous, isotropic, and linearly elastic materials, we are able to derive a set of equations, known as the equations of elasticity, that model the behavior of many structures of practical interest. While known exact analytical solutions are rare, we can take advantage of regular geometric features and restricted applied loads often found in practical structural problems to find simplified sets of equations that are more easily solved. Principal among the simplified sets of equations are those for plane stress, plane strain, and torsion. St. Venant's principle allows us to extend further the validity of known solutions to include the effect of different, but statically equivalent, local loading conditions.

Referring back to the symbolic representation of the equations of elasticity as shown in Figure 1.2.1 and Equations 1.2.1, we shall summarize next the matrix form of the equations for the general case and each of the special cases cited. We shall retain the numbering of the equations as given in the main body of the chapter to assist in finding the detailed derivations and discussions. For the general case of *three-dimensional elasticity,* we have for equilibrium

$$[E]\{\sigma\} = -\{f\}$$

where

$$
\begin{bmatrix}
\dfrac{\partial}{\partial x} & 0 & 0 & \dfrac{\partial}{\partial y} & 0 & \dfrac{\partial}{\partial z} \\[2mm]
0 & \dfrac{\partial}{\partial y} & 0 & \dfrac{\partial}{\partial x} & \dfrac{\partial}{\partial z} & 0 \\[2mm]
0 & 0 & \dfrac{\partial}{\partial z} & 0 & \dfrac{\partial}{\partial y} & \dfrac{\partial}{\partial x}
\end{bmatrix}
\begin{bmatrix}
\sigma_x \\ \sigma_y \\ \sigma_z \\ \tau_{xy} \\ \tau_{yz} \\ \tau_{zx}
\end{bmatrix}
= -
\begin{bmatrix}
f_x \\ f_y \\ f_z
\end{bmatrix}
\qquad (1.2.10)
$$

and for strain–displacement

$$\{\epsilon\} = [D]\{u\}$$

where

$$
\begin{bmatrix}
\epsilon \\ \epsilon_y \\ \epsilon_z \\ \gamma_{xy} \\ \gamma_{yz} \\ \gamma_{zx}
\end{bmatrix}
=
\begin{bmatrix}
\dfrac{\partial}{\partial x} & 0 & 0 \\[2mm]
0 & \dfrac{\partial}{\partial y} & 0 \\[2mm]
0 & 0 & \dfrac{\partial}{\partial z} \\[2mm]
\dfrac{\partial}{\partial y} & \dfrac{\partial}{\partial x} & 0 \\[2mm]
0 & \dfrac{\partial}{\partial z} & \dfrac{\partial}{\partial y} \\[2mm]
\dfrac{\partial}{\partial z} & 0 & \dfrac{\partial}{\partial x}
\end{bmatrix}
\begin{bmatrix}
u \\ v \\ w
\end{bmatrix}
\qquad (1.2.20)
$$

and for stress–strain

$$\{\sigma\} = [G]\{\epsilon\}$$

where

$$
\begin{bmatrix} \sigma_x \\ \sigma_y \\ \sigma_z \\ \tau_{xy} \\ \tau_{yz} \\ \tau_{zx} \end{bmatrix}
\begin{bmatrix}
\lambda + 2G & \lambda & \lambda & 0 & 0 & 0 \\
\lambda & \lambda + 2G & \lambda & 0 & 0 & 0 \\
\lambda & \lambda & \lambda + 2G & 0 & 0 & 0 \\
0 & 0 & 0 & G & 0 & 0 \\
0 & 0 & 0 & 0 & G & 0 \\
0 & 0 & 0 & 0 & 0 & G
\end{bmatrix}
\begin{bmatrix} \epsilon_x \\ \epsilon_y \\ \epsilon_z \\ \gamma_{xy} \\ \gamma_{yz} \\ \gamma_{zx} \end{bmatrix}
\tag{1.2.31}
$$

and where

$$\lambda = \frac{vE}{(1 + v)(1 - 2v)} \tag{1.2.32}$$

The statement of the problem is completed by specifying boundary constraints directly in terms of the displacements $\{u\}$ or on the stresses at the surface in terms of the surface forces

$$
\begin{bmatrix} X_s \\ Y_s \\ Z_s \end{bmatrix} =
\begin{bmatrix}
\sigma_x & \tau_{xy} & \tau_{zx} \\
\tau_{xy} & \sigma_y & \tau_{yz} \\
\tau_{zx} & \tau_{yz} & \sigma_z
\end{bmatrix}
\begin{bmatrix} 1 \\ m \\ n \end{bmatrix}
\tag{1.2.34}
$$

The situation known as *plane stress* occurs when the geometry and loading are such that it is possible to say that

$$\sigma_z = \tau_{zx} = \tau_{yz} = \gamma_{zx} = \gamma_{yz} = 0 \tag{1.3.8}$$

The coupled equations for plane stress are for equilibrium

$$[E]\{\sigma\} = -\{f\}$$

where

$$
\begin{bmatrix}
\dfrac{\partial}{\partial x} & 0 & \dfrac{\partial}{\partial y} \\
0 & \dfrac{\partial}{\partial y} & \dfrac{\partial}{\partial x}
\end{bmatrix}
\begin{bmatrix} \sigma_x \\ \sigma_y \\ \tau_{xy} \end{bmatrix}
= -\begin{bmatrix} f_x \\ f_y \end{bmatrix}
\tag{1.3.9}
$$

and for stress–strain

$$\{\sigma\} = [G]\{\epsilon\}$$

where

$$
\begin{bmatrix} \sigma_x \\ \sigma_y \\ \tau_{xy} \end{bmatrix}
\begin{bmatrix}
\dfrac{E}{1 - v^2} & \dfrac{vE}{1 - v^2} & 0 \\
\dfrac{vE}{1 - v^2} & \dfrac{E}{1 - v^2} & 0 \\
0 & 0 & G
\end{bmatrix}
\begin{bmatrix} \epsilon_x \\ \epsilon_y \\ \gamma_{xy} \end{bmatrix}
\tag{1.3.10}
$$

Sec. 1.4   Conclusions and Summary

55

and for strain–displacement

$$\{\epsilon\} = [D]\{u\}$$

where

$$
\begin{bmatrix} \epsilon_x \\ \epsilon_y \\ \gamma_{xy} \end{bmatrix} =
\begin{bmatrix} \dfrac{\partial}{\partial x} & 0 \\[2ex] 0 & \dfrac{\partial}{\partial y} \\[2ex] \dfrac{\partial}{\partial y} & \dfrac{\partial}{\partial x} \end{bmatrix}
\begin{bmatrix} u \\ v \end{bmatrix}
\tag{1.3.11}
$$

Finally, the two uncoupled equations are

$$\epsilon_z = -\frac{v}{E}(\sigma_x + \sigma_y) = \frac{\partial w}{\partial z} \tag{1.3.12}$$

The boundary conditions for applied load are

$$
\begin{bmatrix} X_s \\ Y_s \end{bmatrix} =
\begin{bmatrix} \sigma_x & \tau_{xy} \\ \tau_{xy} & \sigma_y \end{bmatrix}
\begin{bmatrix} 1 \\ m \end{bmatrix}
\tag{1.3.13}
$$

The situation for *plane strain* occurs when the geometry and loading are such that it is possible to say

$$w = \epsilon_z = \tau_{zx} = \tau_{yz} = \gamma_{zx} = \gamma_{yz} = 0 \tag{1.3.17}$$

The coupled equations for equilibrium and strain–displacement are exactly the same as for plane stress, but for stress-strain

$$\{\sigma\} = [G]\{\epsilon\}$$

where

$$
\begin{bmatrix} \sigma_x \\ \sigma_y \\ \tau_{xy} \end{bmatrix} =
\begin{bmatrix} \lambda + 2G & \lambda & 0 \\ \lambda & \lambda + 2G & 0 \\ 0 & 0 & G \end{bmatrix}
\begin{bmatrix} \epsilon_x \\ \epsilon_y \\ \gamma_{xy} \end{bmatrix}
\tag{1.3.19}
$$

The boundary conditions for the applied load in plane strain are the same as for plane stress; however, in plane strain there is one uncoupled equation that determines σ_z and Z_s:

$$\sigma_z = \lambda(\epsilon_x + \epsilon_y) = Z_s \tag{1.3.21}$$

Further simplifications for slender members in plane stress are possible using St. Venant's principle. We define *equivalent concentrated loads* and *internal stress resultants*. These are dealt with more fully in Chapter 2.

Finally, we derived the *equations of torsion*. The result of assuming a restricted or *constrained* form of the displacements,

$$v = -z\beta = -\theta z x, \qquad w = y\beta = \theta y x \tag{1.3.39}$$

allows us to satisfy equilibrium, Hooke's law, strain–displacement, and the boundary conditions by satisfying the following equations:

$$\tau_{xy} = G\left(\frac{\partial u}{\partial y} - \theta z\right), \qquad \tau_{zx} = G\left(\frac{\partial u}{\partial z} + \theta y\right) \tag{1.3.44}$$

$$\frac{\partial \tau_{xy}}{\partial y} + \frac{\partial \tau_{zx}}{\partial z} = 0 \tag{1.3.45}$$

$$\frac{\partial^2 u}{\partial y^2} + \frac{\partial^2 u}{\partial z^2} = 0 \tag{1.3.46}$$

$$\begin{aligned}
T &= \iint (\tau_{zx}y - \tau_{xy}z)\, dy\, dz \\
&= G \iint \left[\frac{\partial u}{\partial z}y - \frac{\partial u}{\partial y}z + \theta(y^2 + z^2)\right] dy\, dz
\end{aligned} \tag{1.3.47}$$

An alternative form of these equations in terms of a *stress function* is also given.

While it is true that a few exact analytical solutions can be found for the complete and simplified equations, as has been shown, for the most part exact analytical solutions cannot be found for most problems of practical interest. In Chapter 2 we turn to still further simplifications and in subsequent chapters to numerical methods. Both approaches greatly expand the range of problems that can be solved.

PROBLEMS

1. Is a structure with the following stress and body force field in equilibrium? Assume that a, b, ρ and g are constants.

$$\sigma_x = 2ax^2yz \qquad\qquad \tau_{xy} = -axy^2z \qquad f_x = 0$$

$$\sigma_y = \frac{ay^3z}{3} \qquad\qquad \tau_{yz} = bx \qquad\qquad f_y = 0$$

$$\sigma_z = -\rho gz + \frac{ayz^3}{3}, \qquad \tau_{zx} = -axyz^2 \qquad f_z = -\rho g$$

2. Consider the following strain field:

$$\epsilon_x = Ky^2 + Cxy, \qquad \epsilon_y = K(x^2 + y^2), \qquad \epsilon_z = 0$$

$$\gamma_{xy} = Cxy \qquad\qquad \gamma_{yz} = 0 \qquad\qquad \gamma_{zx} = 0$$

(a) Integrate these strain components to find the conditions of C and K such that the displacement field is single valued.

(b) If $K = 0.003$ and

$$\{u(0,\, 0,\, 0)\} = \begin{bmatrix} 1 \\ 2 \\ 0 \end{bmatrix}$$

find the displacement field at $\{u(3,\, 4,\, 0)\}$.

(c) Confirm your conclusions in part (a) by utilizing the compatibility equations.

3. A solid bar with a rectangular cross section is subjected to a uniform axial tension of 100 MPa (megapascals) as shown. No other stresses are present. Find the strain in the lateral directions when $E = 68,950$ MPa and $v = 0.25$.

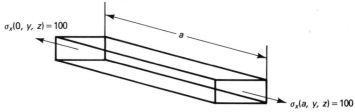

$\sigma_x(0, y, z) = 100$

a

$\sigma_x(a, y, z) = 100$

4. The bar in Problem 3 has a width b and a height c. It is subjected to surface loadings and body forces such that the stress field is

$$\begin{bmatrix} \sigma_x \\ \sigma_y \\ \sigma_z \\ \tau_{xy} \\ \tau_{yz} \\ \tau_{zx} \end{bmatrix} = \begin{bmatrix} y \\ 0 \\ 0 \\ 0 \\ -\dfrac{vz}{1+v} \\ 0 \end{bmatrix}$$

(a) Find the associated strain field.
(b) Give a one-line statement as to why all compatibility conditions are satisfied by inspection, that is, without detailed calculation.
(c) What is the body force field such that equilibrium is satisfied?
(d) What is the surface loading on the face $x = 0$?
(e) What is the surface loading on the face $z = b$?
(f) What conclusions can you draw regarding the relative importance of loadings in parts (d) and (e) if $b \ll c < a$?

5. Consider the following plane stress problem.

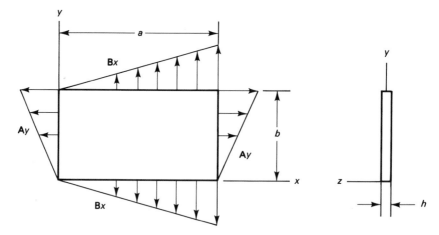

Given the applied loads

$$\sigma_x(a, y) = Ay, \qquad \sigma_y(x, b) = Bx$$

and $\tau_{xy} = 0$ on all surfaces. The geometric constraints and body forces are

$$u(0, 0) = v(0, 0) = v(a, 0) = 0, \qquad f_x(x, y) = f_y(x, y) = 0$$

(a) Show that

$$\sigma_x(x, y) = Ay, \qquad \sigma_y(x, y) = Bx, \qquad \tau_{xy}(x, y) = 0$$

is a solution. What are the values of $\sigma_z(x, y)$, $\tau_{yz}(x, y)$, and $\tau_{zx}(x, y)$?

(b) What are all the strain components? Will there be a z-component of displacement?

(c) Is this a solution to the plane strain problem?

6. Find the approximate stress distribution away from the ends of a slender plate ($a \gg 2c$) that has an end loading given by

$$\sigma_x(0, y) = \sigma_x(a, y) = K\left(1 - \frac{y}{c}\right), \qquad 0 < y < c$$

$$= K\left(1 + \frac{y}{c}\right) \qquad 0 > y > -c$$

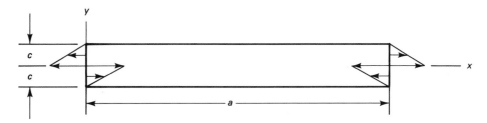

7. The plate in Problem 6 has loads applied to the two ends as follows:

$$\sigma_x(0, y, z) = \sigma_x(a, y, z) = \frac{a}{2}\left(1 + \frac{y}{c}\right)$$

(a) Sketch the loading on the two ends; then find the stresses in the plate. Is this an exact solution?

(b) The end loads are changed to

$$\sigma_x(0, y, z) = \sigma_x(a, y, z) = a, \qquad 0 \le y \le c$$

$$\sigma_x(0, y, z) = \sigma_x(a, y, z) = 0, \qquad -c \le y \le 0$$

Sketch this loading; then find the stresses in the plate. Is this an exact solution? Explain.

8. Show that $u = Cyz$ (C is a constant) is the warping function for a torsional member with an elliptical cross section. Find the torsional constant J. Note that the equation of an ellipse is

$$y^2 + \frac{a^2}{b^2} z^2 = a^2$$

where a and b are the semiaxes.

9. A solid steel shaft 2 meters long with an elliptical cross section is to be used to transmit torque. The shaft may not twist more than $10°$ under a torque of $1000 \text{ N} \cdot \text{m}$. If the major and minor axes of the ellipse are held in the ratio $a/b = 2.0$, what is the smallest cross section that can be used? $E = 2.068 \cdot 10^5$ MPa and $v = 0.3$.

10. Consider a shaft with a triangular cross section oriented with respect to the axes, as shown.

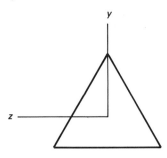

(a) Show that the following is a permissible stress function:

$$\Theta = C[\sqrt{3}\, a(y^2 + z^2) - 2y^3 + 6yz^2 + A]$$

where A, C, and a are constants. Evaluate C in terms of G and υ.

(b) Show that the given stress function is for an equilateral triangle with side a. Find A so that Θ satisfies the boundary conditions.

(c) Using the preceding results, evaluate the stress components and the torsional constant J.

11. Find the torsional constant J and and τ_{max} for the cross section shown. Dimensions are in millimeters and the material is steel.

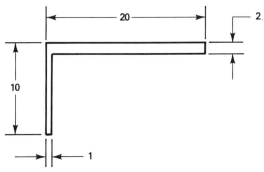

12. Compare two shafts in pure torsion with the same applied torque. One has a closed, hollow, circular cross section as shown in figure (a) and the other has an open cross section with the same geometry and dimensions, except for a slit running the full length of the shaft, as shown in figure (b). Compare the maximum shear stress τ_{max} and rate of twist θ for the two cases.

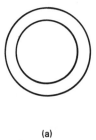

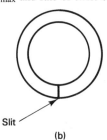

Slit

(a)   (b)

CLASSICAL STRUCTURAL MECHANICS

Slender Bars and Thin Plates

2

2.1 INTRODUCTION

Among the more widely used structures or components of more complex structures are *slender bars* and *thin plates*. A slender bar is a structure in which one dimension is much larger than the other two. These members are often classified according to the kinds of forces that they transmit. If only a tensile axial force is present, they are sometimes called *rods*, but if the axial force is compressive, they are called *columns*. If bending moments and transverse forces are present, they are called *beams,* and when pure torques are applied, they are often called *shafts*. In general, we shall call the member a bar or a beam and mean any of the above and will assume that the restriction to slender bars is implied.

The study of bars grew independently of the theory of elasticity. Based on empirical evidence, certain equations were derived that effectively model the bar's behavior. These equations and their solutions make up the principal topics in most elementary mechanics of materials texts, such as References 1, 2, and 3. In this chapter, we shall revisit some of this material, put it on a firmer theoretical foundation by tying it to the equations of the theory of elasticity in Chapter 1, and extend and generalize some of the methods and conclusions. Among the extensions will be that to bars with arbitrary cross sections and then to bars with a special set of cross sections called *thinwalled*. References that contribute understanding to this subject matter include References 4, 5, and 9.

Next in importance are those structures for which one dimension, usually called the thickness, is small when compared to the other two. This is commonly called the subject of *plates* and *shells*. Generally, plates are flat and shells are curved, although the term *curved plates* is sometimes used when the curvature is modest. This is a vast subject in itself, and so we shall give only a brief introduction to thin, flat plates in this text. References 10 and 11 go much deeper into flat plate theory.

2.2 GENERAL THEORY OF SLENDER BARS

The equations that model the behavior of bars are simplified versions of the equations of elasticity. These are permitted by the nature of the geometry and the loading of the structures. As before, the principal unknowns are stress, strain, and displacement, while the knowns are the geometry, material properties, applied loads, and geometric constraints. What is different are a few simplifying assumptions that permit a more tractable set of governing equations and attendant boundary conditions.

Several things we observed in Section 1.3 converge to help us develop a general theory for slender bars. First, we see that we can deal with equivalent concentrated loads rather than surface tractions in dealing with applied loads on the ends of the bar in determining the stress distributions in most of the bar. Next, if we look carefully at the deflections, we see that in the interior of the bar, the deflection u is uniform across the cross section and varies linearly in the x-direction for axial loads and is uniform in the x-direction and varies linearly in the direction normal to the x-direction for pure moments. This provides a theoretical justification for the common assumption we have studied in elementary mechanics of materials that *plane sections remain plane*. This assumption is also justified from empirical evidence.

The symbolic representation of the quantities and sets of equations used to form a mathematical model for slender bars is given in Figure 2.2.1. In this figure, the column matrices represent the following:

$\{f\}$ distributed applied loads (including body forces)
$\{S\}$ stress resultants
$\{\sigma\}$ stress components
$\{\epsilon\}$ strain components
$\{u\}$ displacement components
$\{u_c\}$ constrained displacements

These quantities are joined together to form the following matrix equations:

$$\{f\} = [E]\{S\}, \qquad \text{equilibrium equations}$$

$$\{S\} = [E_s]\{\sigma\}, \qquad \text{definition of stress resultant}$$

$$\{\sigma\} = [G]\{\epsilon\}, \qquad \text{stress–strain equations} \qquad (2.2.1)$$

$$\{\epsilon\} = [D]\{u\}, \qquad \text{strain–displacement equations}$$

$$\{u\} = [H]\{u_c\} \qquad \text{constrained displacement equations}$$

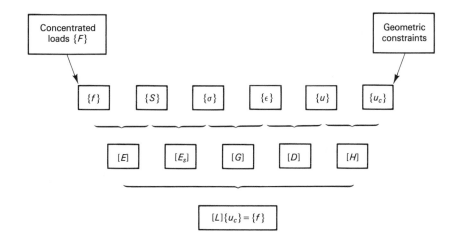

Figure 2.2.1 Symbolic Representation of the Equations for Slender Bars.

Finally, we represent the form of the equations found by combining the preceding set of equations to obtain a relationship between the distributed applied loads and the constrained displacements shown explicitly as

$$[L]\,\{u_c\} = \{f\} \tag{2.2.2}$$

To these equations we can apply directly the boundary conditions in terms of geometric constraints and applied concentrated loads.

The several new terms introduced in this representation are derived and explained in the following paragraphs. Among the items to look for are (1) how the assumption on displacements, called *constrained displacements*, leads to simplified equations, (2) how equilibrium is imposed in terms of stress resultants, (3) how body forces and applied surface forces are expressed in terms of distributed line loads, (4) how equivalent concentrated forces are defined and imposed in the boundary conditions, and (5) how constraints are imposed on the constrained displacements. Taken together with Hooke's law, the strain–displacement equations, and other features from the theory of elasticity, this forms a general theory of slender bars.

In beam theory, we make much use of *equivalent concentrated loads* and *stress resultants* to represent distributed applied loads and distributed internal stresses, respectively. We shall use the notation F_x, F_y, and F_z to designate equivalent concentrated force components in the x-, y-, and z-directions, respectively, and T_x, T_y, and T_z to designate equivalent concentrated moment components about the x-, y-, and z-axes, respectively. Generally, the x-axis lies along the length of the beam, and the location of concentrated loads will be given by an x-coordinate position in parentheses; for example, $F_x(5)$ means a force F_x located at $x = 5$.

Similarly, we shall use the notation S_x, S_y, and S_z to represent the stress resultant force components in the x-, y-, and z-directions, respectively, and M_x, M_y, and M_z to designate stress resultant moment components about the x-, y-, and z-axes, respectively.

For beams with length aligned along the x-axis, the stress resultants are functions of the x-coordinate, and their location will be expressed in functional form. They are defined as follows:

$$S_x(x) = \int \sigma_x(x, y, z)\, dA, \qquad S_y(x) = \int \tau_{xy}(x, y, z)\, dA$$

$$S_z(x) = \int \tau_{zx}(x, y, z)\, dA$$

$$M_y(x) = -\int \sigma_x(x, y, z)\, z\,dA, \qquad M_z(x) = -\int \sigma_x(x, y, z)\, y\,dA \qquad (2.2.3)$$

$$M_x(x) = \int (\tau_{zx}y - \tau_{xy}z)\, dA$$

where A is the cross-sectional area in the y–z plane. Although this form is rarely used, the stress resultants can be put in the form of matrix equations, $\{S\} = [E_s]\{\sigma\}$, in Equations 2.2.1, as

$$\begin{bmatrix} S_x \\ S_y \\ S_z \\ M_x \\ M_y \\ M_z \end{bmatrix} = \begin{bmatrix} \int & 0 & 0 & 0 & 0 & 0 \\ 0 & 0 & 0 & \int & 0 & 0 \\ 0 & 0 & 0 & 0 & 0 & \int \\ 0 & 0 & 0 & -\int z & 0 & \int y \\ -\int z & 0 & 0 & 0 & 0 & 0 \\ -\int y & 0 & 0 & 0 & 0 & 0 \end{bmatrix} \begin{bmatrix} \sigma_x \\ \sigma_y \\ \sigma_z \\ \tau_{xy} \\ \tau_{yz} \\ \tau_{zx} \end{bmatrix} dA \qquad (2.2.4)$$

The sign convention for equivalent concentrated loads and stress resultants needs special attention because these quantities appear to be very similar yet their sign conventions differ. The convention for all concentrated applied forces is that positive force components act in the positive coordinate directions. The positive components of concentrated applied moments act about the axes according to the right-hand rule. The components of stress resultant forces have the same sign convention as stresses; that is, they act in the positive coordinate direction if the surface on which they act has a normal in the positive coordinate direction, and they act in the negative coordinate direction if the surface on which they act has a normal in the negative coordinate direction. The widely accepted convention for positive components of the stress resultant moments is that a positive normal stress at a positive coordinate location on the cross section will contribute a negative moment. If this is confusing now, it may clear up when examples are given.

We shall also make use of distributed loads along the length of the bar, denoted by f_x, f_y, and f_z, which have the units of force per unit length and are applied in the x-, y-, and z-directions, respectively, and one distributed moment per unit length about the axis along the length of the beam, usually the x-axis, denoted by t_x. Distributed moments about the other two axes are so rare in nature that we do not bother to define them. All the preceding distributed loads are functions of the coordinate along the length of the beam. The lateral loads are positive in the positive coordinate directions, and the moment is positive by the right-hand rule. These distributed loads are formed from both the body forces and surface forces applied to the bar. They will be explained and discussed further at appropriate places in the text.

The justification for using the same symbols for distributed line loads in bar theory as for body forces in the theory of elasticity is based, in part, on the fact that there are just so many convenient symbols and not enough to go around and, in part, on the fact that in beam theory body forces (units of force per unit volume) are converted to line loads (units of force per unit length) by being multiplied by the cross-sectional area of the bar and, therefore, never appear explicitly in their original form. These line loads are combined with the line loads resulting from surfaces forces; thus, the distributed line loads include the body forces, and separate symbols representing body forces and line loads are not needed.

2.2.1 Slender Bars with Axial Loads

In Section 1.3.2, we dealt with a long, narrow strip of a thin plate in plane stress with a purely axial and symmetrical end loading, and with the help of St. Venant's principle we found a uniform stress in the interior regardless of the distribution of the end loading. This is where we first introduced the idea of equivalent concentrated loads and stress resultants. Now let us generalize this to a slender bar of constant but arbitrary cross section, as shown in Figure 2.2.2.

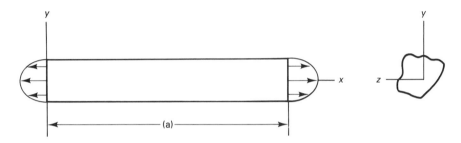

Figure 2.2.2   Slender Bar with Arbitrary Cross Section.

Let us suppose that equal but opposite distributed loads are applied to the ends and that the only nonzero components of the equivalent concentrated loads are the x-components of force. Furthermore, since the stresses on the end surfaces must be equal to the distributed applied load at the end surface, we can say

$$F_x(0) = -\int \sigma_x(0, y, z)\, dA, \qquad F_x(a) = \int \sigma_x(a, y, z)\, dA \qquad (2.2.5)$$

As noted previously, all other components of the equivalent concentrated end loads are zero. From equilibrium, we may say

$$S_x(x) = F_x(a) = F \qquad (2.2.6)$$

and all the other stress resultants are zero. From St. Venant's principle, we may assume that

$$\sigma_x = \frac{S_x(x)}{A} = \frac{F}{A} \qquad (2.2.7)$$

where A is the cross-sectional area. Since there are no other applied loads, it is assumed that there is no opportunity for any other stresses to develop in the bar. Such assumptions are supported by the study in Chapter 1 and are verified by experiment. To complete this problem, from the strain–displacement and stress–strain equations we note that

$$\epsilon_x = \frac{\partial u_0}{\partial x} = \frac{\sigma_x}{E} = \frac{S_x(x)}{EA} \qquad (2.2.8)$$

or

$$EA\frac{\partial u_0}{\partial x} = S_x(x) \qquad (2.2.9)$$

where it is recognized that $u(x, y, z)$ is reduced to $u_0(x)$. For the simple loading shown, this may be integrated to obtain

$$EAu_0(x) = Fx + A_1 \qquad (2.2.10)$$

Thus, one geometric constraint is needed to let us evaluate the constant of integration. For example, if $u_0(0) = 0$, a condition that fixes the left end of the bar and prevents rigid body motion, then $A_1 = 0$.

Actual applied loads on slender bars are usually much more complicated than the simple end loads acting only in the x-direction that we have studied so far in this chapter; for example, there can be applied axial loads distributed along the length of the bar as shown in Figure 2.2.3.

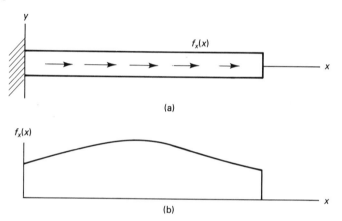

Figure 2.2.3 Axial Loading on a Bar.

These loads are shown as forces per unit length along the bar. They might arise physically, for example, as applied tractions on the outer surface of the bar. These tractions, which have units of force per unit area, can be converted to loads, which have units of force per unit length, by integration. The resulting load then is a function only of the coordinate x, and an example distribution of that load is shown in Figure 2.2.3(b). There can also be an applied equivalent concentrated end load on the bar or concentrated loads at points along the bar, but these are not shown in the figure.

To develop equations that model the behavior of this bar, we consider the equilibrium of an element as shown in Figure 2.2.4. Using a Taylor's series expansion and neglecting the higher terms, we have

$$S_x(x + dx) = S_x(x) + \frac{dS_x}{dx} dx \qquad (2.2.11)$$

From a summation of forces in the x-direction, we have

$$S_x(x) + \frac{dS_x}{dx} dx - S_x(x) + f_x(x) dx = 0 \qquad (2.2.12)$$

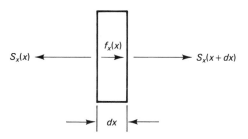

Figure 2.2.4 Equilibrium of an Axially Loaded Bar Element.

which reduces to

$$\frac{dS_x}{dx} = -f_x(x) \tag{2.2.13}$$

This is the form of the equilibrium equation in Equation 2.2.1 appropriate to this simple case.

We now take Equation 2.2.9, which, strictly speaking, has been shown to be correct for a uniform bar with a concentrated end load and, assume that it holds for a distributed axial loading and for a changing cross-sectional area. After differentiating Equation 2.2.9 once with respect to x and combining the result with Equation 2.2.13, we obtain

$$(EAu_0')' = S_x' = -f_x(x) \tag{2.2.14}$$

where primes denote differentiation with respect to x. From this equation, we may obtain the deflection of the bar by direct integration. This is the form of Equation 2.2.2 applicable to axially loaded bars. Empirical evidence verifies that this equation can be used when the beam has a modest change in cross-sectional area along its length, that is, when $A(x)$. Of course, when EA is constant, the equation becomes

$$EAu_0'' = -f_x(x) \tag{2.2.15}$$

Both Equations 2.2.14 and 2.2.15 are second-order equations that will generate two constants when integrated; therefore, two boundary conditions must be specified to complete the statement of the problem. How these equations may be used to obtain the stress and deflection in an actual case is shown in the next example.

Example 2.2.1

If the bar shown in Figure 2.2.3 has a constant cross section, a length a, and is loaded by a uniform distributed axial load, f_0, and a concentrated load, F, at $x = a$, find the stress in and the deflection of the bar.

Solution: The governing equation for the deflection of a uniform bar is

$$EAu_0'' = -f_0 \tag{a}$$

Integrating twice, we have

$$EAu_0(x) = -\frac{f_0 x^2}{2} + A_1 x + A_2 \tag{b}$$

We need two boundary conditions to evaluate the two constants of integration. One of these is readily seen to be the restraint at the left end, from which we obtain

$$u_0(0) = \frac{A_2}{EA} = 0 \longrightarrow A_2 = 0 \tag{c}$$

The other boundary condition can be found at the right end. There we have an applied end load, or, from Equation 2.2.6,

$$S_x(a) = EAu_0'(a) = -f_0a + A_1 = F \tag{d}$$

or

$$A_1 = F + f_0a \tag{e}$$

from which we have

$$u_0(x) = \frac{1}{EA}\left[-\frac{f_0x^2}{2} + (F + f_0a)\,x\right] \tag{f}$$

Finally, the stress may be obtained from

$$\sigma_x(x) = \frac{S_x(x)}{A} = \frac{EAu_0'(x)}{A} = \frac{f_0(a - x) + F}{A} \tag{g}$$

Note that all other stresses and deflections are zero.

2.2.2 Slender Bars with Torsional Loads

Next we consider a slender bar with a purely torsional loading. Once again we turn to Chapter 1, this time to Section 1.3.3, for guidance. There we discovered that the deflection of a long, slender bar with torsional end load, T, at $x = a$ is given in terms of an angular displacement, β, by

$$\beta(x) = \theta x = \frac{Tx}{GJ} \tag{2.2.16}$$

where the constraint $\beta(0) = 0$ is implied. We may recognize from moment equilibrium about the x-axis that for this case

$$M_x(x) = T \tag{2.2.17}$$

and after differentiating and rearranging, that

$$GJ\beta' = M_x(x) \tag{2.2.18}$$

We wish to generalize this to the case of distributed applied moment and other boundary conditions.

Consider an element of a bar with a distributed applied torsional load as shown in Figure 2.2.5(a) and an element in equilibrium in Figure 2.2.5(b).

Using a Taylor's series expansion and neglecting higher-order terms, we have, for the torsional stress resultant,

$$M_x(x + dx) = M_x(x) + \frac{dM_x}{dx}\,dx \tag{2.2.19}$$

From a summation of moments about the x-axis, we have

$$M_x(x) + \frac{dM_x}{dx}\,dx - M_x(x) + t_x(x)\,dx = 0 \tag{2.2.20}$$

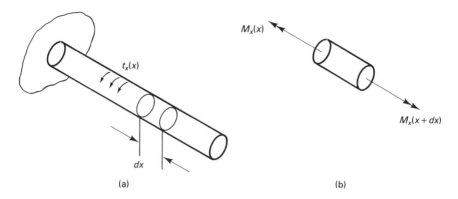

Figure 2.2.5 Bar with Distributed Torsional Load.

which reduces to

$$\frac{dM_x}{dx} = -t_x(x) \tag{2.2.21}$$

If we now assume that Equation 2.2.18, which was derived for the case of constant stress resultant and constant cross section, still applies for the case of variable stress resultant and cross section, we can differentiate it once and combine the result with Equation 2.2.21 to obtain

$$(GJ\beta')' = M_x' = -t_x(x) \tag{2.2.22}$$

When GJ is a constant, we have

$$GJ\beta'' = -t_x(x) \tag{2.2.23}$$

Boundary conditions are imposed at both ends in terms of constraint or applied torque. Mathematically, this equation and its boundary conditions are identical to the case of distributed axial load as described in Section 2.2.1. The stress in the cross section, however, is not a simple matter, as we have seen in Chapter 1, but may be found for certain cross sections by the methods described there. We will add to the available solutions for stress in torsion later in this chapter.

2.2.3 Slender Bars in Bending and Shear

Next we are inspired by the plane stress analysis of the long, narrow strip with an applied moment discussed in Section 1.3.2. Let us generalize this for arbitrary cross section. We start by assuming the end load distribution is such that there are equivalent concentrated moment components, but no net transverse, axial, or torsional loads. Using the fact that the distributed end loads are equal to the stresses at the ends, we may say that the only nonzero equivalent concentrated loads, defined by the right-hand rule, are

$$\begin{aligned} T_y(0) &= -\int \sigma_x(0, y, z)z \, dA, & T_y(a) &= \int \sigma_x(a, y, z)z \, dA \\ T_z(0) &= \int \sigma_x(0, y, z)y \, dA & T_z(a) &= -\int \sigma_x(a, y, z)y \, dA \end{aligned} \tag{2.2.24}$$

Let us assume further that the beam is constrained from rigid body motion, say, by restraints at the ends. Exactly how it must be restrained will become clear shortly. In any case, we can imagine that under the given load it will deform. In Figure 2.2.6, we show a portion of the beam in both its deformed and undeformed position, greatly exaggerated in amplitude, as we would observe it when looking along the z-axis at its projection on the x-y plane.

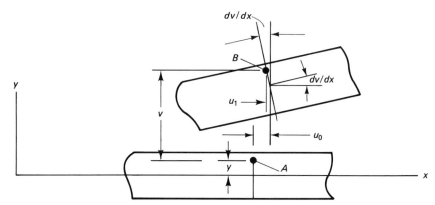

Figure 2.2.6 Beam Deformation in Bending.

We are particularly interested in the movement of a point in the body from its undeformed position A to its final deformed position B. The point at A lies a distance x along the x-axis and a distance y above it. After deformation, the same physical point, now at B, may lie a greater distance along and above the x-axis, depending on the deformation. In the theory of beams, we take advantage of the fact that the relative deformation of points in the cross section is necessarily small, because the maximum cross section dimensions are small, and assume that $v(x)$ only. Compare this assumed form with the actual deformations for the special cases in Examples 1.3.1 and 1.3.2. In other words, in beam theory, the Poisson's ratio effect is ignored. As a result, every point on the cross section deforms the same amount in the y-direction.

Since plane cross sections are assumed to remain plane, we have for small deformations a translation along the x-axis, u_0, of the whole cross section and another deformation in the x-direction, u_1, due to rotation about the z-axis, where

$$u_1 = -y \frac{dv}{dx} \tag{2.2.25}$$

This may be seen in Figure 2.2.6; however, we are observing only part of the rotation of the cross section. Another deformation in the x-direction, u_2, occurs from rotation about the y-axis, where

$$u_2 = -z \frac{dw}{dx} \tag{2.2.26}$$

and it is assumed also that $w(x)$ only. To be consistent, we should add subscripts and denote $v_0(x)$ and $w_0(x)$, but this is not usually done. Thus, the total x-direction deflection of a point in the beam is

$$u(x, y, z) = u_0(x) - y\frac{dv}{dx} - z\frac{dw}{dx} = u_0 - yv' - zw' \qquad (2.2.27)$$

where primes are used to denote differentiation with respect to x. This is the deformation pattern referred to as *plane sections remain plane*. This is the basis for the definition of the constrained displacement matrix, $\{u\} = [H]\{u_c\}$ in Equations 2.2.1. Its components for this case are

$$\begin{bmatrix} u \\ v \\ w \end{bmatrix} = \begin{bmatrix} 1 & -y\frac{d}{dx} & -z\frac{d}{dx} \\ 0 & 1 & 0 \\ 0 & 0 & 1 \end{bmatrix} \begin{bmatrix} u_0(x) \\ v(x) \\ w(x) \end{bmatrix} \qquad (2.2.28)$$

When the preceding expression for the displacements is inserted in the strain–displacement equations, $\{\epsilon\} = [D]\{u\}$ in Equations 2.2.1, the only strain associated with the preceding displacement is

$$\epsilon_x = \frac{\partial u}{\partial x} = u_0' - yv'' - zw'' \qquad (2.2.29)$$

All other strain components are zero.

The term constrained displacement is not common in the literature. It has been adopted here to provide a convenient distinction between the true displacements of the theory of elasticity and the restricted displacements of beam and plate theory. Note that in the preceding notation we use u_0 to denote the x-displacement for $y = z = 0$ and u to denote x-displacement for any value of x, y, and z. Also, v and w denote y- and z-displacements for any value of x, y, and z; however, $v(x, y, z) = v(x)$ and $w(x, y, z) = w(x)$ in beam theory.

Since no surface forces are acting in the y- and z-directions, it is reasonable to assume that there are no y- or z-components of stress, nor are there any shear stresses; therefore, from Hooke's law, $\{\sigma\} = [G]\{\epsilon\}$ in Equations 2.2.1, the one stress component is

$$\sigma_x = E\epsilon_x = E(u_0' - yv'' - zw'') \qquad (2.2.30)$$

The stress resultant for the axial force is

$$S_x(x) = \int \sigma_x(x, y, z) \, dA = E[u_0' \int dA - v'' \int y \, dA - w'' \int z \, dA] \qquad (2.2.31)$$

This can be greatly simplified by carefully choosing the x-axis to pass through the centroid of the cross-sectional area. If the x-axis is centroidal, then

$$\int y \, dA = 0, \qquad \int z \, dA = 0 \qquad (2.2.32)$$

and

$$S_x(x) = EAu_0' \qquad (2.2.33)$$

This may be seen to be the same result obtained in Equation 2.2.9. Note that the axial displacement depends only on the axial stress resultant force. We say that the equation for u_0 is uncoupled from those for v and w. Had we included torsional effects, we would have concluded that Equation 2.2.18 holds in this case and is also uncoupled from v, w, and u_0.

The stress resultants for the moments are

$$M_y(x) = -\int \sigma_x(x, y, z)z \, dA = E[-u_0' \int z \, dA + v'' \int yz \, dA + w'' \int z^2 \, dA]$$

$$= E[v'' I_{yz} + w'' I_{yy}]$$

$$M_z(x) = -\int \sigma_x(x, y, z)y \, dA = E[-u_0' \int y \, dA + v'' \int y^2 \, dA + w'' \int zy \, dA] \qquad (2.2.34)$$

$$= E[v'' I_{zz} + w'' I_{yz}]$$

where the appropriate integrals have been identified as the area moments of inertia and the product of inertia. We note that the lateral displacements $v(x)$ and $w(x)$ depend only on the stress resultant moments and are independent of the axial load and torsional load. In any case, in the absence of an axial applied load, $S_x(x) = 0$, and therefore $u_0(x) = 0$; and in the absence of an applied torque, $M_x(x) = 0$, and therefore $\beta(x) = 0$.

It is useful at this point to consider Equations 2.2.34 as two equations in terms of two unknowns and solve for v'' and w'' in terms of the moments. If we do this, we obtain

$$v''(x) = \frac{1}{E} \frac{I_{yy}M_z - I_{yz}M_y}{I_{yy}I_{zz} - I_{yz}^2}, \qquad w''(x) = -\frac{1}{E} \frac{I_{yz}M_z - I_{zz}M_y}{I_{yy}I_{zz} - I_{yz}^2} \qquad (2.2.35)$$

and the stress is found by substituting Equations 2.2.35 into Equation 2.2.30, from which we obtain

$$\sigma_x = -\left(\frac{I_{yy}M_z - I_{yz}M_y}{I_{yy}I_{zz} - I_{yz}^2}\right)y + \left(\frac{I_{yz}M_z - I_{zz}M_y}{I_{yy}I_{zz} - I_{yz}^2}\right)z + \frac{S_x}{A} \qquad (2.2.36)$$

In the special case where the product of inertia is zero, that is,

$$I_{yz} = 0 \qquad (2.2.37)$$

the preceding equations simplify to

$$v''(x) = \frac{M_z}{EI_{zz}}, \qquad w''(x) = \frac{M_y}{EI_{yy}}$$

$$\sigma_x(x, y, z) = -\frac{M_z y}{I_{zz}} - \frac{M_y z}{I_{yy}} + \frac{S_x}{A} \qquad (2.2.38)$$

These are the familiar equations of elementary beam theory, which are studied in elementary mechanics of materials.

To solve Equations 2.2.35 for the deflections and Equations 2.2.36 for the stress or, alternatively, when the product of inertia is zero, Equations 2.2.38 for the deflections and stress, we must find the values of the internal stress resultants in terms of the applied loads with the aid of equilibrium. Once we have commited ourselves to the displacement pattern of Equation 2.2.28, in general, we cannot hope to satisfy the equations of equilibrium as presented in Chapter 1. We must be content to satisfy equilibrium in an approximate way in terms of the stress resultants. That this turns out to be an accurate solution to many practical problems is one of the most fortunate occurrences in structural analysis. How this is done is shown in the next few paragraphs.

In Figure 2.2.7, we have a differential element of the beam showing the internal stress resultants acting on that element in the x-y plane.

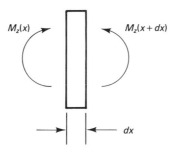

$M_z(x)$ $M_z(x + dx)$

Figure 2.2.7 Equilibrium of a Beam Element in Pure Bending.

Using the Taylor's series expansion, we recognize that

$$M_z(x + dx) = M_z(x) + \frac{dM_z}{dx}\,dx \tag{2.2.39}$$

Similarly, although not pictured, about the y-axis, we have

$$M_y(x + dx) = M_y(x) + \frac{dM_y}{dx}\,dx \tag{2.2.40}$$

Then, summing moments about the z-axis and about the y-axis, we have

$$\frac{dM_y}{dx} = 0, \qquad \frac{dM_z}{dx} = 0 \tag{2.2.41}$$

This is the form the equilibrium equations, $\{f\} = [E]\{S\}$ in Equations 2.2.1, take in this case. The applied loads are all concentrated loads, which are introduced via the boundary conditions. This can best be seen by example.

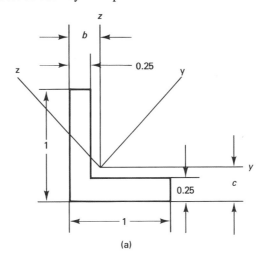

(a)

Example 2.2.2

A cantilever beam, that is, a beam fixed in a wall at one end and free at the other end, has a length a and the cross section shown in figure (a). All dimensions are in the same units whether English or metric. The following are the geometric constraints for a cantilever beam:

$$v(0) = v'(0) = w(0) = w'(0) = 0 \tag{a}$$

The applied loads on the ends of the beam are such that

$$T_y(a) = M \tag{b}$$

and all other equivalent concentrated loads are zero. Find the stresses and the deflections of the beam.

Solution: First, we may note that in the absence of axial loads $u_0(x) = 0$, and in the absence of torsional loads, $\beta(x) = 0$. Consequently, the axial and torsional stresses are also zero. These quantities may be found from the uncoupled equations for axial and torsional deflections and stresses given in Sections 2.2.1 and 2.2.2.

To find the bending stresses and deflections, we must find the centroid of the cross section and then the moments and product of inertia with respect to that centroid. If necessary, we can refer back to an elementary mechanics book or an engineering handbook for methods of calculating these quantities. In the figure, the location of the centroid is given by b and c. Their values to three significant figures are

$$b = c = 0.339 \tag{c}$$

The centroidal moments and product of inertia with respect to the y-z axes are

$$I_{yy} = I_{zz} = 0.0369, \qquad I_{yz} = -0.0201 \tag{d}$$

From Equation 2.2.36, the stress in the beam is

$$\sigma_x = \frac{0.0201M}{(0.0369)^2 - (0.0201)^2} y + \frac{0.0369M}{(0.0369)^2 - (0.0201)^2} z$$
$$= 21.0My + 38.5Mz \tag{e}$$

All other components of stress are zero. The deflections are found from Equations 2.2.35, which for this case are

$$v''(x) = \frac{1}{E} \frac{-0.0201M}{(0.0369)^2 - (0.0201)^2} = -\frac{21.0M}{E}$$

$$w''(x) = \frac{1}{E} \frac{-0.0369M}{(0.0369)^2 - (0.0201)^2} = -\frac{38.5M}{E} \tag{f}$$

Integrating twice, we get

$$v(x) = -\frac{21.0Mx^2}{2E} + B_1 x + B_2, \qquad w(x) = -\frac{38.5Mx^2}{2E} + C_1 x + C_2 \tag{g}$$

and from the four boundary conditions, we find the four constants of integration:

$$v(0) = B_2 = 0, \qquad v'(0) = B_1 = 0$$
$$w(0) = C_2 = 0 \qquad w'(0) = C_1 = 0 \tag{h}$$

Thus, the deflections are

$$v(x) = -\frac{21.0Mx^2}{2E}, \qquad w(x) = -\frac{38.5Mx^2}{2E} \tag{i}$$

Note that a consequence of a nonzero product of inertia is that a moment in the x-z plane about the y-axis produces a deflection $v(x)$ out of the plane.

For an alternative way to solve this problem, find the principal axes of inertia and use Equations 2.2.38. For this case, we can see from figure (a) that the principal axes are at 45° to the original axes from the symmetry of the cross section. These axes are shown on the figure as the y-z axes. The values of the principal moments of inertia are

$$I_{yy} = 0.0570, \qquad I_{zz} = 0.0168 \tag{j}$$

We must also find the components of the applied moment with respect to the y-z axes. They are

$$M_y = -M \sin 45°, \qquad M_z = -M \cos 45° \tag{k}$$

From Equation 2.3.38, the stress is

$$\sigma_x = \frac{0.707M}{0.0168E} y + \frac{0.707M}{0.0570E} z = \frac{42.1M}{E} y + \frac{12.4M}{E} z \tag{l}$$

and the deflections are given by (where roman v and w are used to denote deflections with respect to the y-z axes)

$$v''(x) = -\frac{0.707M}{0.0168E} = -\frac{42.08M}{E}$$

$$w''(x) = -\frac{0.707M}{0.0570E} = -\frac{12.4M}{E} \tag{m}$$

After integrating and applying the boundary conditions,

$$v(x) = -\frac{21.04Mx^2}{E}, \qquad w(x) = -\frac{6.2Mx^2}{E} \tag{n}$$

The vector sum of the deflections in equations (i) and (n) are identical.

In many practical beam problems, applied loads include transverse, or lateral, forces, that is, forces normal to the x-direction on the ends of the bar and on the outer surfaces. In Figure 2.2.8, we show a general case of transverse loading divided into components in the y- and z-directions. These are shown as forces per unit length of the beam. If, for example, the original load is a force per unit area over the upper surface of the beam, we would integrate across the width of the beam to produce the force per unit

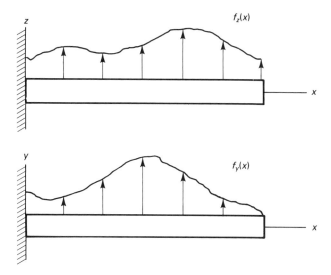

Figure 2.2.8 Two Components of Distributed Applied Loads.

length shown. The load could as easily be applied as a normal surface load on the bottom face of the beam or as surface shear loads on the sides. However the loads are applied, they are resolved into transverse loads per unit length and are assumed to act on the center line of the beam, unless otherwise specified.

The usual schematic diagram for a cantilever beam, implying the beam is built into a rigid wall, is shown in the figure. In this case there also can be applied end loads on the free end with components in all three coordinate directions, although they are not shown in the figure.

The transmission of transverse shear forces by beams is always accompanied by bending and may be accompanied by torsion as well. For purposes of this section, we shall assume that whatever the stress distribution present it produces no net twisting about the x axis; that is, there are no torsional loads. The significance of this assumption will become apparent later when we study the thin-walled beam. We shall also exclude any axial loads. We shall introduce combined loading in a later section. The relationships between the stress resultants for a beam with lateral loads are found by using the equilibrium of a small element of the beam. Such an element is shown in Figure 2.2.9.

The element in Figure 2.2.9(a) is projected on the x-z plane and the element in Figure 2.2.9(b) is in the x-y plane. Using a Taylor's series expansion and neglecting the higher terms, we have

$$S_y(x + dx) = S_y(x) + \frac{dS_y}{dx}\, dx, \qquad S_z(x + dx) = S_z(x) + \frac{dS_z}{dx}\, dx$$

$$M_y(x + dx) = M_y(x) + \frac{dM_y}{dx}\, dx, \qquad M_z(x + dx) = M_z(x) + \frac{dM_z}{dx}\, dx \tag{2.2.42}$$

From equilibrium of forces in the z-direction, we have

$$S_z(x + dx) = S_z(x) + \frac{dS_z}{dx}\, dx - S_z + f_z(x)\, dx = 0 \tag{2.2.43}$$

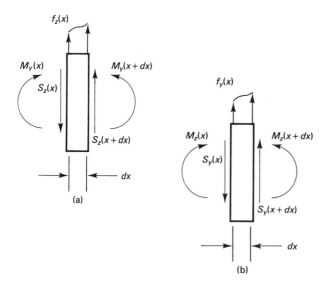

Figure 2.2.9 Equilibrium of a Beam Element with Distributed Loads.

Classical Structural Mechanics   Chap. 2

which reduces to

$$\frac{dS_z}{dx} = -f_z(x) \tag{2.2.44}$$

and, similarly, for forces in the y-direction,

$$\frac{dS_y}{dx} = -f_y(x) \tag{2.2.45}$$

From equilibrium of moments about the y-axis,

$$M_y + \frac{dM_y}{dx}\,dx - M_y + S_z\,dx + f_z\,dx\,\frac{dx}{2} = 0 \tag{2.2.46}$$

which reduces to

$$\frac{dM_y}{dx} = -S_z \tag{2.2.47}$$

and, similarly, for moments about the z-axis,

$$\frac{dM_z}{dx} = -S_y \tag{2.2.48}$$

If we differentiate Equations 2.2.47 and 2.2.48 once and combine with Equations 2.2.44 and 2.2.45, we get

$$\frac{d^2M_y}{dx^2} = -\frac{dS_z}{dx} = f_z(x), \qquad \frac{d^2M_z}{dx^2} = -\frac{dS_y}{dx} = f_y(x) \tag{2.2.49}$$

Equations 2.2.44, 2.2.45, and 2.2.49 are the equilibrium equations, $[E]\{S\} = \{f\}$ in Equations 2.2.1, for this particular case; however, it is not convenient to put them in the form of a single matrix equation. The following matrix form is more useful:

$$-\begin{bmatrix} \dfrac{d}{dx} & 0 \\ 0 & \dfrac{d}{dx} \end{bmatrix}\begin{bmatrix} S_y \\ S_z \end{bmatrix} = \begin{bmatrix} f_y \\ f_z \end{bmatrix}, \qquad \begin{bmatrix} \dfrac{d^2}{dx^2} & 0 \\ 0 & \dfrac{d^2}{dx^2} \end{bmatrix}\begin{bmatrix} M_z \\ M_y \end{bmatrix} = \begin{bmatrix} f_y \\ f_z \end{bmatrix} \tag{2.2.50}$$

The assumption that plane sections remain plane, which is seen to be valid for pure bending, at least in the interior of the beam, is not strictly true when shear forces are present in the interior of the beam. Shear stresses produce shear deformations, which cause the cross section to warp. For slender bars, however, it can be shown that the amount of warping is small and can often be neglected. It is common, then, to retain the assumption that plane sections remain plane and to use the normal stress distribution obtained for pure bending when shear stresses are present.

The consequences of the preceding assumption are very important. For one thing, it enables us to calculate the normal stresses by Equation 2.2.36 or 2.3.38 where now the moments in those equations are not the pure moments described there, but are moments that result from the transverse applied loads. These moments are, in general, functions of x. Furthermore, this new meaning of the moments applies to the deflection equations given in Equations 2.2.35 or 2.2.38. We can, in fact, combine the deflection equations with the equilibrium equations in terms of the stress resultants and come up with two very useful equations. First, differentiate Equations 2.2.35 once more with respect to x. If the beam has a constant cross section and, therefore, the moments of inertia and product of inertia are independent of x, we have

$$v'''(x) = \frac{1}{E} \frac{I_{yy}M_z' - I_{yz}M_y'}{I_{yy}I_{zz} - I_{yz}^2}, \qquad w'''(x) = -\frac{1}{E} \frac{I_{yz}M_z' - I_{zz}M_y'}{I_{yy}I_{zz} - I_{yz}^2} \qquad (2.2.51)$$

From Equations 2.2.47 and 2.2.48, these may be written as

$$v'''(x) = -\frac{1}{E} \frac{I_{yy}S_y - I_{yz}S_z}{I_{yy}I_{zz} - I_{yz}^2}, \qquad w'''(x) = \frac{1}{E} \frac{I_{yz}S_y - I_{zz}S_z}{I_{yy}I_{zz} - I_{yz}^2} \qquad (2.2.52)$$

Now differentiate with respect to x once more:

$$v''''(x) = -\frac{1}{E} \frac{I_{yy}S_y' - I_{yz}S_z'}{I_{yy}I_{zz} - I_{yz}^2}, \qquad w''''(x) = \frac{1}{E} \frac{I_{yz}S_y' - I_{zz}S_z'}{I_{yy}I_{zz} - I_{yz}^2} \qquad (2.2.53)$$

From Equations 2.2.44 and 2.2.45, these may be written as

$$v''''(x) = \frac{1}{E} \frac{I_{yy}f_y - I_{yz}f_z}{I_{yy}I_{zz} - I_{yz}^2}, \qquad w''''(x) = -\frac{1}{E} \frac{I_{yz}f_y - I_{zz}f_z}{I_{yy}I_{zz} - I_{yz}^2} \qquad (2.2.54)$$

This is the form of the equations given symbolically by Equation 2.2.2 for this case. In a given problem, all quantities to the right of the equal sign in Equation 2.2.54 are known constants or functions of x. To obtain the deflection, we must integrate these equations four times. For each equation, we get four constants of integration. To find the value of these constants, we must have four boundary conditions for each equation. Boundary conditions are discussed in greater detail in the next section.

Now let us simplify the equations for the case where the product of inertia of the cross section is zero. Equations 2.2.52, after some rearrangement of terms, reduce to

$$EI_{zz}v''' = -S_y(x), \qquad EI_{yy}w''' = -S_z(x) \qquad (2.2.55)$$

and Equations 2.2.54 reduce to

$$EI_{zz}v'''' = f_y(x), \qquad EI_{yy}w'''' = f_z(x) \qquad (2.2.56)$$

Since it is always possible to find the principal axes of inertia and adopt a coordinate system for which the product of inertia is zero, these, rather than Equations 2.2.44, are the equations that we will work with in most examples.

Before going on, let us return to Equations 2.2.35 and remove the requirement of a constant cross section. If the beam cross section varies with x, the moments and product of inertia become functions of x. If that variation is modest, these equations are still a reasonable model of the behavior of the actual system. In such a case the differentiation with respect to x must be done accordingly. We shall not carry this out for the general case, but only for when the product of inertia is zero. Then Equations 2.2.55 are replaced by

$$(EI_{zz}v'')' = -S_y(x), \qquad (EI_{yy}w'')' = -S_z(x) \qquad (2.2.57)$$

and Equations 2.2.56 are replaced by

$$(EI_{zz}v'')'' = f_y(x), \qquad (EI_{yy}w'')'' = f_z(x) \qquad (2.2.58)$$

Note that the lateral distributed forces will produce normal stresses in the y- and z-directions on the surface and in the interior of the beam; however, they are ignored in beam theory because they will always be small compared to the normal stresses generated in the x-direction by these very same lateral loads when the bar is slender.

It is time for another example. In this and most subsequent examples, we shall assume that the principal axes of inertia will be used. For any beam, the cross section axes can always be given the appropriate orientation.

Example 2.2.3

The cantilever beam in Example 2.2.2 has a uniform distributed load, f_0, applied in the z-direction. Find the deflection and stress of the beam.

Solution: When the load is referred to the principal axes,

$$f_y = \frac{\sqrt{2}}{2} f_0, \qquad f_z = \frac{\sqrt{2}}{2} f_0 \tag{a}$$

The equations to be solved are

$$EI_{zz} v'''' = \frac{\sqrt{2}}{2} f_0, \qquad EI_{yy} w'''' = \frac{\sqrt{2}}{2} f_0 \tag{b}$$

Integrating four times,

$$v(x) = \frac{\sqrt{2}}{48} \frac{f_0 x^4}{EI_{zz}} + \frac{B_1 x^3}{6} + \frac{B_2 x^2}{2} + B_3 x + B_4$$

$$w(x) = \frac{\sqrt{2}}{48} \frac{f_0 x^4}{EI_{yy}} + \frac{C_1 x^3}{6} + \frac{C_2 x^2}{2} + C_3 x + C_4 \tag{c}$$

The boundary conditions are

$$v(0) = 0 \qquad\qquad v'(0) = 0$$
$$EI_{zz} v''(a) = T_z(a) = 0, \qquad EI_{zz} v'''(a) = F_y(a) = 0$$

$$w(0) = 0 \qquad\qquad w'(0) = 0$$
$$EI_{yy} w''(a) = T_y(a) = 0 \qquad EI_{yy} w'''(a) = F_z(a) = 0 \tag{d}$$

Upon substitution of equations (c) into the boundary conditions for $v(x)$, we get

$$v(0) = B_4 = 0, \qquad v'(0) = B_3 = 0$$

$$v''(a) = \frac{\sqrt{2} f_0 a^2}{4EI_{zz}} + B_1 a + B_2 = 0$$

$$v'''(a) = \frac{\sqrt{2} f_0 a}{2EI_{zz}} + B_1 = 0 \tag{e}$$

from which

$$B_1 = -\frac{\sqrt{2} f_0 a}{2EI_{zz}}, \qquad B_2 = \frac{\sqrt{2} f_0 a^2}{4EI_{zz}} \tag{f}$$

and

$$v(x) = \frac{\sqrt{2} f_0}{2EI_{zz}} \left(\frac{x^4}{24} - \frac{x^3 a}{6} + \frac{x^2 a^2}{4} \right) \tag{g}$$

Similarly,

$$w(x) = \frac{\sqrt{2} f_0}{2EI_{yy}} \left(\frac{x^4}{24} - \frac{x^3 a}{6} + \frac{x^2 a^2}{4} \right) \tag{h}$$

The normal stress is found using Equation 2.2.38:

$$\sigma_x(x) = -\frac{M_z y}{I_{zz}} - \frac{M_y z}{I_{yy}} = -\frac{\sqrt{2} f_0}{2} \left(\frac{x^2}{2} - xa + \frac{a^2}{2} \right) \left(\frac{y}{I_{zz}} + \frac{z}{I_{yy}} \right) \tag{i}$$

So far we have found ways to find the deflection of a beam under rather general loading, and from these we can find the normal stresses. We cannot avoid the fact, however, that shear stresses are also present in the beam. These shear stresses are directly related to the stress resultants shown in Equation 2.2.57. Unfortunately, there are no simple general solutions for the shear stresses in beams except for particular shapes of the cross section. For rectangular cross sections, for example, we are able to find the shear stresses readily.

To see how this is done, let us look at the normal stress on a section of a beam loaded by a distributed transverse force as shown in Figure 2.2.10(a). The beam section has a height $2c$, a depth h, and a length dx.

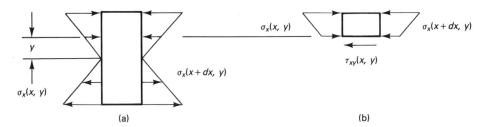

Figure 2.2.10 Element of a Beam in Bending and Shear.

In Figure 2.2.10(b), we show a portion of the section that appears above the coordinate y. The unbalanced normal force must be in equilibrium with the shear force or

$$\int \sigma_x(x + dx, y) h \, dy - \int \sigma_x(x, y) h \, dy - \tau_{xy} h \, dx = 0 \tag{2.2.59}$$

When it is noted that

$$\sigma_x(x, y) = -\frac{M_z y}{I_{zz}}$$

$$\sigma_x(x + dx, y) = -\left(M_z + \frac{dM_z}{dx} dx \right) \frac{y}{I_{zz}} = -(M_z - S_y \, dx) \frac{y}{I_{zz}} \tag{2.2.60}$$

we can obtain the following from Equation 2.2.59:

$$\tau_{xy} = \frac{S_y}{I_{zz}} \int y \, dy = \frac{1}{2} \frac{S_y}{I_{zz}} (c^2 - y^2) \tag{2.2.61}$$

Since the shear stress at a point, in this case the point x, y, z, is the same in two perpendicular directions, the value of the shear stress given in Equation 2.2.61 is the value on a cross section perpendicular to the x-axis. A similar result is obtained for τ_{zx}. Because the shear stresses are directly related to S_y and S_z, these particular stress resultants are called *shear forces*. After some further discussion of applied loads and boundary conditions, we shall find the shear stresses in some example beams.

2.2.4 Beam Applied Loads and Boundary Conditions

In the theory of elasticity, discussed in Chapter 1, the boundary conditions at each and every point on the surface of the body are expressed either in terms of applied surface tractions or geometric constraints. In beam theory, because of the use of stress resultants in satisfying equilibrium, some of the applied surface loads appear in the governing equations while others appear in the boundary conditions. Distributed lateral loads, for example, appear in the governing equations. As mentioned before, in beam theory no distinction is made if these loads are applied to the upper surface, the lower surface, or both. For example, the loads shown in Figure 2.2.11(a) and (b) are identical as far as beam theory is concerned and may be made up of applied surfaces loads or body forces formulated as forces per unit length. These loads would appear to cause a surface stress, σ_y, on the upper surface in Figure 2.2.11(a) and on the lower surface in Figure 2.2.11(b). In fact, such surface stresses are ignored as small compared to the value of σ_x generated by these loads. Furthermore, the body forces, which are applied loads in terms of force per unit volume, are explicit elements of the governing equations in the theory of elasticity, but are configured as distributed applied loads in beam theory in units of force per unit length and added to any other applied loads acting on the beam.

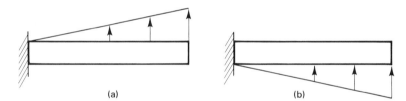

(a)   (b)

Figure 2.2.11   Equivalent Beam Lateral Loads.

Lateral loads are defined as acting in such a way that no twisting moment is generated. For simple cross sections, including all those that are symmetrical about the z- and y-axes, this means that the lateral loads act through the center line of the beam, the x-axis, which is the loci of centroids of the cross sections. Lateral loads that act off this line can be replaced by a statically equivalent load acting on this line plus a torsional moment about it. For nonsymmetrical cross sections, this line of action may not be the centroidal axis, as we shall see later in this chapter.

On the other hand, equivalent concentrated loads acting at the ends of the beam or concentrated loads at intermediate points appear in the boundary conditions, not in the governing equation. In general, boundary conditions consist of geometric constraints, applied concentrated loads, or elastic constraints. In reality, the geometric constraint equations used in the mathematical modeling of structures are idealizations of the physical constraints found in real structures.

In Figure 2.2.12, we show several beams with a variety of boundary conditions and applied loads. These figures also illustrate several of the schematic conventions for representing idealized boundary conditions. Let us assume that for the beams in Figure

2.2.12 the product of inertia is zero, the cross section is uniform, the length is a, and all loads are in the x-y plane; thus, the deflections are governed by Equations 2.2.56. We can see immediately that $w(x) = 0$. The one remaining equation to solve is

$$EI_{zz}v'''' = f_y \qquad (2.2.62)$$

Integrating this equation four times produces four constants of integration, which determines that four boundary conditions are needed to complete the statement of a problem.

In Figure 2.2.12(a), we have the usual schematic representation of a cantilever beam. Such a beam has two idealized geometric constraints at the left end, where it is said to be built into a wall, and two applied load conditions at the right end, where it is free of geometric constraint. The four boundary conditions, required by the four constants of integration when the governing equation is solved, are

$$v(0) = 0, \qquad v'(0) = 0, \qquad EIv''(a) = 0, \qquad EIv'''(a) = -F \qquad (2.2.63)$$

The first says that there is no deflection at $x = 0$; the second, that the slope is zero at $x = 0$; the third, from Equation 2.2.38, that there is no applied moment at $x = a$; and the fourth, from Equation 2.2.55, that the stress resultant in shear at $x = a$, $S_y(a)$, is equal to the applied load at $x = a$.

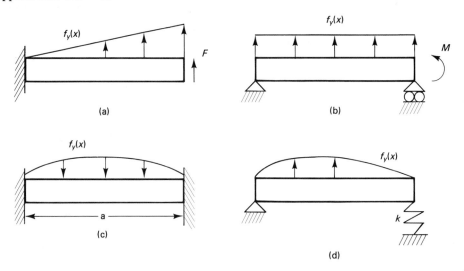

Figure 2.2.12 Some Applied Loads and Boundary Conditions.

For the beam shown in Figure 2.2.12(b), we have the standard schematic representation of the *simply supported* beam. The two triangular supports represent pinned or hinged restraint; that is, no lateral deflection is permitted at the supports, but rotation is allowed. The roller depicted at the right end support allows for axial displacement. This is not important here because the axial displacement equation is completely uncoupled from the lateral displacements. In Chapter 5 we shall take up a situation where this is not so, and the movable support will be necessary. Note that, while the pivot point is shown on the underside of the beam, it is assumed to be acting on the center line, and the point is assumed to be hinged so that the beam cannot lift off the pivot. This happens to be the schematic convention used in most publications. The governing equation is the same as for the cantilever beam, given by Equation 2.2.62, but the boundary conditions are now

$$v(0) = 0, \qquad EI_{zz}v''(0) = 0, \qquad v(a) = 0, \qquad EI_{zz}v''(a) = M \qquad (2.2.64)$$

The third example, Figure 2.2.12(c), is the *clamped* or *fixed end* beam. Its boundary conditions are all geometric constraints, or

$$v(0) = 0, \qquad v'(0) = 0, \qquad v(a) = 0, \qquad v'(a) = 0 \qquad (2.2.65)$$

Before we go on to the next case in Figure 2.2.12, let us do an example problem.

Example 2.2.4

A beam with a constant rectangular cross section is loaded and constrained as shown in figure (a). Find the deflections and stresses in the beam.

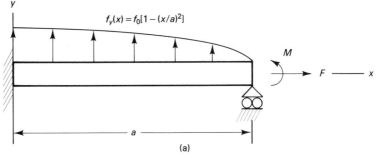

$$f_y(x) = f_0[1 - (x/a)^2]$$

(a)

Solution: First, let us solve for the axial deflection and stress, which are uncoupled from the lateral deflections and stresses. The governing equation for the axial deflection is

$$EAu_0'' = 0 \qquad (a)$$

and the boundary conditions are

$$u_0(0) = 0, \qquad EAu_0'(a) = F \qquad (b)$$

Integrating twice and applying the boundary conditions, we get

$$u_0'(x) = A_1 \longrightarrow EAu_0'(a) = EAA_1 = F \longrightarrow A_1 = \frac{F}{EA} \qquad (c)$$

$$u_0(x) = A_1x + A_2 \longrightarrow u_0(0) = A_2 = 0$$

and, therefore, the deflection is

$$u_0(x) = \frac{Fx}{EA} \qquad (d)$$

and the stress is

$$\sigma_x(x) = \frac{F}{A} \qquad (e)$$

Now we can turn to the displacements and stresses from the lateral loads. The governing equation is

$$EI_{zz}v'''' = f_0\left(1 - \frac{x^2}{a^2}\right) \qquad (f)$$

The boundary conditions are

$$v(0) = 0, \qquad v'(0) = 0, \qquad v(a) = 0, \qquad EI_{zz}v''(a) = M \qquad \text{(g)}$$

Integrating four times,

$$EI_{zz}v''' = f_0\left(x - \frac{x^3}{3a^2}\right) + B_1$$

$$EI_{zz}v'' = f_0\left(\frac{x^2}{2} - \frac{x^4}{12a^2}\right) + B_1 x + B_2$$

$$EI_{zz}v' = f_0\left(\frac{x^3}{6} - \frac{x^5}{60a^2}\right) + B_1\frac{x^2}{2} + B_2 x + B_3 \qquad \text{(h)}$$

$$EI_{zz}v = f_0\left(\frac{x^4}{24} - \frac{x^6}{360a^2}\right) + B_1\frac{x^3}{6} + B_2\frac{x^2}{2} + B_3 x + B_4$$

and we find the constants of integration from the boundary conditions:

$$EI_{zz}v(0) = B_4 = 0, \qquad EI_{zz}v'(0) = B_3 = 0$$

$$EI_{zz}v(a) = f_0\left(\frac{a^4}{24} - \frac{a^6}{360a^2}\right) + B_1\frac{a^3}{6} + B_2\frac{a^2}{2} = 0 \qquad \text{(i)}$$

$$EI_{zz}v''(a) = f_0\left(\frac{a^2}{2} - \frac{a^4}{12a^2}\right) + B_1 a + B_2 = M$$

From equations (i),

$$B_1 = \frac{3M}{2a} - \frac{183f_0 a}{360}, \qquad B_2 = -\frac{M}{2} + \frac{11f_0 a^2}{120} \qquad \text{(j)}$$

and so

$$EI_{zz}v(x) = f_0\left(\frac{x^4}{24} - \frac{x^6}{360a^2}\right) + B_1\frac{x^3}{6} + B_2\frac{x^2}{2}$$

$$= -\frac{f_0}{360a^2}x^6 + \frac{f_0}{24}x^4 + \left(\frac{M}{4a} - \frac{183f_0 a}{2160}\right)x^3 \qquad \text{(k)}$$

$$+ \left(-\frac{M}{4} + \frac{11f_0 a^2}{240}\right)x^2$$

The normal stress resulting from the lateral loads is added to that from the axial loads and is given by Equation 2.2.38. The moment used in the stress formula is the second derivative of equation (k), or

$$M_z(x) = EI_{zz}v''(x) = -\frac{f_0}{12a^2}x^4 + \frac{f_0}{2}x^2 + \left(\frac{3M}{2a} - \frac{183f_0 a}{360}\right)x$$

$$+ \left(-\frac{M}{2} + \frac{11f_0 a^2}{120}\right) \qquad \text{(l)}$$

and so the total normal stress is

$$\sigma_x = \frac{F}{A} - \left[-\frac{f_0}{12a^2}x^4 + \frac{f_0}{2}x^2 + \left(\frac{3M}{2a} - \frac{183f_0 a}{360}\right)x \right.$$

$$\left. + \left(-\frac{M}{2} + \frac{11f_0 a^2}{120}\right)\right]\frac{y}{I_{zz}} \qquad \text{(m)}$$

Since this beam has a rectangular cross section, we can also find the shear stress. It is given by Equation 2.2.61, where the shear force is given by the first of equations (h), or

$$S_y(x) = -EI_{zz}v''' = -f_0\left(x - \frac{x^3}{3a^2}\right) - \frac{3M}{2a} + \frac{183}{360}f_0a \tag{n}$$

and so the shear stress is

$$\tau_{xy} = \left[-f_0\left(x - \frac{x^3}{3a^2}\right) - \frac{3M}{2a} + \frac{183}{360}f_0a\right]\left(\frac{c^2 - y^2}{2I_{zz}}\right) \tag{o}$$

The final example, Figure 2.2.12(d), shows a beam with an *elastic, or spring,* constraint where the elastic behavior of the spring is represented by a *spring constant, k,* with units of force/length. Springs are used to represent, well, springs, or other elastic members. We can all realize that the fixed and simply supported boundary conditions are idealizations. For a cantilever beam built into a wall, for example, the wall itself is an elastic body. In many cases, the wall is so much more rigid than the beam that the idealization of no displacement and no rotation at the wall is very accurate. In other cases it may be necessary to model the wall as an elastic member. This may be done by replacing the wall with a translational and a rotational spring. By making the spring constants very stiff we approximate the idealized wall.

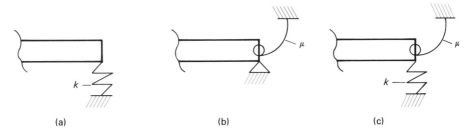

Figure 2.2.13. Boundary Conditions with Springs.

Three possible conditions of elastic support are shown schematically in Figure 2.2.13. The elastic behavior of a spring is given by the *spring constant.* For a linear lateral or *extensional spring,* the spring constant will be denoted by k, and it has the units of force per unit displacement; thus, the force in an extensional spring is equal to the spring constant times its extensional displacement. When the spring is fixed on one end and attached to a beam on the other, the extensional displacement of the spring is the lateral displacement of the beam. For the beam in Figure 2.2.13(a), the force in the spring is

$$F_e(a) = kv(a) \tag{2.2.66}$$

Similarly, for a linear *rotational spring,* the spring constant will be denoted by μ, and it has the units of moment per radian of rotation; thus, the moment in a rotational spring is equal to the spring constant times the rotational displacement. When the rotational spring is fixed at one end and fastened to a beam on the other, the rotational displacement is the first derivative of the displacement of the beam. For the beam in Figure 2.2.13(b), the moment in the spring is

$$M_e(a) = \mu v'(a) \tag{2.2.67}$$

To find the boundary conditions imposed on a beam by the presence of springs, we turn to the free-body diagram. Consider a very thin slice of each beam in Figure 2.2.13 at the end where the spring is attached. Free bodies for each of these cases are given in Figure 2.2.14, where the internal stress resultants and the forces and moments resulting from spring displacement are shown.

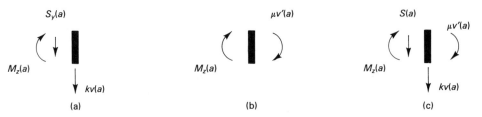

Figure 2.2.14   Free Bodies at the Beam Boundary.

Summation of forces and moments for the free body in Figure 2.2.14(a) will give us the boundary conditions

$$-S_y(a) - kv(a) = 0 \longrightarrow EI_{zz}v'''(a) - kv(a) = 0$$

$$M_z(a) = 0 \longrightarrow EI_{zz}v''(a) = 0$$

(2.2.68)

For the free body in Figure 2.2.14(b), we have the boundary conditions

$$M_z(a) + \mu v'(a) = 0 \longrightarrow EI_{zz}v''(a) + \mu v'(a) = 0$$

$$v(a) = 0$$

(2.2.69)

and for the free body in Figure 2.2.14(c)

$$-S_y(a) - kv(a) = 0 \longrightarrow EI_{zz}v'''(a) - kv(a) = 0$$

$$M_z(a) + \mu v'(a) = 0 \longrightarrow EI_{zz}v''(a) + \mu v'(a) = 0$$

(2.2.70)

It will be helpful to look at an example of a beam with elastic supports.

Example 2.2.5

Find the deflections and stresses in the beam shown in figures (a) and (b).

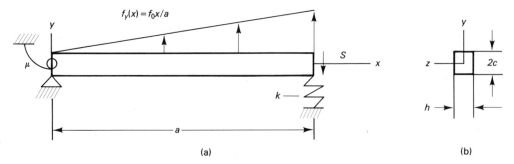

Solution: The governing equation is

$$EI_{zz}v''''(x) = \frac{f_0 x}{a}$$

(a)

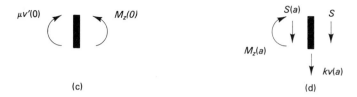

(c) (d)

and the boundary conditions can be obtained from free bodies at each end of the beam as shown in figures (c) and (d). For the left end

$$v(0) = 0, \qquad M_z(0) - \mu v'(0) = EI_{zz}v''(0) - \mu v'(0) = 0 \qquad \text{(b)}$$

and for the right end

$$-S_y(a) - kv(a) - S = EI_{zz}v'''(a) - kv(a) - S = 0$$

$$M_z(a) = EI_{zz}v''(a) = 0 \qquad \text{(c)}$$

Integrating equation (a) four times, we get

$$EI_{zz}v'''(x) = \frac{f_0}{a}\frac{x^2}{2} + B_1$$

$$EI_{zz}v''(x) = \frac{f_0}{a}\frac{x^3}{6} + B_1x + B_2$$

$$EI_{zz}v'(x) = \frac{f_0}{a}\frac{x^4}{24} + B_1\frac{x^2}{2} + B_2x + B_3 \qquad \text{(d)}$$

$$EI_{zz}v(x) = \frac{f_0}{a}\frac{x^5}{120} + B_1\frac{x^3}{6} + B_2\frac{x^2}{2} + B_3x + B_4$$

From the boundary conditions, we have

$$EI_{zz}v(0) = B_4 = 0$$

$$EI_{zz}v''(0) - \mu v'(0) = B_2 - \frac{\mu B_3}{EI_{zz}} = 0$$

$$EI_{zz}v''(a) = \frac{f_0a^2}{6} + B_1a + B_2 = 0 \qquad \text{(e)}$$

$$EI_{zz}v'''(a) = \frac{f_0a}{2} + B_1 - \frac{k}{EI_{zz}}\left(\frac{f_0a^4}{120} + B_1\frac{a^3}{6} + B_2\frac{a^2}{2} + B_3a \right) = S$$

From the second and third of these, we get

$$B_2 = -\frac{f_0a^2}{6} - B_1a, \qquad B_3 = B_2\frac{EI_{zz}}{\mu} = \left(-\frac{f_0a^2}{6} - B_1a \right)\frac{EI_{zz}}{\mu} \qquad \text{(f)}$$

and combining this with the fourth, we have

$$\frac{f_0a}{2} + B_1 - \frac{k}{EI_{zz}}\left[\frac{f_0a^4}{120} + B_1\frac{a^3}{6} + \left(-\frac{f_0a^2}{6} - B_1a \right)\frac{a^2}{2} \right.$$

$$\left. + \left(-\frac{f_0a^2}{6} - B_1a \right)\frac{EI_{zz}}{\mu}a \right] = S \qquad \text{(g)}$$

Solving for B_1, we get

$$B_1 = \frac{S - (f_0 a/2) - (9f_0 a^4 k/120EI_{zz}) - (f_0 a^3 k/6\mu)}{1 + (a^2 k/\mu) + (a^3 k/3EI_{zz})} \qquad \text{(h)}$$

and then B_2 and B_3 may be found from equations (f). All the constants are now known, and the deflection is given by the fourth of equations (d).

The normal stress and shear stress are found using Equations 2.2.38 and 2.2.61 and the appropriate expressions for the bending moment and shear force obtained from differentiating the deflection. Thus

$$\sigma_x = -\frac{M_z y}{I_{zz}} = -EI_{zz} v'' \frac{y}{I_{zz}} = -\left(\frac{f_0}{6a} x^3 + B_1 x + B_2\right)\frac{y}{I_{zz}} \qquad \text{(i)}$$

and

$$\tau_{xy} = \frac{S_y}{2I_{zz}}(c^2 - y^2) = \frac{EI_{zz} v'''}{2I_{zz}}(c^2 - y^2)$$

$$= -\frac{[(f_0/2a)x^2] + B_1}{2I_{zz}}(c^2 - y^2) \qquad \text{(j)}$$

So far, we have carefully introduced constraints and concentrated applied loads only at the ends of the beams where they, very considerately, find a place in the problem via the boundary conditions. Furthermore, the distributed loads have been simple continuous functions over the whole length of the beam, and the cross-sectional properties have been constant. In many real problems, the constraints and concentrated applied loads may not be at the ends, and the distributed loads and section properties may be discontinuous, as shown, for example, in Figure 2.2.15.

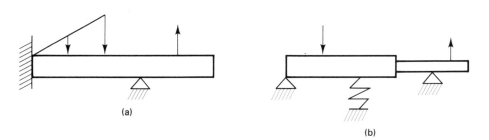

(a)

(b)

Figure 2.2.15 Beams with More Complex Loading and Restraint.

The solution of such beams by the method we have just discussed, while straightforward, puts a severe strain on our patience. Using this method, we would have to divide the beam into segments at each discontinuity, constraint, or concentrated load and write a separate differential equation for each segment. The beam in Figure 2.2.15(a), for example, would be divided into four segments and the beam in Figure 2.2.13(b) into six segments. Integrating each segment four times we would get 16 constants of integration in the first case and 24 in the second. Sixteen and twenty-four appropriate boundary conditions would have to be found in each respective case. This would take an inordinate amount of time to solve, especially when it is known that there are better ways. We shall investigate some of those better ways in Chapter 3.

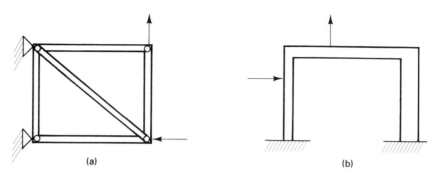

Figure 2.2.16 Trusses and Frames.

Axially and laterally loaded beam segments are combined to form structures commonly known as *trusses* and *frames*. In Figure 2.2.16(a), we have a structure known as a *pin-jointed truss*, which is composed of segments that carry only axial internal forces, and in Figure 2.2.16(b), we have a structure known as a frame composed of beam segments that can transmit shear and bending as well as axial forces. The preceding equations of beam theory can be used to solve these types of problems; however, they also present difficulties that are better overcome by other methods of analysis, which will be presented in Chapters 3 and 4.

In spite of the difficulties encountered in solving the beam equations for the more complicated cases, the equations we have just dealt with are among the most valuable in all of structural analysis. They provide us with exact analytical solutions for a class of structures of great practical interest. Not only are these solutions valuable in their own right, but they are, perhaps, even more valuable in checking and confirming a number of approximate methods that are beginning to dominate practical structural analysis. We shall use these solutions for this purpose in Chapters 3 and 4.

2.3 SHEAR STRESS IN THIN-WALLED BEAMS

We have developed a general theory for beams under axial, torsional, and transverse loading for arbitrary cross-sectional shape that provides excellent accuracy in modeling the deflections and the normal stresses. Unfortunately, no simple general equations exist for the shear stresses in beams of arbitrary cross section. We have found the shear stresses due to transverse shear loads in a beam with a rectangular cross section in Equation 2.2.61 and for that we are grateful. And in Chapter 1 we were able to find the shear stresses due to torsional loads for a few cross sections. The rest of the good news is that if the cross section is thin walled we can find the shear stress in a beam with what is otherwise an arbitrary cross section for both transverse shear and torsion. Two categories of thin-walled beams are shown in Figure 2.3.1, the (a) open and (b) closed sections. The difference in the behavior in shear and torsion of these two types of cross sections provides an interesting study in the next two sections.

When the equations in this chapter for bending and shear were derived, we had explicitly stated that no twist, or torsion, was present. We shall now learn that this has important consequences on how the lateral load can be applied. The fact is that the shear

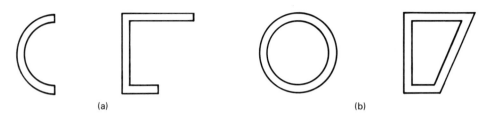

(a)   (b)

Figure 2.3.1 Examples of Thin-Walled Cross Sections

stress resultant must pass through a certain point in the cross section if there is to be no twist of the section. This point is called the *shear center,* and, in general, it does not coincide with the centroid of the cross section. To develop the location of this point, we shall first find the shear stress equations for transverse shear without twist and observe what conditions must be satisfied for it to be true.

It is assumed that both the inner and outer surfaces of the walls of the beam are free of stress. The key to finding the shear stress in thin-walled beams is found in the observation that the shear stress can have no component normal to the boundary at the boundary. This observation was used to advantage in finding the torsional stresses in Chapter 1. Here we carry it further by noting that, since the inner and outer surfaces of the thin wall are near each other, no appreciable component of shear stress normal to the boundary can occur anywhere on the cross section. We shall assume, therefore, that the shear stress is parallel to the center line of the thin wall and is uniform across the thickness of the wall. Since the shear stress is uniform across the thickness, it is often convenient to work with a force per unit length called the *shear flow* and denoted by q, which is defined as the shear stress times the thickness.

We proceed by considering the equilibrium of a small element of the thin-walled section. It is convenient to use a coordinate s that coincides with the center line of the thin wall. The element, along with its state of stress, is shown in Figure 2.3.2.

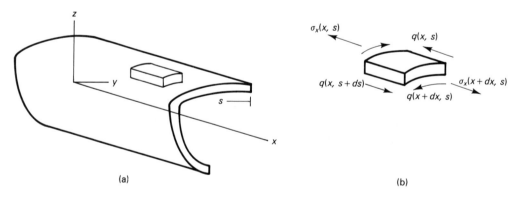

(a)   (b)

Figure 2.3.2 An Element of a Thin-Walled Beam.

From a Taylor's series expansion, we get

$$q(x, s + ds) = q(x, s) + \frac{\partial q}{\partial s} \, ds, \qquad \sigma_x(x + dx, s) = \sigma_x(x, s) + \frac{\partial \sigma_x}{\partial x} \, dx \quad (2.3.1)$$

and from the summation of forces in the x-direction, we get

$$\left(q(x,\,s)+\frac{\partial q}{\partial s}\,ds\right)dx-q\,dx+\left(\sigma_x+\frac{\partial \sigma_x}{\partial x}\,dx\right)h\,ds-\sigma_x h\,ds=0 \qquad (2.3.2)$$

which reduces to

$$\frac{\partial q}{\partial s}=-h\frac{\partial \sigma_x}{\partial x} \qquad (2.3.3)$$

As stated earlier, it is a common assumption that the normal stresses obtained for pure bending are reasonably accurate for beams when shear stresses are present; thus, we may introduce Equation 2.2.36 or Equation 2.2.38 into the preceding expression. For convenience, we can leave out the uncoupled axial force contribution to the normal stress since it can be added back at any time it is needed. For any given value of x, the result is

$$\frac{dq}{ds}=\frac{h}{I_{yy}I_{zz}-I_{yz}{}^2}\,[(I_{yy}M_z{}'-I_{yz}M_y{}')y-(I_{yz}M_z{}'-I_{zz}M_y{}')z] \qquad (2.3.4)$$

or

$$\frac{dq}{ds}=\frac{h}{I_{yy}I_{zz}-I_{yz}{}^2}\,[(-I_{yy}S_y+I_{yz}S_z)y+(I_{yz}S_y-I_{zz}S_z)z] \qquad (2.3.5)$$

and in the special case where the product of inertia is zero,

$$\frac{dq}{ds}=-h\left(\frac{S_y}{I_{zz}}y+\frac{S_z}{I_{yy}}z\right) \qquad (2.3.6)$$

This may be integrated along the path s to obtain the shear flow. This can be shown in line integral form, as follows:

$$q=-\int\left(\frac{S_y}{I_{zz}}y+\frac{S_z}{I_{yy}}z\right)h\,ds+C \qquad (2.3.7)$$

where C is the constant of integration shown explicitly. It is a property of line integrals that the constant of integration can be assigned the value of the shear flow at some point on the line. For example, we can let $C=q(0)$; that is, C is the value of the integral at the origin of the coordinate s. The lower limit of the integral can then be expressed as $s=0$ and the upper limit as the variable value of the coordinate s. The integral is then written

$$q=q_0-\int_0^s\left(\frac{S_y y}{I_{zz}}+\frac{S_z z}{I_{yy}}\right)h\,ds \qquad (2.3.8)$$

where $q(0)=q_0$. The ways for determining q_0 for closed sections are different from those for open sections. We shall discuss these in the next two sections. Before we do that, let us examine the phenomenon of shear center.

For the preceding analysis to be correct, there must be no twist of the beam. It is argued that for this to be true the applied shear load must pass through the same point in the cross section as the resultant of the shear flow produced by Equation 2.3.8. This can be confirmed by looking at the equilibrium of a free-body section of a beam with a shear load at one end and an internal shear flow at the other, as shown in Figure 2.3.3. An open section is shown in the figure, but the following argument is equally true for a closed section.

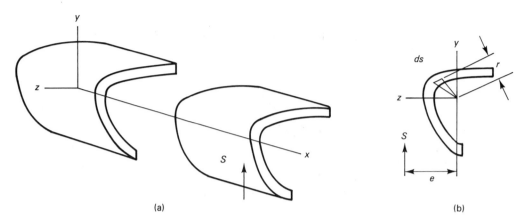

Figure 2.3.3   Shear Load and Shear Flow through the Shear Center.

(a)   (b)

Unless the moment of the shear load about any point in the y-z plane is the same as the moment of the shear flow about the same point in the y-z plane, the section of the beam shown as a free body would be subjected to twist. The statement that these two moments are the same is

$$S_y e_z = \int qr \, ds \qquad (2.3.9)$$

A similar examination of the case where a shear force is present in the z-direction will provide us with the equation

$$S_z e_y = \int qr \, ds \qquad (2.3.10)$$

From these equations, we can find the shear center in any specific case.

We may note that the shear center will always lie on a plane of symmetry if one is present, because the preceding two integrals will have terms on each side of the plane of symmetry, which will cancel each other.

2.3.1 Shear and Torsion in Thin-Walled Open Sections

In open sections, we can always define the origin of the s-coordinate at a free edge of the section. Since no shear load is applied to the edge in the x-direction, there can be no shear stress at that edge in the s-direction and, therefore, $q_0 = 0$.

To complete the integration, usually we must express the coordinate s in terms of y and z, or in some cases in terms of polar coordinates, r and φ. The use of this equation for finding shear flow in practical problems is best shown by example. Two examples for open sections follow.

Example 2.3.1

Find the shear stress in a beam with a narrow rectangular cross section and a shear force applied only in the y-direction, as shown in figure (a).

Solution: We already know the answer to this. It is given in Equation 2.2.61. This gives us a chance to check the validity of the solution obtained from the integral form given in Equation 2.3.8. In this case, the integral is

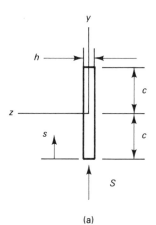

(a)

$$q = q_0 - \int_0^s \frac{S_y y}{I_{zz}} h \, ds = -\int_{-c}^y \frac{S_y y}{I_{zz}} h \, dy \qquad \text{(a)}$$

where we note that $q_0 = 0$ and, since $s = y + c$, $ds = dy$. Note the change in limits to reflect the change in coordinate from s to y. Upon integration, we have

$$q = -\frac{S h y^2}{2 I_{zz}} \bigg|_{-c}^y \qquad \text{(b)}$$

which becomes

$$q = \frac{S h}{2 I_{zz}} (c^2 - y^2) = \tau_{xy} h \qquad \text{(c)}$$

and which agrees with Equation 2.2.61.

To find the shear center, we shall apply Equation 2.3.9 using the centroid of the cross section as the origin. Since the shear flow q acts on a straight line through the origin, the distance r is zero, and therefore the value of the integral is zero. Thus $e_z = 0$. This confirms that the shear center will lie on a plane of symmetry in this special case.

Example 2.3.2

Find the shear flow in a beam with the cross section shown in figure (a).

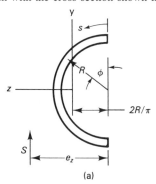

(a)

Solution: The integral equation for this case can be converted to more convenient polar coordinates, r and φ. Note that, while the x-axis must be at the centroid of the cross section, shown in the figure at a distance $2R/\pi$ from the center of the radius, for the beam equations to apply, the coordinate y and the differential ds can both be put in terms of R and φ, which have their origin offset from the centroid, without any loss of generality. This happens because z is not an explicit variable in the equation, and the y-coordinate is not affected by the z-shift in the origin of the axes. In terms of R and φ, we have

$$y = R \cos \varphi \tag{a}$$

and since $s = R\varphi$, we have

$$ds = R \, d\varphi \tag{b}$$

so that our equation for shear flow becomes

$$q = q_0 - \int_0^s \frac{S_y y}{I_{zz}} h \, ds = -\frac{S h R^2}{I_{zz}} \int_0^\varphi \cos \varphi \, d\varphi \tag{c}$$

Note the limits of integration from $\varphi = 0$ to a variable point φ along the cross section. Upon integration, we have

$$q = -\frac{S h R^2}{I_{zz}} \sin \varphi = -\frac{2S}{\pi R} \sin \varphi \tag{d}$$

since $I_{zz} = \pi R^3 h/2$. To find the shear center, we shall take the origin of the r-φ coordinates as the point about which moments are found. When we shift to r-φ coordinates and note that the shear flow along the curved path is always at a distance R, Equation 2.3.9 becomes

$$S_y e_z = \int_0^\pi q R^2 \, d\varphi \tag{e}$$

or, when we insert the value of q from equation (d),

$$S_y e_z = \int_0^\pi \frac{S_y h R^4}{I_{zz}} \sin \varphi \, d\varphi \tag{f}$$

which reduces to

$$\epsilon_z = \int_0^\pi \left(\frac{h R^4}{I_{zz}} \right) \sin \varphi \, d\varphi = \frac{2 h R^4}{I_{zz}} = \frac{4R}{\pi} \tag{g}$$

We may note that the shear center lies a distance $2R/\pi$ to the left of the centroid, which places it off the cross section.

When the shear force does not act through the shear center, the beam will twist. The question is, how much? We can find the answer in Section 1.3.3. There we examined the torsional deformation and stress for a thin-walled open section under pure torsion. We can combine what we learned there with the shear flow equations given in this section.

Consider a beam with a thin-walled open section in which the shear stress resultant is not through the shear center. To adapt it to the analyses just discussed, we resolve the shear force into a statically equivalent shear force through the shear center and a torsional moment about the shear center. The effective shear stress for the shear force through the shear center is given by Equations 2.3.8, and the effective shear stress of the torsional moment is given by Equation 1.3.77. To see how this all ties together, we had best look at an example.

Example 2.3.3

A cantilever beam has a uniform loading that is acting along the centroidal x-axis, as shown in figure (a), and has the same cross section as the beam in Example 2.3.2, as shown in figure (b). Find the shear stresses.

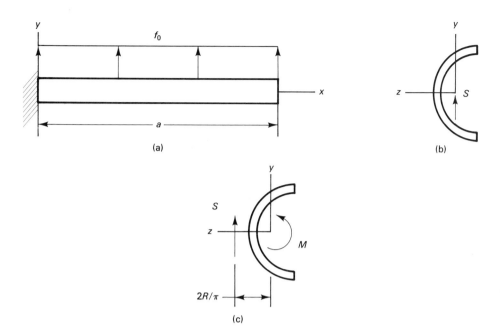

(a)

(b)

(c)

Solution: We integrate the governing equation for bending once to obtain the shear force at any cross section.

$$EI_{zz}v'''' = f_0 \longrightarrow EI_{zz}v''' = f_0 x + B_1 = -S_y(x) \tag{a}$$

From the boundary condition on shear at the free end, we conclude that $B_1 = -f_0 a$; thus,

$$S_y(x) = f_0(a - x) \tag{b}$$

Since S_y acts through the centroid, which is not the shear center, we resolve S_y into a force through the shear center of the same magnitude, or $S = S_y$, and a moment about the shear center of $M = 2SR/\pi$, as shown in figure (c). This is a statically equivalent set of loads. We can now find the shear flow for a force through the shear center with Equation 2.3.8, but this has already been done in Example 2.3.2, or

$$q = \frac{2S}{\pi R} \sin \varphi = -\frac{2f_0(a - x)}{\pi R} \sin \varphi \tag{c}$$

The shear stress due to the torsional load is given by Equation 1.3.77, which for this case is

$$\tau_{xs} = -\frac{2M}{J} n = -\frac{4SR}{\pi J} n = -\frac{4f_0(a - x)R}{\pi J} n \tag{d}$$

where n is measured normal to the coordinate s and the torsional constant, from Equation 1.3.76, for this case is

$$J = \frac{\pi R h^3}{3} \tag{e}$$

where h is the wall thickness. The total shear stress at any point on any cross section is the sum of equations (c), where $\tau_{xs} = q/h$, and (d).

The bending deflection and normal stress are obtained from further integration of equation (a), and the twist from Equation 2.2.23, which for this case is

$$GJ\beta'' = -t_x = \frac{f_0 2R}{\pi} \qquad (f)$$

Calculating the shear flow and the shear center can be confusing because of the amount of mathematical bookkeeping necessary for the more complicated cross sections. We shall give another example, which should help to clarify the process.

Example 2.3.4

Find the total shear stress in the beam with the cross section shown in figure (a) when the shear force acts through the center of the vertical web.

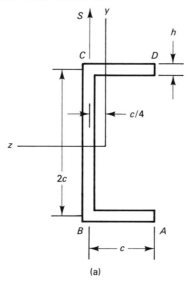

(a)

Solution: Let us put the origin of the coordinate s at point A and have s increase in the direction of BCD. The equation for the shear flow due to transverse shear force is

$$q = q_0 - \int \frac{S_y y}{I_{zz}} h \, ds \qquad (a)$$

which, for this case, may be stated in three parts. In the region AB, because $y = -c$ and $ds = dz$, it becomes

$$q_{AB} = q_A + \int_{-0.75c}^{z} \left(\frac{Sc}{I_{zz}}\right) h \, dz = \frac{Sch}{I_{zz}}(z + 0.75c) \qquad (b)$$

where we note that $q_A = 0$. From this, we may note that

$$q_B = \frac{Sc^2 h}{I_{zz}} \qquad (c)$$

In the region BC, because $ds = dy$, the shear flow becomes

$$q_{BC} = q_B - \int_{-c}^{y} \left(\frac{S_y y}{I_{zz}}\right) h \, dy = \frac{Sc^2 h}{I_{zz}} - \frac{Sh}{2I_{zz}}(y^2 - c^2) \qquad (d)$$

and we may note that

$$q_C = q_A = \frac{Sc^2 h}{I_{zz}} \qquad (e)$$

Finally, in the region CD, because $y = c$ and $ds = -dz$, the shear flow becomes

$$q_{CD} = q_C + \int_{0.25c}^{z} \left(\frac{S_y c}{I_{zz}}\right) h \, dz = \frac{Sc^2 h}{I_{zz}} - \frac{Sch}{I_{zz}} (z - 0.25c)$$

$$= \frac{Sch}{I_{zz}} (0.75c + z) \tag{f}$$

We can check to see if $q_D = 0$, since D is at a free edge and, sure enough, it does.

To find the shear center, let us take moments about the point B since this immediately eliminates q_{AB} and q_{BC} since there is no moment arm. Thus,

$$S_y e_z = \int qr \, ds = \int_{0.25c}^{-0.75c} q_{CD} \, r \, ds$$

$$= \int_{0.25c}^{-0.75c} \frac{Sch}{I_{zz}} (0.75c + z) = \frac{Sc^4 h}{I_{zz}} \tag{g}$$

and so the shear center is found on the z-axis (plane of symmetry), and to the left of the point B an amount

$$e_z = \frac{c^4 h}{I_{zz}} \tag{h}$$

Since the load is not through the shear center, it must be resolved into a load through the shear center and a moment about it. That moment will produce torsion in the beam and is seen to have the value

$$M_x = \frac{Sc^4 h}{I_{zz}} \tag{i}$$

and the shear stress and rate of twist due to torsion are

$$\tau_{xs} = -\frac{2M_x}{J} z = -\frac{2Sc^4 h}{I_{zz} J} n \tag{j}$$

$$\theta = \frac{M_x}{GJ} = \frac{Sc^4 h}{I_{zz} GJ} \tag{k}$$

where $J = 4ch^3/3$ and n is interpreted as the normal distance from the center line of the thin-walled section.

2.3.2 Torsion and Shear of Thin-Walled Closed Sections

The torsion of a closed section like those shown in Figure 2.3.1 (b) is more easily handled. The assumption of uniform shear stress between the edges of the thin walls is supported by the stress distribution we found in Example 1.3.7 for the hollow circular cross section. If the inner and outer radii are nearly the same, the stresses at the inner and outer edges are nearly the same. A close approximation would be to replace that distribution with a constant shear stress with a value equal to that on the center line between the two edges.

This may be extended to closed, thin-walled beams of arbitrary cross section, as shown in Figure 2.3.4. A small element of the beam shown in Figure 2.3.4(a) is enlarged and shown with shear flow in Figure 2.3.4(b). If the beam is loaded with a pure torque, T, which from equilibrium is equal to the stress resultant torsional moment, M_x, the shear flow will be the same on all cross sections; that is,

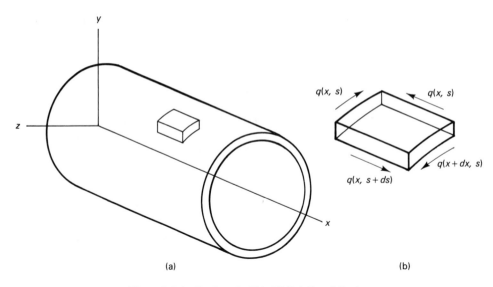

Figure 2.3.4 Torsion of a Thin-Walled Closed Section.

$$q(x, s) = q(x + dx, s) = q(s) \tag{2.3.11}$$

or q is independent of x. Using the Taylor series expansion we have

$$q(s + ds) = q(s) + \frac{dq}{ds} \, ds \tag{2.3.12}$$

and for equilibrium in the s-direction

$$q(s) + \frac{dq}{ds} \, ds - q(s) = 0 \tag{2.3.13}$$

or

$$\frac{dq}{ds} = 0 \longrightarrow q = \text{constant} \tag{2.3.14}$$

The value of this constant shear flow may be determined by equating the torsional stress resultant to the torque resulting from the shear flow:

$$M_x = \oint qr \, ds = q \oint r \, ds \tag{2.3.15}$$

where r is the moment arm of each element of force $q \, ds$, as shown in Figure 2.3.5.

The line integral in Equation 2.3.15 can be changed to an area integral by noting that $r \, ds/2$ is an element of area enclosed by the triangle of base ds and height r, as shown in Figure 2.3.5. The integral then becomes

$$M_x = 2q \int dA = 2Aq \longrightarrow q = \frac{M_x}{2A} \tag{2.3.16}$$

The area is the total area enclosed by the center line of the closed cross section. Note that it is not the area of the material in the cross section.

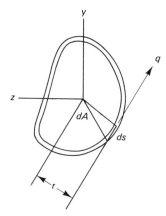

Figure 2.3.5 Arbitrary Cross Section in Torsion.

The relationship between twist and torque can be obtained from the strain–displacement equations. In terms of the x-s coordinates,

$$\gamma_{xs} = \frac{du}{ds} + \frac{du_s}{dx} = \frac{\tau_{xs}}{G} = \frac{q}{Gh} \tag{2.3.17}$$

where u_s is the displacement component in the s-direction. Analogous to Equation 1.3.39, we have

$$u_s = r\beta \longrightarrow \frac{du_s}{dx} = r\frac{d\beta}{dx} = \theta r \tag{2.3.18}$$

and therefore

$$q = Gh\left(\frac{du}{ds} + \theta r\right) \tag{2.3.19}$$

Since q and θ are constants, this becomes

$$\frac{du}{ds} = \frac{q}{Gh} - \theta r \tag{2.3.20}$$

or

$$\oint du = \frac{1}{G}\oint\frac{q}{h}\,ds - \theta\oint r\,ds \tag{2.3.21}$$

The left-hand integral is zero since the path around the thin-walled section is closed. We know also that

$$\oint r\,ds = 2A \tag{2.3.22}$$

where A is the enclosed area of the cross section. Finally, this gives us

$$\theta = \frac{1}{2AG}\oint\frac{q}{h}\,ds \tag{2.3.23}$$

When the value of q for the pure torsion case is inserted, we get

$$\theta = \frac{M_x}{4A^2G} \oint \frac{ds}{h} = \frac{M_x}{GJ} \qquad (2.3.24)$$

where

$$J = \frac{4A^2}{\oint ds/h} \qquad (2.3.25)$$

Now we need some examples of torsion of closed sections. In the next example, we also compare an open section with a closed section.

Example 2.3.5

Find the shear stress and rate of twist for beams loaded with a pure torsional moment, $M_x = T$, and with the cross sections shown. The beam in figure (a) has a slit running the full length of the beam; thus the cross section is an open section, while the one in figure (b) is a closed section. Compare the two in their ability to carry the same applied torque.

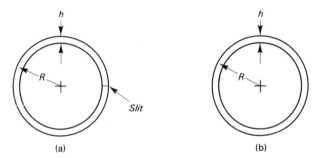

(a)   (b)

Solution: For the open section in figure (a), the rate of twist θ is given by Equation 1.3.75, which for this case is

$$\theta = \frac{M_x}{GJ} = \frac{3M_x}{2G\pi Rh^3} \qquad (a)$$

The shear stress is given by Equation 1.3.77:

$$\tau_{xs} = -2G\theta n = -\frac{3M_x}{\pi Rh^3}n \qquad (b)$$

where n is interpreted as the normal distance from the center line of the thin wall. The largest values of shear stress would occur on the outer and inner edges where $n = \pm h/2$, or

$$(\tau_{xs})_{max} = \pm\frac{3M_x}{2\pi Rh^2} \qquad (c)$$

The rate of twist for the closed section in figure (b) is given by Equation 2.3.24, which for this case is

$$\theta = \frac{M_x}{GJ} = \frac{M_x \oint ds/h}{4GA^2} = \frac{M_x\pi R}{2GA^2h} = \frac{M_x}{2G\pi R^3h} \qquad (d)$$

The shear stress is given by Equation 2.3.16, which for this case is

$$\tau_{xs} = \frac{q}{h} = \frac{M_x}{2Ah} = \frac{M_x}{2\pi R^2 h} \tag{e}$$

The effect of the same applied torque on each section can be shown by setting up the ratios of the twist and the shear stress for the two sections. Using o and c as subscripts for the open and closed sections, respectively, we have

$$\frac{\theta_o}{\theta_c} = \frac{3M_x}{2G\pi Rh^3} \frac{2G\pi R^3 h}{M_x} = 3\frac{R^2}{h^2} \tag{f}$$

Since $R/h \gg 1$, we see that the open section is very weak in torsion and, in fact, should not be used when torsional loads are of concern.

The ratio of stresses is also informative. Using the maximum value of stress for the open section, we have

$$\frac{(\tau_{xs})_o}{(\tau_{xs})_c} = \frac{3M_x h}{2\pi Rh^3} \frac{2\pi R^2 h}{M_x} = 3\frac{R}{h} \tag{g}$$

The maximum shear stress in torsion from the same applied load is also much larger for the open section, reinforcing our conclusion that open sections should be avoided when torsional loads are to be carried.

Now consider a beam with a closed cross section, such as that shown in Figure 2.3.4, and with a transverse shear load. Generally, the transverse load will not act through the shear center, and combined bending, transverse shear, and torsion will be present. We can start with Equation 2.3.8, but in the case of a closed section there is no convenient starting point for the s-coordinate where the shear flow is known; thus, q_0 is unknown.

After finding the shear flow in the closed section with the Equation 2.3.8, we use our definition of the shear center and the equation for the twist of a thin-walled closed section to find the value of q_0. Going back to Equation 2.3.23, this time we use the shear flow obtained from Equation 2.3.8 to find the twist. If the applied load is through the shear center, the twist is zero, or

$$\theta = \frac{1}{2AG} \oint \frac{q}{h} ds = 0 \longrightarrow \oint \frac{q}{h} ds = 0 \tag{2.3.26}$$

This equation is used to evaluate q_0, and once it is known the shear center can be found using Equations 2.3.9 and 2.3.10.

An example will do so much better than more words.

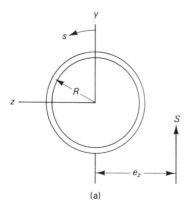

(a)

Example 2.3.6

Find the shear flow and the shear center for a beam with the circular, thin-walled cross section shown in figure (a), on the bottom of page 101.

Solution: Let us choose the origin of the s-coordinate at the top of the section and go counterclockwise. From Equation 2.3.26, we get

$$q = q_0 - \frac{S_y h}{I_{zz}} \oint y \, ds = q_0 - \frac{SR^2 h}{I_{zz}} \oint \cos \varphi \, d\varphi = q_0 - \frac{SR^2 h}{I_{zz}} \sin \varphi \qquad (a)$$

since

$$y = R \cos \varphi, \qquad ds = R \, d\varphi \qquad (b)$$

If the section does not twist,

$$\oint \frac{q}{h} \, ds = \oint \left(q_0 - \frac{SR^2 h}{I_{zz}} \sin \varphi \right) \left(\frac{R}{h} \right) d\varphi = 0 \qquad (c)$$

from which we obtain

$$q_0 = 0 \qquad (d)$$

The shear center is located by

$$S_y e_z = \oint qr \, ds = \oint \left(q_0 - \frac{SR^2 h}{I_{zz}} \sin \varphi \right) R^2 d\varphi \qquad (e)$$

from which

$$e_z = 0 \qquad (f)$$

which is not surprising. It could have been anticipated by the symmetry of the cross section. The shear center always lies on a plane of symmetry if there is one.

Example 2.3.7

Find the shear flow in the beam with the cross section shown. The transverse shear force is applied on a line through the center line of the vertical web of the beam.

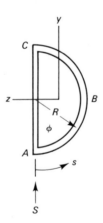

Solution: After letting $y = -R \cos \varphi$ and $ds = R \, d\varphi$, the shear flow in the curved part due to transverse shear is

$$q_{ABC} = q_0 - \frac{S_y h}{I_{zz}} \int_0^s y \, ds = q_A + \frac{Sh}{I_{zz}} \int_0^\varphi R^2 \cos \varphi \, d\varphi$$

$$= q_A + \frac{Sh}{I_{zz}} R^2 \sin \varphi \tag{a}$$

It may be noted that $q_C = q_A$. Then, along the vertical web, since $ds = -dy$,

$$q_{CA} = q_0 + \frac{S_{yh}}{I_{zz}} \int_0^s y \, ds = q_A + \frac{Sh}{2I_{zz}} (y^2 - R^2) \tag{b}$$

To find the value of q_A, assume the section does not twist, or

$$\oint \frac{q}{h} \, ds = \int_0^\pi \left(q_A + \frac{Sh}{I_{zz}} R^2 \sin \varphi \right) \left(\frac{R}{h} \right) d\varphi + \int_R^{-R} \left(q_A + \frac{Sh}{2I_{zz}} (y^2 - R^2) \right) \left(\frac{-dy}{h} \right)$$

$$= \frac{R\pi + 2R}{h} q_A + \frac{4SR^3}{3I_{zz}} = 0 \tag{c}$$

or

$$q_A = -\frac{4ShR^2}{3(\pi + 2)I_{zz}} \tag{d}$$

To find the shear center, let

$$S_y e_z = \oint qr \, ds = \int_0^\pi \left(q_A + \frac{Sh}{I_{zz}} R^2 \sin \varphi \right) R^2 \, d\varphi$$

$$= \frac{ShR^4}{I_{zz}} \left[\frac{4\varphi}{3(\pi + 2)} + \cos \varphi \right] \Big|_0^\pi \tag{e}$$

$$= -\frac{ShR^4}{I_{zz}} \frac{2(\pi + 6)}{3(\pi + 2)}$$

and therefore the shear center lies to the right of the point A as given by

$$e_z = -\frac{hR^4}{I_{zz}} \frac{2(\pi + 6)}{3(\pi + 2)} \tag{f}$$

Finally, find the shear flow due to the moment about the shear center:

$$q = \frac{M_x}{2A} = -\frac{S_y e_z}{2A} = -\frac{ShR^4}{2AI_{zz}} \frac{2(\pi + 6)}{3(\pi + 2)} \tag{g}$$

The shear flow analysis can be extended to more complex structures; for example, the multicelled, thin-walled beam depicted in Figure 2.3.6.

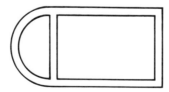

Figure 2.3.6 Cross Section of Multicelled, Thin-Walled Beam.

Space does not permit a development of the multicelled analysis here, but a brief review is given in Appendix F.

Sec. 2.3   Shear Stress in Thin-Walled Beams   **103**

2.3.3 Stiffened, Thin-Walled Beams

Open or closed sections with stiffeners such as those depicted in Figure 2.3.7(a) are a familiar structural type that can be analyzed by thin-walled shear flow theory.

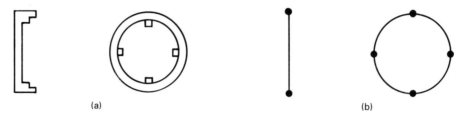

(a)   (b)

Figure 2.3.7   Real and Idealized Cross Sections of Stiffened, Thin-Walled Beams.

Idealizations of these beams are shown in Figure 2.3.7(b). Actually, there are two idealizations depending on the relative amount of material in the stiffener and in the thin wall. If the area of the stiffeners is relatively large compared to the area of the wall between stiffeners, it is often assumed that all the bending is carried by the stiffeners and all the shear by the walls. In this idealization, the area of the stiffeners is assumed to act at a point. Often the adjacent wall areas are added to the stiffener areas. The moment of inertia of the cross section now becomes a sum of the form

$$I_{zz} = \Sigma \, y_i^2 A_i \tag{2.3.27}$$

and the integral in Equation 2.3.8 for the shear flow reduces to

$$q = q_0 - \Sigma \, \frac{S_y}{I_{zz}} y_i A_i \tag{2.3.28}$$

How these are applied will now be shown by example.

Example 2.3.8

Find the shear flow and shear center for a stiffened, thin-walled beam with the cross section shown in figure (a).

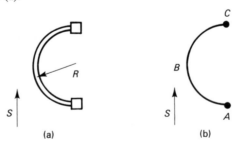

(a)   (b)

Solution: The idealization is shown in figure (b). Let us assume that the area of the thin wall, which is equal to πRh, is small compared to the area of the stiffeners. Then let the area of the stiffeners in the idealized section be equal to the area of the real stiffeners plus one-half the area of the wall. We shall denote the area of the lower idealized stiffener as A_L and the upper as A_U, and let $A_L = A_U = A$.

We handle the summation much like we do the integration. We start with a free edge to the right of the lower stiffener at point A where the shear flow is zero; therefore, $q_A = 0$.

With the first summation over the lower stiffener, we find the shear flow in the web:

$$q_B = q_A - \Sigma \frac{S_y}{I_{zz}} y_i A_i = \frac{S}{I_{zz}} RA = \frac{SRA}{2R^2A} = \frac{S}{2R} \tag{a}$$

This is the constant value of the shear flow in the web. We can check to see that it goes to zero again when the summation passes the upper stiffener.

$$q_C = q_B - \Sigma \frac{S_y}{I_{zz}} y_i A_i = \frac{S}{2R} - \frac{S}{I_{zz}} RA = 0 \tag{b}$$

To find the shear center, we use Equation 2.3.9, which for this case is

$$Se_z = \int qr\, ds = \int_0^{\pi R} \frac{S}{2R} R\, ds = \frac{\pi RS}{2} \tag{c}$$

Therefore,

$$e_z = \frac{\pi R}{2} \tag{d}$$

Example 2.3.9

Find the shear flow and the shear center for the box beam with the shear load shown in figure (a). Such a configuration is often the central load-carrying member of an aircraft wing. The dimensions are such that the area of the two front stiffeners is twice that of the other four. All wall thicknesses are the same.

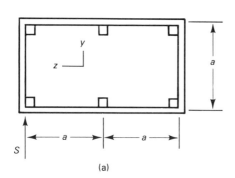

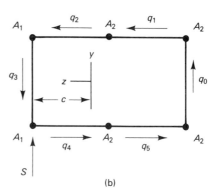

(a)

(b)

Solution: The idealization is shown in figure (b), where $A_1 = 2A_2$. Part of the web area is added to the stiffener area to make up the idealized stiffener areas. Assuming that the web thicknesses do not contribute to the moments of area or the moments of inertia, we can find the centroid location, c, and the moment of inertia, I_{zz}, with respect to the centroid. From symmetry, the centroid is on the z-axis, and the horizontal location measured from the front web is

$$c = \frac{2(aA_2 + 2aA_2)}{2(A_1 + 2A_2)} = \frac{3a}{4} \tag{a}$$

The moment of inertia is

$$I_{zz} = 2(A_1 + 2A_2)\frac{a^2}{4} = 2a^2A_2 \tag{b}$$

Let us start with the right vertical web where we have identified the shear flow as q_0, and move counterclockwise using Equation 2.3.28.

$$q = q_0 - \Sigma \frac{S_y}{I_{zz}} y_i A_i \tag{c}$$

Then

$$q_1 = q_0 - \frac{S}{I_{zz}} \frac{a}{2} A_2 = q_0 - \frac{S}{4a}$$

$$q_2 = q_0 - \frac{S}{4A} - \frac{S}{I_{zz}} \frac{a}{2} A_2 = q_0 - \frac{S}{2a}$$

$$q_3 = q_0 - \frac{S}{2A} - \frac{S}{I_{zz}} \frac{a}{2} A_1 = q_0 - \frac{S}{a}$$

$$q_4 = q_0 - \frac{S}{4A} + \frac{S}{I_{zz}} \frac{a}{2} A_1 = q_0 - \frac{S}{2a} \tag{d}$$

$$q_5 = q_0 - \frac{S}{2a} + \frac{S}{I_{zz}} \frac{a}{2} A_2 = q_0 - \frac{S}{4a}$$

$$q_6 = q_0 - \frac{S}{4a} + \frac{S}{I_{zz}} \frac{a}{2} A_2 = q_0$$

The last equation in the list is a check to see if we get back the value we started with.

To find the value of q_0, let us invoke the condition of no twist:

$$\oint \frac{q}{h} ds = \frac{1}{h} (q_0 + q_1 + q_2 + q_3 + q_4 + q_5)a$$

$$= \frac{a}{h} \left[q_0 + 2\left(q_0 - \frac{S}{4a}\right) + 2\left(q_0 - \frac{S}{2a}\right) + q_0 - \frac{S}{a} \right] \tag{e}$$

$$= \frac{a}{h} \left(6q_0 - \frac{5s}{2a}\right) = 0$$

or

$$q_0 = \frac{5S}{12a} \tag{f}$$

Thus, the shear flows all around the section are

$$q_0 = \frac{5S}{12a}, \qquad q_1 = \frac{S}{6a}, \qquad q_2 = -\frac{S}{12a}$$

$$q_3 = -\frac{7S}{12a} \qquad q_4 = -\frac{S}{12a} \qquad q_5 = \frac{S}{6a} \tag{g}$$

To find the shear center, take moments about the lower-left corner. That way we include only q_0, q_1, and q_2.

$$Se_z = \int qr \, ds = q_0 2a^2 + q_1 a^2 + q_2 a = \frac{11Sa}{12} \tag{h}$$

or

$$e_z = -\frac{11a}{12} \tag{i}$$

The torsional moment about the shear center is then $M_x = -S(11a/12)$, and the torsional shear flow, q_T, is

$$q_T = \frac{M_x}{2A} = -\frac{11Sa}{12}\frac{1}{4a^2} = -\frac{11s}{48a} \tag{j}$$

When the two shear flows are added and the notation for the total, $q_{it} = q_i + q_T$, is used, we get

$$q_{0t} = \frac{3S}{16a} \qquad q_{1t} = \frac{S}{16a} \qquad q_{2t} = -\frac{5S}{16a}$$

$$q_{3t} = -\frac{13S}{16a}, \qquad q_{4t} = -\frac{5S}{16a}, \qquad q_{5t} = -\frac{S}{16a} \tag{k}$$

This idealization is very useful, but may be too much of an approximation when the web area is an appreciable amount of the total area of the cross section. It is possible to retain the influence of the web thickness and also introduce concentrated stiffeners. To do this, we combine features of the two types of analysis discussed previously. While integrating along a web, we use the form

$$q = q_0 - \frac{S_y h}{I_{zz}} \int z \, ds \tag{2.3.29}$$

and when integrating over a stiffener, we use the form

$$q = q_0 - \Sigma \frac{S_y}{I_{zz}} y_i A_i \tag{2.3.30}$$

This is best described by an example.

Example 2.3.10

Reconsider the problem in Example 2.3.8 accounting for both the web and concentrated stiffeners. The cross-sectional shape is repeated in figure (a).

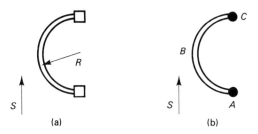

(a)   (b)

Solution: The idealization is shown in figure (b). The area of each idealized stiffener is just the area of the stiffener. We start, as in Example 2.3.8, by noting that $q_A = 0$ and by summing over the lower stiffener and integrating over the web:

$$q_B = q_A - \Sigma \frac{S_y y_i A_i}{I_{zz}} - \frac{S_y h}{I_{zz}} \int_0^s y \, ds = \frac{SRA}{I_{zz}} - \frac{ShR^2}{I_{zz}} \int_0^\varphi \cos\varphi \, d\varphi$$

$$= \frac{SRA}{I_{zz}} - \frac{ShR^2}{I_{zz}} \sin\varphi \tag{a}$$

where

$$I_{zz} = \int y^2 \, dA = 2AR^2 + R^2h \int_0^\pi \cos\varphi \, d\varphi = 2AR^2 + \frac{\pi R^3 h}{2} \tag{b}$$

We shall return to the subject of beams in Chapter 3. For now, we shall take up the subject of thin flat plates.

2.4 THIN PLATES IN BENDING

Thin flat plates are structures for which one dimension, the thickness, is small compared to the other two and whose mid-surface is planar. The classical theory of thin flat plates is a two-dimensional generalization of beam theory. The assumptions of beam theory are adapted to the differing geometry, but are easily recognized in the process. As in beam theory, we retain all the assumptions about material properties, small deflections, and so on, characteristic of the linear theory of elasticity and add geometrical and loading simplifications suitable for this structural element.

Consider a thin flat plate of constant thickness, as shown in Figure 2.4.1. A rectangular plate is shown, but any shape in the x-y plane can be used. The midplane of the plate coincides with the x-y plane; thus the upper and lower surfaces are at $\pm h/2$. Transverse loads are applied as forces or pressures perpendicular to the midplane and give rise to bending and shear stresses analogous to bending and shear stresses in beams. Edge loads may also be applied in the plane of the plate, giving rise to midplane stresses analogous to axial stresses in beams. We shall assume for now that the in-plane loads and deflections are negligibly small and consider only transverse loads and deflections until Chapter 5.

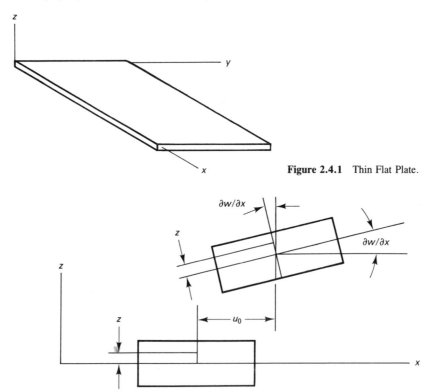

Figure 2.4.1 Thin Flat Plate.

Figure 2.4.2 Plate Bending Deformation.

Let us look at an element of the plate in both the unloaded and undeflected position and in the loaded and deflected position, as we have done before for the beam. From Figure 2.4.2, analogous to the beam derivation (see Figure 2.2.2), we have

$$u = u_0 - z \frac{\partial w}{\partial x} \tag{2.4.1}$$

and from a similar figure drawn in the y-z plane, we would see that

$$v = v_0 - z \frac{\partial w}{\partial y} \tag{2.4.2}$$

This is based on the plate equivalent of plane sections remaining plane, which may be stated as straight line elements originally normal to the middle surface remain normal and straight after deformation.

Since we are neglecting the in-plane deformations (and the in-plane loads), we have the following strain components:

$$\epsilon_x = -z \frac{\partial^2 w}{\partial x^2}, \qquad \epsilon_y = -z \frac{\partial^2 w}{\partial y^2}, \qquad \gamma_{xy} = -2z \frac{\partial^2 w}{\partial x \partial y} \tag{2.4.3}$$

All the other strain components are zero, and the other strain–displacement relations need not be considered.

It is further assumed that the stress component in the z-direction is small and can be neglected in the stress–strain equations. Pressures on the upper and lower surface of the plate will certainly generate stresses at those surfaces, but, just as in beam theory, these stresses are far smaller than the bending stresses these same loads generate; hence, they are neglected. Thus, from Hooke's law,

$$\epsilon_x = \frac{1}{E}(\sigma_x - \upsilon \sigma_y), \qquad \epsilon_y = \frac{1}{E}(\sigma_y - \upsilon \sigma_x), \qquad \gamma_{xy} = \frac{1}{G}\tau_{xy} \tag{2.4.4}$$

It follows that

$$\sigma_x = \frac{E}{1 - \upsilon^2}(\epsilon_x + \upsilon \epsilon_y) = -z \frac{E}{1 - \upsilon^2}\left(\frac{\partial^2 w}{\partial x^2} + \upsilon \frac{\partial^2 w}{\partial y^2}\right)$$

$$\sigma_y = \frac{E}{1 - \upsilon^2}(\epsilon_y + \upsilon \epsilon_x) = -z \frac{E}{1 - \upsilon^2}\left(\frac{\partial^2 w}{\partial y^2} + \upsilon \frac{\partial^2 w}{\partial x^2}\right) \tag{2.4.5}$$

$$\tau_{xy} = G\gamma_{xy} = -z \frac{E}{1 + \upsilon}\frac{\partial^2 w}{\partial x \partial y}k$$

In dealing with equilibrium, it is convenient to work with stress resultants in terms of distributed moments rather than stresses, which are defined as follows:

$$M_x = \int_{-h/2}^{h/2} \sigma_x z \, dz = -D\left(\frac{\partial^2 w}{\partial x^2} + \upsilon \frac{\partial^2 w}{\partial y^2}\right)$$

$$M_y = \int_{-h/2}^{h/2} \sigma_y z \, dz = -D\left(\frac{\partial^2 w}{\partial y^2} + \upsilon \frac{\partial^2 w}{\partial x^2}\right) \tag{2.4.6}$$

$$M_{xy} = \int_{-h/2}^{h/2} \tau_{xy} z \, dz = -D(1 - \upsilon)\frac{\partial^2 w}{\partial x \partial y}$$

where

$$D = \frac{Eh^3}{12(1 - v^2)} \qquad (2.4.7)$$

These are stress resultants with units of force per unit length, rather than just force as was found in beam theory. Note the differences in definition and sign convention between plates and beams for M_x and M_y; that is, compare Equations 2.4.6 with 2.2.3 and note that M_{xy} is unique to plate theory and is often called the *twisting moment*. The use of the same or similar notation to mean different things in beam and plate theory may be confusing, but both are widely adopted conventions, which we shall not try to overthrow. Generally, we can keep them straight within the context of the problem being solved.

In addition, we define distributed shear forces

$$Q_x = \int_{-h/2}^{h/2} \tau_{zx}\, dz, \qquad Q_y = \int_{-h/2}^{h/2} \tau_{yz}\, dz \qquad (2.4.8)$$

These shear forces, the applied moments, and the applied transverse loads are used together to establish equilibrium. Consider the equilibrium of a small element, $dx\, dy$, of the plate as shown in Figure 2.4.3. Recognizing that

$$Q_x + dQ_x = Q_x + \frac{\partial Q_x}{\partial x}\, dx \qquad (2.4.9)$$

and that similar relations hold for the other quantities, we can satisfy equilibrium by summing forces on the element in the z-direction and summing

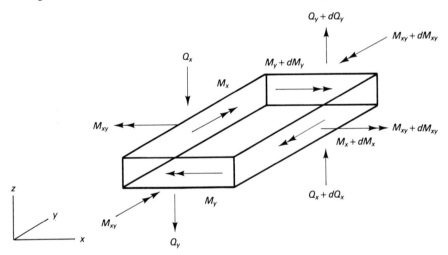

Figure 2.4.3 Plate Element in Bending and Shear.

moments about the x and y axes, which will give us

$$\frac{\partial Q_x}{\partial x} + \frac{\partial Q_y}{\partial y} + p = 0$$

$$\frac{\partial M_{xy}}{\partial x} - \frac{\partial M_y}{\partial y} + Q_y = 0, \qquad \frac{\partial M_{xy}}{\partial y} - \frac{\partial M_x}{\partial x} + Q_x = 0 \qquad (2.4.10)$$

By combining the equilibrium equations for moments with those that define the stress resultants in terms of the deflections, we obtain

$$Q_x = \frac{\partial M_x}{\partial x} - \frac{\partial M_{xy}}{\partial y} = -D \frac{\partial}{\partial x}\left(\frac{\partial^2 w}{\partial x^2} + v \frac{\partial^2 w}{\partial y^2}\right) - D(1 - v) \frac{\partial}{\partial y}\left(\frac{\partial^2 w}{\partial x \partial y}\right)$$

$$Q_y = \frac{\partial M_y}{\partial y} - \frac{\partial M_{xy}}{\partial x} = -D \frac{\partial}{\partial y}\left(\frac{\partial^2 w}{\partial y^2} + v \frac{\partial^2 w}{\partial x^2}\right) - D(1 - v) \frac{\partial}{\partial x}\left(\frac{\partial^2 w}{\partial x \partial y}\right)$$

(2.4.11)

For plates of constant thickness, these reduce to

$$Q_x = -D \frac{\partial}{\partial x}\left(\frac{\partial^2 w}{\partial x^2} + v \frac{\partial^2 w}{\partial y^2}\right)$$

$$Q_y = -D \frac{\partial}{\partial y}\left(\frac{\partial^2 w}{\partial y^2} + v \frac{\partial^2 w}{\partial x^2}\right)$$

(2.4.12)

If these expressions are substituted into the equilibrium equation for transverse forces in Equations 2.4.10, we obtain the differential equation that governs the bending deflection of a thin flat uniform plate for small deflections:

$$\nabla^4 w = \frac{\partial^4 w}{\partial x^4} + 2 \frac{\partial^4 w}{\partial x^2 \partial y^2} + \frac{\partial^4 w}{\partial y^4} = \frac{p}{D}$$

(2.4.13)

All that is needed to complete the formulation of the mathematical model for the plate is to define appropriate boundary conditions. We shall restrict our discussion to some of the more common conditions applied to rectangular plates.

(a) *Simple support.* When an edge is restrained from lateral deflection but is free to rotate, it is called simply supported. Consider that the edge at $x = a$ is so supported; then

$$w(a, y) = 0, \qquad M_x(a, y) = -D\left(\frac{\partial^2 w(a, y)}{\partial x^2} + \frac{\partial^2 w(a, y)}{\partial y^2}\right) = 0 \qquad (2.4.14)$$

But since $w(a, y) = 0$, it follows that $\partial^2 w(a, y)/\partial y^2 = 0$, and this reduces to

$$w(a, y) = 0, \qquad \frac{\partial^2 w(a, y)}{\partial x^2} = 0 \qquad (2.4.15)$$

Similar conclusions can be reached for other simply supported edges.

(b) *Fixed or clamped edges.* If the edge at $x = a$ is restrained in both lateral deflection and rotation,

$$w(a, y) = 0, \qquad \frac{\partial w(a, y)}{\partial x} = 0 \qquad (2.4.16)$$

and similarly for other fixed edges.

(c) *Free edge.* At first it would appear that if the edge at $x = a$ is stress free, the boundary conditions would be

$$M_x(a, y) = M_{xy}(a, y) = Q_x(a, y) = 0 \qquad (2.4.17)$$

It is pointed out in References 10 and 11 that Kirchhoff has proved that the preceding three conditions are too many and that just two conditions are sufficient to completely determine the deflection w. These two conditions are given without proof as follows:

$$Q_x(a, y) = -\frac{\partial M_{xy}(a, y)}{\partial y} = -\frac{\partial^3 w(a, y)}{\partial x^3} - (2 - v)\frac{\partial^3 w(a, y)}{\partial x\, \partial y^2} = 0$$

(2.4.18)

$$M_{xy}(a, y) = -\frac{\partial^2 w(a, y)}{\partial x^2} - \frac{\partial^2 w(a, y)}{\partial y^2} = 0$$

We shall let the subject of boundary conditions rest at this point and briefly discuss methods of solution. No general solution to Equation 2.4.13 has been found. As with the partial differential equations studied earlier, exact analytical solutions have been found in only a few cases where loading and boundaries are simple. We shall examine only two such cases to provide a basis for evaluating some approximate methods, which will be introduced in Chapter 3.

First, consider a strip of unit width of a plate that is infinite in the $\pm y$ directions and for which the edges at $x = 0$ and $x = a$ are constrained to remain straight. If p is a function of x only, then w is a function of x only. Equation 2.4.13 then reduces to

$$D\frac{d^4 w}{dx^4} = p(x)$$

(2.4.19)

This is the same as the beam equation with $D = Eh^3/12(1 - v^2)$ replacing EI. D is seen to be equal to EI per unit width of the plate but with the presence of the factor $1/(1 - v^2)$.

The reason for this factor can be seen by considering the stress–strain relations, taking into account the usual plate assumption $\sigma_z = 0$. Then

$$\epsilon_x = \frac{1}{E}(\sigma_x - v\sigma_y), \qquad \epsilon_y = \frac{1}{E}(\sigma_y - v\sigma_x)$$

(2.4.20)

Since the plate bends only in the x-direction, it can be presumed that $\epsilon_y = 0$; thus,

$$\sigma_y = v\sigma_x$$

(2.4.21)

and

$$\sigma_x = \frac{E\epsilon_x}{1 - v^2}$$

(2.4.22)

The corresponding relation for a beam does not contain the factor $1/(1 - v^2)$ because the narrow cross section is free to strain laterally.

Next let us consider the special case of a rectangular plate of length a and width b with simple supports on all four edges. In addition, give it the special set of sinusoidal loadings so that Equation 2.4.13 becomes

$$D\nabla^4 w = p(x, y) = a_{ij} \sin\frac{i\pi x}{a} \sin\frac{j\pi y}{b}$$

(2.4.23)

where i and j are integers and a_{ij} is a set of known constants associated with each pair of integers. The boundary conditions are

$$w(0, y) = w(a, y) = \frac{\partial^2 w(0, y)}{\partial x^2} = \frac{\partial^2 w(a, y)}{\partial x^2} = 0$$

$$w(x, 0) = w(x, b) = \frac{\partial^2 w(x, 0)}{\partial y^2} = \frac{\partial^2 w(x, b)}{\partial y^2} = 0$$

(2.4.24)

A solution is suggested of the form

$$w(x, y) = A_{ij} \sin \frac{i\pi x}{a} \sin \frac{j\pi y}{b}$$

(2.4.25)

Note that the boundary conditions are identically satisfied by this form. Upon substitution into Equation 2.4.23, we obtain

$$A_{ij}\left(\frac{i^4\pi^4}{a^4} + 2\frac{i^2 j^2 \pi^4}{a^2 b^2} + \frac{j^4 \pi^4}{b^4}\right) = \frac{a_{ij}}{D}$$

(2.4.26)

Given an a_{ij}, we can find an A_{ij} from Equation 2.4.26 that satisfies Equation 2.4.23.

Now i and j are integers that may take on any value. For each set of values, we have a solution. From the principle of superposition, a solution can be formed from the sum of all the individual solutions, or

$$p(x, y) = \Sigma\Sigma a_{ij} \sin \frac{i\pi x}{a} \sin \frac{j\pi y}{b}$$

(2.4.27)

and

$$w(x, y) = \Sigma\Sigma A_{ij} \sin \frac{i\pi x}{a} \sin \frac{j\pi y}{b}$$

(2.4.28)

It is interesting to note that an arbitrary load can be represented by the series summation of Equation 2.4.26. To show this, multiply both sides of Equation 2.4.27 by $\sin r\pi x/a \sin s\pi y/b$ and integrate from 0 to a on x and from 0 to b on y, or

$$\int\int p(x, y)\sin \frac{r\pi x}{a} \sin \frac{s\pi y}{b} dx\, dy$$

$$= \Sigma\Sigma\, a_{ij}\int\int \sin \frac{r\pi x}{a} \sin \frac{s\pi y}{b} \sin \frac{i\pi x}{a} \sin \frac{j\pi y}{b} dx\, dy$$

(2.4.29)

Since

$$\int_0^a \sin \frac{i\pi x}{a} \sin \frac{r\pi x}{a} dx = 0, \qquad i \neq r$$

$$= \frac{a}{2} \qquad i = r$$

$$\int_0^b \sin \frac{j\pi y}{b} \sin \frac{s\pi y}{b} dy = 0 \qquad j \neq s$$

$$= \frac{b}{2} \qquad j = s$$

(2.4.30)

it follows that

$$a_{ij} = \frac{4}{ab} \int_0^a \int_0^b p(x, y) \sin \frac{i\pi x}{a} \sin \frac{j\pi y}{b} \, dx \, dy \qquad (2.4.31)$$

Thus, for any loading, $p(x, y)$, the coefficients a_{ij} may be found. Such a series is known as a *Fourier series* and is widely used in the mathematical analysis of structures.

Example 2.4.1

Find the deflection of a rectangular plate with simple supported edges that has a uniform applied load, p_0.

Solution: First, find a_{ij}.

$$a_{ij} = \frac{4p_0}{ab} \int_0^a \sin \frac{i\pi x}{a} \, dx \int_0^b \sin \frac{j\pi y}{b} \, dy = 0, \qquad i \text{ and/or } j \text{ even}$$

$$\qquad (a)$$

$$= \frac{16p_0}{\pi^2 ij}, \qquad i \text{ and } j \text{ odd}$$

Thus, for i and j odd,

$$w(x, y) = \frac{16p_0}{\pi^6 D} \sum\sum \frac{1}{ij[(i^2/a^2) + (j^2/b^2)]^2} \sin \frac{i\pi x}{a} \sin \frac{j\pi y}{b} \qquad (b)$$

It is tempting at this point to try to find other solutions; however, in the practical analysis of plates, other methods are more successful. We shall return to the analysis of plates when these other methods have been established. We have taken this brief time because the derivation of the plate equation is so analogous to the beam and because the relationships established will be useful in the later developments of plate theory by other methods.

2.5 CONCLUSIONS AND SUMMARY

The simplifying assumptions that lead to beam and plate theory have been proved in practice as some of the most powerful tools in all of engineering analysis. A principal mathematical advantage of beam theory over the theory of elasticity is that the problems have been reduced from partial to ordinary differential equations. The methods are particularly adept at providing deflections and normal stresses, but are less able to provide shear stresses. Part of that difficulty is overcome with the additional feature of thin-walled cross sections, which, fortunately, are widely used in aerospace and automotive applications and other fields as well.

Referring back to the symbolic representation of the equations of beam and plate theory in Figure 2.2.1 and Equations 2.2.1 and 2.2.2, we shall summarize here the more important of these equations for the several general and special cases cited. We shall retain the numbering of the equations as given in the main body of the chapter to assist in finding the detailed derivations and discussions.

From the *general beam theory*, we obtain the equation for the axial deformation:

$$(EAu_0')' = -f_x(x) \qquad (2.2.14)$$

Once the deflections are found, we can calculate the internal stress resultant

$$S_x(x) = EA\frac{\partial u_0}{\partial x} \tag{2.2.8}$$

and find the stress

$$\sigma_x = \frac{S_x(x)}{A} \tag{2.2.7}$$

For the torsional deformation,

$$(GJ\beta')' = -t_x(x) \tag{2.2.22}$$

Once this equation is solved, the internal stress resultant may be found from

$$M_x(x) = GJ\beta' \tag{2.2.18}$$

and the stresses by the methods of Section 1.3.3 or Section 2.3, as appropriate. Of special interest is the torsional stress in a thin-walled closed section given by

$$q = \frac{M_x}{2A} \tag{2.3.16}$$

and the rate of twist by

$$\theta = \frac{M_x}{GJ} \tag{2.3.24}$$

where

$$J = \frac{4A^2}{\int (ds/h)} \tag{2.2.25}$$

This J would be used in Equation 2.2.23 to find the torsional deformation of a beam with a thin-walled, closed section.

For the bending deformation,

$$(EI_{zz}v'')'' = f_y(x), \qquad (EI_{yy}w'')'' = f_z(x) \tag{2.2.58}$$

Collectively, these are the forms of Equation 2.2.2 for a beam in bending and shear. Once the deflections are found from these equations, the stresses may be determined by working back through the strain–displacement and stress–strain equations. The normal stress in beams can be found from the deflections by

$$M_z = EI_{zz}v'', \qquad M_y = EI_{yy}w''$$

$$\sigma_x(x, y, z) = -\frac{M_z}{I_{zz}}y - \frac{M_y}{I_{yy}}z \tag{2.2.38}$$

The shear stress from transverse loads are generally cross section dependent and are much harder to find. Transverse shear stresses have been found for rectangular cross sections. For *thin-walled* cross sections, some general shear stress solutions are possible. For these sections, we have found that the *shear flow* due to a transverse shear force is given by

$$q = q_0 - \int_0^s \left(\frac{S_y y}{I_{zz}} + \frac{S_z z}{I_{yy}}\right) h \, ds \tag{2.3.8}$$

and the *shear center*, a point through which the shear force must pass for the equation to be valid, is found from

$$S_y e_z = \int qr\, ds \tag{2.3.9}$$

$$S_z e_y = \int qr\, ds \tag{2.3.10}$$

The equation for the deflection of *thin plates* is

$$\nabla^4 w = \frac{\partial^4 w}{\partial x^4} + 2\frac{\partial^4 w}{\partial x^2\, \partial y^2} + \frac{\partial^4 w}{\partial y^4} = \frac{p}{D} \tag{2.4.13}$$

This is the form of Equation 2.2.2 for plates. The stresses may be determined by working back through the strain–displacement and stress–strain equations. Once the deflections are found, the stresses are obtained from

$$\sigma_x = \frac{E}{1 - v^2}(\epsilon_x + v\epsilon_y) = -z\frac{E}{1 - v^2}\left(\frac{\partial^2 w}{\partial x^2} + v\frac{\partial^2 w}{\partial y^2}\right)$$

$$\sigma_y = \frac{E}{1 - v^2}(\epsilon_y + v\epsilon_x) = -z\frac{E}{1 - v^2}\left(\frac{\partial^2 w}{\partial y^2} + v\frac{\partial^2 w}{\partial x^2}\right) \tag{2.4.5}$$

$$\tau_{xy} = G\gamma_{xy} = -z\frac{E}{1 + v}\frac{\partial^2 w}{\partial x \partial y}$$

Even the power of general beam and thin plate theory is limited, because as loading and restraints become more complex so do the solutions. Still more powerful methods are needed to extend solutions to more complex beams and plates, to more complex geometries and loading not covered by beam and plate theory, and to built-up combinations of structures. We shall find some of these methods in Chapter 3.

PROBLEMS

1. A solid bar with a circular cross section is acted on by three simultaneous loadings: (1) a body force m with units of N/mm^3, (2) a surface traction τ with units of N/mm^2 acting on the cylindrical surface of the bar parallel to the surface, and (3) a surface traction σ with units N/mm^2 acting normal to the end surface at $x = a$. The bar is fixed at $x = 0$. Find the displacements and stresses in the bar.

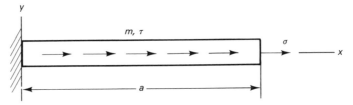

2. The bar in Problem 1 is subjected to a distributed applied torque

$$t_x(x) = t_0 a\left(1 - \frac{x}{a}\right)$$

where t_0 is a constant. Find the displacements and stresses in the bar.

3. A cantilever beam has the cross section shown. A pure moment M_z is applied at the free end ($M_y = 0$). Find the moments and product of inertia of the cross section. Find the deflections and stresses in the beam.

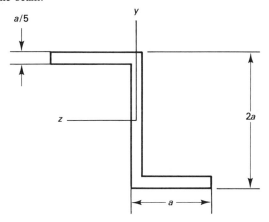

4. A uniform beam with the same cross section as Example 2.2.2 has the loading and boundary support shown. Find the deflection w in the z-coordinate direction and also the deflection v in the y-coordinate direction. Note that the only loading is in the y-coordinate direction.

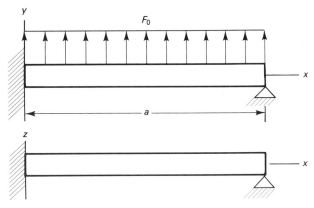

5. A uniform beam has the cross section, loading, and boundary support shown. Find the deflections and stresses in the beam. At what value of x is the normal stress the largest? The shear stress?

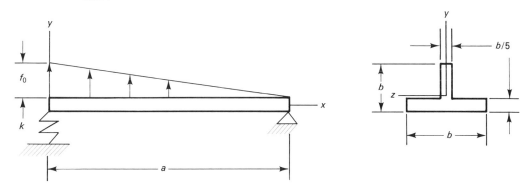

6. Consider the cantilever beam with a square cross section and with a rigid plate attached to the free end so that a load can be applied off center as shown. The load F is applied at the point A.

(a) Find all three displacement components.

(b) Find all six strain components.

(c) Find all six stress components.

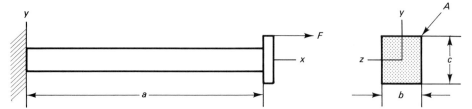

7. Find the shear flow and shear center for a beam with the cross section shown. Note that the origin of the y-z axes is not at the centroid. Why not?

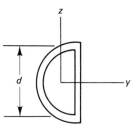

8. A slender, thin-walled tube with the cross section shown is loaded in pure torque. If $d = 50$ mm, the wall thickness is 3 mm, and the torque is 10,000 N-mm, what is the stress in the tube and the rate of twist? What is the total angle of twist of one end relative to the other end if the length of the tube is 500 mm?

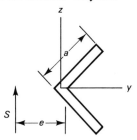

9. Two beams have the cross sections shown and are loaded by equal shear forces as shown. The two cross sections are identical except (a) is a closed section and (b) is an open section because of a slit running the full length of the beam, as shown. In both the wall thickness is h. Find the shear stress in each cross section.

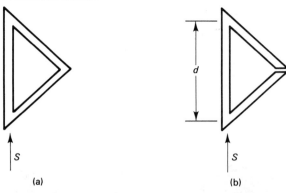

10. Consider the thin-walled beam with stiffeners in figure (a) and two different idealizations in figures (b) and (c). In (c), only the spar caps and stringers support bending stresses, while in (b) both the spar caps and stringer do. All dimensions are in millimeters. The two spar caps each have an area of 500 mm^2, and the two stringers each have half that area. Find the shear stress for cases (b) and (c).

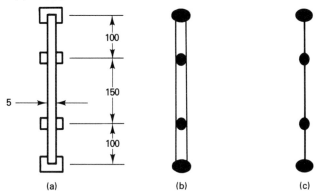

(a)   (b)   (c)

11. Spar caps are added to the beam in Problem 8 as shown. The area of each spar cross section is $A = dh$. Find the shear stress in each cross section. Consider the case where the spars carry all the bending stress, as well as the case where they do not.

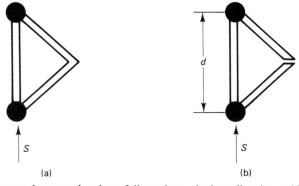

(a)   (b)

12. A simply supported rectangular plate of dimensions a in the x-direction and b in the y-direction has the following loading.

$$p(x, y) = p_0\left(1 - \frac{x}{a}\right)\left(1 - \frac{y}{b}\right)$$

What are the deflections and the stresses in the plate?

WORK AND ENERGY METHODS

Introduction to the Finite Element Method

3

3.1 INTRODUCTION

Powerful methods based on work and energy can be developed as alternatives to the differential equation methods discussed in the first two chapters. The foundation of these methods is the *principle of virtual work*. From this principle, we can establish the equilibrium of the solid body, and in this sense it replaces Newton's laws in the formulation of a set of equations for finding the deformations, strains, and stresses.

In this chapter, after we develop the principle of virtual work and some closely related energy principles, we apply them to problems in static structural mechanics. At first we consider some simple traditional approaches, including the *Rayleigh-Ritz method*, and then we apply the most modern form, which is known as the *finite element method (FEM)* or *finite element analysis (FEA)*. The approach to problem solving known as the finite element method has been developed in the last three decades to become the most powerful and widely used tool in the analysis of solid bodies under load.

The material in this chapter may be found in many books. References 4 and 5 deal with the work and energy principles in much the same context as we do here. References 13 through 17 are some of the more recent books on the finite element method. All these references contain many additional references, including some of the original sources.

3.2 PRINCIPLE OF VIRTUAL WORK

To show the place of the principle of virtual work in the grand plan, we return to the symbolic representation of the equations of solid mechanics, shown in Figure 1.2.1, and adapt it to this new approach. As noted in Figure 3.2.1, we retain exactly the same representation and definition of the stresses, strains, and deformations as we have used in the previous two chapters, and we retain Hooke's law and the strain–displacement equations as before. We define equilibrium quite differently, however, without the use of the differential equations of equilibrium; in fact, equilibrium is stated in the form of an integral equation, $\delta W = 0$. And all the applied loads and the geometric constraints are contained explicitly in the integral equation, not as boundary conditions.

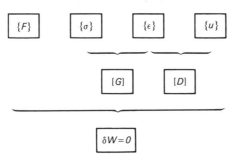

Figure 3.2.1 Symbolic Representation of the Virtual Work-Based Equations of Solid Mechanics.

In this symbolic representation, body forces are lumped with other applied forces, including both distributed and concentrated applied loads, in the matrix $\{F\}$. Thus, the equations to be solved are now

$$\{\sigma\} = [G]\{\epsilon\} \qquad \text{stress–strain equations}$$
$$\{\epsilon\} = [D]\{u\} \qquad \text{strain–displacement equations}$$

(3.2.1)

These are joined with the applied loads and geometric constraints to satisfy equilibrium in the symbolic form

$$\delta W = 0 \qquad (3.2.2)$$

Equilibrium stated in the form of Equation 3.2.2 will provide us with some powerful generalizations but with few specific solutions of practical problems. We can turn to an approximate method of solving the equations known as the Rayleigh-Ritz method. To do this, we add two new terms to the symbolic representation as shown in Figure 3.2.2. The new terms are

$$\{q\} \quad \text{generalized displacements}$$

$$\{Q\} \quad \text{generalized forces}$$

and this requires two new equations:

$$\{u\} = [\Omega]\{q\} \qquad \text{admissible function equations}$$
$$\{Q\} = [A]\{F\} \qquad \text{generalized force equations}$$

(3.2.3)

The preceding equations are combined into a new set of linear algebraic equations of the form

$$[K]\{q\} = \{Q\} \tag{3.2.4}$$

The meaning of these new terms and new equations will all be explained in due course.

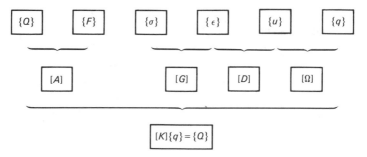

Figure 3.2.2 Symbolic Representation of the Rayleigh-Ritz Form of the Equations of Solid Mechanics.

The principle of virtual work we are about to consider in detail is so simple in form yet so powerful in concept that we must proceed carefully or risk misunderstanding vital points. We begin, therefore, with some illustrations of virtual work as it applies to a particle under load. This helps to fix some of the major concepts of virtual work in a very simple setting. We follow with the application to a general three-dimensional body and then specialize that for the case of the slender bar under axial load.

3.2.1 Virtual Work for a Particle

Consider a single particle that is acted on by a set of n concurrent forces as shown in Figure 3.2.3. Now imagine that the particle is displaced an amount δa and the forces move with it, but with no change in the magnitude or the direction of any of the forces. It is important to note that we are not speaking of the actual displacement of the particle caused by the action of the applied forces. We are speaking of an imagined displacement in which the forces are imagined to behave as stated. For this reason, the displacement δa is referred to as a *virtual displacement* to distinguish it from the actual displacement. Throughout this chapter, the symbol δ before any quantity will denote it to be a virtual or imaginary quantity.

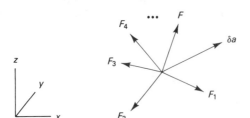

Figure 3.2.3 A Particle in Equilibrium.

The work imagined to be done on the particle during the virtual displacement is the sum of all the components of force in the direction of δa times δa. This is called the *virtual work*. If we divide the virtual displacement into components δu, δv, δw in the x-, y-, z-directions, we have for the virtual work

$$\delta W = \Sigma \, F_{ix} \, \delta u + \Sigma \, F_{iy} \, \delta v + \Sigma \, F_{iz} \, \delta w \qquad (3.2.5)$$

where F_{ix} is the component of the ith force in the x-direction and similarly for the y- and z-directions. We may note that if the particle is in equilibrium

$$\Sigma \, F_{ix} = 0, \qquad \Sigma \, F_{iy} = 0, \qquad \Sigma \, F_{iz} = 0 \qquad (3.2.6)$$

and therefore

$$\delta W = 0 \qquad (3.2.7)$$

In words, if a particle is in equilibrium under the action of a set of concurrent forces, the total virtual work done by the forces during an arbitrary virtual displacement is zero. Since δa is arbitrary, Equation 3.2.7 is true for all virtual displacements, and it is possible to reverse the preceding statement to define equilibrium as follows:

A particle is in equilibrium if and only if the virtual work of all the forces acting on the particle is zero during an arbitrary virtual displacement.

Thus, Equation 3.2.7. becomes a condition that establishes or defines equilibrium, rather than a result of equilibrium.

Let us consider a slightly more complicated case of a particle. Let the particle be a point mass attached to a support by a linear spring with a spring constant k. Have acting on it a body force F_b, as well as other applied forces, as shown in Figure 3.2.4. For convenience of illustration, let us restrict all quantities to the x-y plane and have the x-direction coincide with the direction of the spring force. Now imagine the particle to have achieved a static equilibrium position under a spring extension u and then to be given a virtual displacement with components δu and δv. The virtual work is

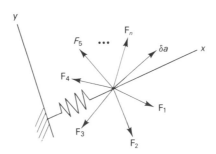

Figure 3.2.4 A Particle Restrained by a Spring.

$$\delta W = \Sigma \, (F_{ix} \, \delta u + F_{iy} \, \delta v) + F_{bx} \, \delta u + F_{by} \, \delta v - (ku) \, \delta u = 0 \qquad (3.2.8)$$

where the unknown force in the spring is ku. By regrouping terms, we have

$$\delta W = (\Sigma \, F_{ix} + F_{bx} - ku) \, \delta u + (\Sigma \, F_{iy} + F_{by}) \, \delta v = 0 \qquad (3.2.9)$$

and, therefore, for equilibrium

$$\Sigma \, F_{ix} + F_{bx} - ku = 0, \qquad \Sigma \, F_{iy} + F_{by} = 0 \qquad (3.2.10)$$

These are the familiar equations of equilibrium, which can also be obtained by applying Newton's laws, but we just got them using the principle of virtual work. And they can be used to find the unknown displacement u.

We can use this example to identify important features of the general statement of virtual work, which we shall establish for a solid body in the next section. Note that there are three states of the particle: (I) an original state where there are no loads and the spring is unstrained, (II) a final equilibrium state after the loads are applied and the spring has been stretched an amount u, and (III) an imaginary state where a virtual displacement δu is present. Note further that the true displacement of the particle is the difference in displacement between state I and II and that the virtual displacement is the imagined difference between states II and III. Note also that the condition that the forces do not change during the virtual displacement applies to the internal spring force ku, which is clearly displacement dependent, as well as to the external applied forces.

It is convenient to identify certain terms in the expression for virtual work by separate notation, as follows:

$$\delta W = \delta W_e + \delta W_i = 0 \tag{3.2.11}$$

where δW_e is the virtual work of the *external forces* and can be broken further into

$$\delta W_e = \delta W_s + \delta W_b \tag{3.2.12}$$

where δW_s is the virtual work of the *surface forces* and δW_b of the *body forces*. The other term δW_i is the virtual work of the *internal forces*, in this case of the forces in the spring. For the preceding particle,

$$\delta W_s = \Sigma \,(F_{ix} \,\delta u + F_{iy} \,\delta v), \qquad \delta W_b = F_{bx} \,\delta u + F_{by} \,\delta v$$

$$\delta W_i = -(ku) \,\delta u \tag{3.2.13}$$

These illustrations should help in gaining an understanding of the more complex problem of virtual work in a three-dimensional solid body.

3.2.2 Virtual Work for a Deformable Body

Consider a general solid body as shown in Figure 3.2.5. It is shown in two dimensions for convenience of illustration, but think of it as a general three-dimensional body. It may be acted on by body forces that have units of force per unit volume and surface forces with units of force per unit area. Distributed line loads and concentrated loads may also be present. It may also be geometrically constrained so that no rigid body motions can occur by distributed surface constraints or line or point constraints. In short, it is exactly the same body considered in Section 1.2. It is important to note that we can specify the load or the constraint at each and every point on the surface, but we cannot specify both. Everywhere on the surface where the constraint is not specified, the load must be specified, and everywhere the load is not explicitly given, it is assumed to be specified as zero.

As steps in developing the principle of virtual work, we first identify three states of the body. In state I, constraints are in place, but the body is unloaded. To keep things simple, we consider the case where the constraints all impose zero displacements and so the body is undeformed and unstressed. This can be generalized to include nonzero constraint displacements, but we do not wish to introduce too many new concepts all at once.

State II is the true state of loading and deformation of the body; that is, the true loads are applied to the body defined in state I and it is allowed to deform to its true state of

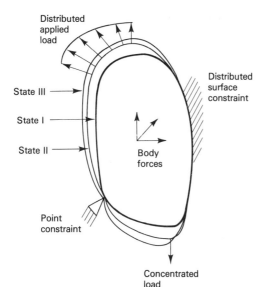

Distributed applied load

State III

State I

State II

Body forces

Distributed surface constraint

Point constraint

Concentrated load

Figure 3.2.5 Three States of a Deformable Body.

deformation. It is assumed that the deformation takes place slowly so that no dynamic or thermal effects are introduced; thus, in state II the body is in static equilibrium. Our goal is to find the true state of stress, strain, and deformation of the body as it exists in state II. This was our goal in Chapters 1 and 2, as well.

Finally, state III is the imaginary or virtual state of the body obtained by assigning virtual displacements. It is in this state of the body that the principle of virtual work provides us with a suitable statement of equilibrium.

Note well that we take as given that the geometric constraints imposed in state I can never be violated by the true displacements in state II or the virtual displacements in state III. Furthermore, the material of the body is assumed to be distributed continuously, and both the components of deformation u, v, and w and the components of the virtual displacements δu, δv, and δw are differentiable functions of the coordinates x, y, and z. Functions that have all the properties described are called *kinematically admissible functions* or sometimes *geometrically admissible functions*. Thus, any kinematically admissible function can be used as a virtual displacement.

The principle of virtual work is as fundamental as Newton's laws and as such is equally valid for establishing equilibrium. For a deformable body it may be stated as follows:

A deformable solid body is in equilibrium if and only if the total virtual work is zero for every kinematically admissible virtual displacement.

It is helpful in relating the stress, strain, and deformation components to the concepts of this principle to go through a derivation of the mathematical statement of the principle using the differential equations of equilibrium we derived in Chapter 1 with Newton's laws. Because this derivation is a bit involved and would interrupt the rhythm of the current discussion, it has been placed in Appendix G and only the results are presented here.

The virtual work of a deformable solid body is divided into two parts, as follows:

$$\delta W = \delta W_e + \delta W_i \qquad (3.2.14)$$

where the external virtual work is given by

$$\delta W_e = \delta W_s + \delta W_b \qquad (3.2.15)$$

The external virtual work of surface forces is given by

$$\delta W_s = \int (X_s \, \delta u + Y_s \, \delta v + Z_s \, \delta w) \, dA \qquad (3.2.16)$$

where the integration is carried out over that portion of the surface where surface forces are prescribed. Similarly, the external virtual work of body forces is given by

$$\delta W_b = \int (f_x \, \delta u + f_y \, \delta v + f_z \, \delta w) \, dV \qquad (3.2.17)$$

where the integration is carried out over the volume of the body. Concentrated and distributed line forces on the surface are not explicitly presented in this statement. They can be assumed to be special forms of Equation 3.2.16 and no loss of generality is incurred. How they are introduced in practical applications will be given later.

The internal virtual work (as shown in Appendix G) is given by

$$\delta W_i = -\delta U \qquad (3.2.18)$$

where

$$\delta U = \int (\sigma_x \, \delta\epsilon_x + \sigma_y \, \delta\epsilon_y + \sigma_z \, \delta\epsilon_z + \tau_{xy} \, \delta\gamma_{xy} + \sigma_{yz} \, \delta\gamma_{yz} + \tau_{zx} \, \delta\gamma_{zx}) \, dV \qquad (3.2.19)$$

In summary, the statement for virtual work is

$$\delta W = \delta W_s + \delta W_b - \delta U$$

$$= \int (X_s \, \delta u + Y_s \, \delta v + Z_s \, \delta w) \, dA + \int (f_x \, \delta u + f_y \, \delta v + f_z \, \delta w) \, dV \qquad (3.2.20)$$

$$- \int (\sigma_x \, \delta\epsilon_x + \sigma_y \, \delta\epsilon_y + \sigma_z \, \delta\epsilon_z + \tau_{xy} \, \delta\gamma_{xy} + \sigma_{yz} \, \delta\gamma_{yz} + \tau_{zx} \, \delta\gamma_{zx}) \, dV$$

All this can be put in matrix form using the matrix definitions for stress, strain, displacement, body force, Hooke's law, strain–displacement equations, and applied surface loads. We shall designate the transpose of a matrix with the superscript T. Thus, when we write, for example, $\{\delta u\}^T$, we shall have the transpose of the column matrix $\{\delta u\}$ or a row matrix. In matrix form

$$\delta W = \int \{\delta u\}^T \{F_s\} \, dA + \int \{\delta u\}^T \{f\} \, dV - \int \{\delta\epsilon\}^T \{\sigma\} \, dV = 0 \qquad (3.2.21)$$

So far the symbol δ has been a prefix on the symbols for work, displacement, and strain to characterize them as virtual quantities. We shall now recognize it as an operator so that, for example,

$$\delta\epsilon_x = \delta \frac{\partial u}{\partial x} = \frac{\partial \delta u}{\partial x} = \frac{\partial}{\partial x} (\delta u) \qquad (3.2.22)$$

Now we introduce the matrix form of the strain–displacement and stress–strain equations from Equations 3.2.1. This allows us to write

$$\delta U = \int \{\delta\epsilon\}^T \{\sigma\} \, dV = \int ([D] \, \{\delta u\})^T [G][D]\{u\} \, dV \qquad (3.2.23)$$

The virtual work may now be written

$$\delta W = \delta W_s + \delta W_b - \delta U$$

$$= \int \{\delta u\}^T \{F_s\} \, dA + \int \{\delta u\}^T \{f\} \, dV - \int ([D] \{\delta u\})^T [G][D]\{u\} \, dV = 0 \tag{3.2.24}$$

The problem is to find $\{u\}$ so that the value of $\delta W = 0$.

A branch of mathematics called the *calculus of variations* is designed to solve this type of problem. If we exercise that mathematics on the full three-dimensional solid body, we shall discover that the value of $\{u\}$ that causes the virtual work to be zero is precisely the value that satisfies the differential equations of elasticity given in Chapter 1. Lest we appear to be going in circles, remember that virtual work is an independent and equivalent statement of equilibrium, and we should expect it to conclude the same as Newton's laws.

Fortunately, there are circumstances in which the preceding integrals can be evaluated without going through the formal exercise of the calculus of variations. This approach is sometimes called the *direct method* of the calculus of variations. This method can be used with the simplified equations of elasticity and, in particular, with slender bars, so let us prepare for the next section by formulating virtual work for slender bars.

3.2.3 Virtual Work for Slender Bars

First, consider a slender bar with axial loading only. The only nonzero component of displacement is u_0 and the only virtual displacement needed is δu_0. In this case, the strain-displacement matrix and the Hooke's law matrix each contain only one term. Since one-element matrices are simply scalars, we have

$$\epsilon_x = \frac{\partial u_0}{\partial x}, \qquad \sigma_x = E\epsilon_x \tag{3.2.25}$$

Also, if we include the body force with the applied load, as is the usual practice, the external work comes from $f_x(x)$, which is the applied axial load per unit length along the x-axis, and from any concentrated loads present. Remember that concentrated loads and distributed line loads were not explicitly represented in the external work term in the virtual work for the full three-dimensional solid given previously. This affords us an opportunity to show how these are introduced into the external work term.

For the bar, the external virtual work may be stated to be

$$\delta W_e = \int \delta u_0 f_x(x) \, dx + \Sigma \, \delta u_{0i} F_{xi} \tag{3.2.26}$$

where F_{xi} is the concentrated load in the x-direction at the point x_i, and δu_{0i} is the virtual displacement at that point. Thus $\delta u_{0i} = \delta u_0(x_i)$. The statement for virtual work becomes

$$\delta W = \delta W_e - \delta U = \int \delta u_0 f_x(x) \, dx + \Sigma \, \delta u_{0i} F_{xi} - \int \{\delta \epsilon\}^T \{\sigma\} \, dV$$

$$= \int \delta u_0 f_x(x) \, dx + \Sigma \, \delta u_{0i} F_{xi} - \int \frac{\partial \delta u_0}{\partial x} E \frac{\partial u_0}{\partial x} A \, dx = 0 \tag{3.2.27}$$

where A is the cross-sectional area of the bar.

The secret for getting a solution to a practical problem out of virtual work is to find a way to evaluate the integral for the internal work despite the fact that it contains the unknown deflection components. The next example shows a way to evaluate the integral for a very simple structural system consisting of an axially loaded bar, but this cannot necessarily be generalized to more complex systems. Nevertheless, it shows virtual work can be used to solve problems directly.

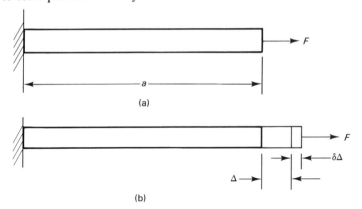

(a)

(b)

Example 3.2.1

Use virtual work to find the deflections, strains, and stresses in a bar fixed at one end and with a concentrated load on the other as shown in figure (a).

Solution: We know that the beam will elongate and we can express the deflection of the free end by Δ, as shown in figure (b), and we can add a virtual displacement $\delta\Delta$. Given Equation 3.2.27, we can recognize that the external virtual work caused by the one concentrated force is simply

$$\delta W_e = \delta\Delta \, F \tag{a}$$

To put the internal virtual work in a form we can use, we also recognize that for this problem we can represent the strain and the virtual strain by

$$\epsilon_x = \frac{\Delta}{a} \longrightarrow \delta\epsilon_x = \frac{\delta\Delta}{a} \tag{b}$$

The stress is now

$$\sigma_x = E\epsilon_x = \frac{E\Delta}{a} \tag{c}$$

so that the virtual work becomes

$$\delta W = \delta\Delta F - \int_0^a \frac{\delta\Delta}{a} E \frac{\Delta}{a} A \, dx = 0 \tag{d}$$

Since everything inside the integral is a constant, all we have to do is integrate dx from 0 to a and we have

$$\delta W = \delta\Delta F - \frac{\delta\Delta}{a} \frac{E\Delta}{a} Aa = \delta\Delta\left(F - \frac{EA\Delta}{a}\right) = 0 \tag{e}$$

and, since $\delta\Delta$ is arbitrary,

$$F - \frac{EA\Delta}{a} = 0 \longrightarrow \Delta = \frac{Fa}{EA} \tag{f}$$

Also,

$$\epsilon_x = \frac{F}{EA}, \qquad \sigma_x = \frac{F}{A} \tag{g}$$

The key to this solution is that it is possible to formulate the unknown displacements, strains, and stresses in such a way that the evaluation of the virtual work integral could be carried out explicitly. This approach can be extended to assemblies of axial loaded bars, or trusses, to beams and assemblies of beams, or frames, and to some other structures. This is not often done any more because a more convenient form of solution has been developed. We shall be considering this other method later in this chapter. For now, we shall complete the formulation of the virtual work expressions for the slender bar in anticipation of developing this other method.

The virtual work expression for a bar in pure torsion is very like the one for axial loading. We note from Section 1.3.3 that for a circular cross section in cylindrical coordinates

$$\gamma = \frac{\tau}{G} = r \frac{\partial \beta}{\partial x} \longrightarrow \delta \gamma = r \frac{\partial \delta \beta}{\partial x} \tag{3.2.28}$$

The virtual work becomes

$$\delta W = \delta W_e - \delta U = \int \delta \beta t_x(x)\, dx + \Sigma\, \delta \beta_i T_{xi} - \int \{\delta \epsilon\}^T \{\sigma\}\, dV$$

$$= \int \delta \beta t_x(x)\, dx + \Sigma\, \delta \beta_i T_{xi} - \int \frac{\partial \delta \beta}{\partial x}\, G\, \frac{\partial \beta}{\partial x}\, (\int r^2\, dA)\, dx \tag{3.2.29}$$

$$= \int \delta \beta t_x(x)\, dx + \Sigma\, \delta \beta_i T_{xi} - \int \frac{\partial \delta \beta}{\partial x}\, GJ\, \frac{\partial \beta}{\partial x}\, dx$$

where $J = \int r^2\, dA$, the polar moment of inertia. For noncircular cross sections, the appropriate torsional constant is used for J.

Next we develop the expression for the virtual work in a beam with transverse loading. For convenience, let us simplify the problem by considering $I_{yz} = 0$, no axial loads, and transverse loads only in x-y plane. We may remember that the deformation pattern assumed in beam theory, often stated as plane sections remain plane, does not account for shear deformation. Therefore, the equations for beam displacement and beam normal stress do not include evidence of the presence of shear stress. This leads to little error in the resulting equations because the true shear deformation compared to the deformation of bending is very small for beams. It can be shown that neglecting the shear contribution to the internal work is exactly equivalent to what we did in neglecting shear deformation in deriving the differential equations for beams, and so we neglect it in what follows. We shall give further justification of this neglect in a later section. It follows that only the one displacement component v is needed, and it is seen that the matrices reduce to scalars:

$$\epsilon_x = -y\, \frac{\partial^2 v}{\partial x^2}, \qquad \sigma_x = E\epsilon_x \tag{3.2.30}$$

Concentrated loads for beams include both transverse forces and moments, so the external virtual work is

$$\delta W_e = \int \delta v f_y(x) \, dx + \Sigma \, F_{yi} \, \delta v_i + \Sigma \, M_{zj}(\delta v_j)' \qquad (3.2.31)$$

where $f_y(x)$ is the applied transverse load per unit length along the x-axis, F_{yi} is a concentrated transverse load at the point x_i, v_i is $v(x_i)$, M_{zj} is a concentrated moment, and $(\delta v_j)'$ is $\delta v'(x_j)$. We recognize here that the first derivative of the deflection is, for small deflections, the angle through which a concentrated moment moves to produce work.

The statement for virtual work becomes

$$\delta W = \int \delta v f_y(x) \, dx + \Sigma \, F_{yi} \, \delta v_i + \Sigma \, M_{zj}(\delta v_j)' - \int \{\delta \epsilon\}^T \{\sigma\} \, dV$$

$$= \int \delta v f_y(x) \, dx + \Sigma \, F_{yi} \, \delta v_i + \Sigma \, M_{zj}(\delta v_j)' - \int \frac{\partial^2 \delta v}{\partial x^2} \, E \, \frac{\partial^2 v}{\partial x^2} \, (\int y^2 \, dA) \, dx \qquad (3.2.32)$$

$$= \int \delta v f_y(x) \, dx + \Sigma \, F_{yi} \, \delta v_i + \Sigma \, M_{zj}(\delta v_j)' - \int \frac{\partial^2 \delta v}{\partial x^2} \, EI_{zz} \, \frac{\partial^2 v}{\partial x^2} \, dx$$

We shall now add one more feature to the beam equation. As we noted in Section 2.2.4, springs are often added to beams, particularly to represent flexible boundary conditions. The effect of springs may be added to the virtual work expression. First, we note that springs are structural members, and when work is done on the structure, they contribute to the internal work. Two standard forms of linear elastic springs are of primary interest. One contributes a force from lateral elongation and the other from rotation. These are depicted in Figure 2.2.13. The force in a lateral spring and the moment in a rotational spring, as given in Section 2.2.4, are

$$F_e = kv(x_m), \qquad M_e = \mu v'(x_n) \qquad (3.2.33)$$

where m and n are indexes indicating the values of x where the springs are placed. Using the notation $v_m = v(x_m)$ and $v'_n = v'(x_n)$, we can write the internal virtual work for several springs attached to a beam as

$$\delta \Delta U = \Sigma \, k_m v_m \, \delta v_m + \Sigma \, \mu_n v_n' \, \delta v_n' \qquad (3.2.34)$$

This term is added, or perhaps we should say subtracted, from Equation 3.2.32 when springs are present.

We could at this point solve some problems with beam bending for beams and frames with discrete applied forces and moments in a manner analogous to the preceding example. Distributed loads, however, present difficulties in attempting to use this method. We are after a more general approach that does deal nicely with distributed as well as discrete loads, so we shall now ask how we can use virtual work in problems where the selection of discrete values is not so obvious. This leads us to an important and powerful method for problem solving, which deserves a section of its own.

3.2.4 Rayleigh-Ritz Method

Suppose it is possible to represent the distributed displacement field $\{u\}$ of any solid body by an expression of the form

$$\{u\} = [\Omega]\{q\} \qquad (3.2.35)$$

where $\{\Omega\}$ is a set of known kinematically admissible functions of the coordinates and $\{q\}$ is a set of constants. This idea is not foreign to our experience. It is common mathematical practice to represent a known function by a series of other known functions as, for example, in Fourier series analysis. In our problem, at this point, however, both $\{u\}$ and $\{q\}$ are unknown quantities. Nevertheless, if this is possible, and we shall show in a moment that it is, we can say

$$\{\delta u\} = [\Omega]\{\delta q\} \tag{3.2.36}$$

and the expression for virtual work in Equation 3.2.24 may be written

$$\delta W = \int \{\delta u\}^T \{F_s\} \, dA + \int \{\delta u\}^T \{f\} \, dV - \int ([D] \{\delta u\})^T [G][D]\{u\} \, dV$$

$$= \int \{\delta q\}^T [\Omega]^T \{F_s\} \, dA + \int \{\delta q\}^T [\Omega]^T \{f\} \, dV \tag{3.2.37}$$

$$- \int ([D][\Omega]\{\delta q\})^T [G][D][\Omega]\{q\} \, dV = 0$$

Since we can interpret $\{\delta q\}$ as a set of arbitrary discrete virtual displacements and since they are constants, we can factor it from the integrals and write the virtual work as

$$\delta W = \{\delta q\}^T (\int [\Omega]^T \{F_s\} \, dA + \int [\Omega]^T \{f\} \, dV$$

$$- (\int ([D][\Omega])^T [G][D][\Omega] \, dV)\{q\}) = 0 \tag{3.2.38}$$

Since $\{q\}$ is also a constant, we factored it from the third integral. A common notation is to define

$$[B] = [D][\Omega] \tag{3.2.39}$$

and when this is combined with the fact that since $\{\delta q\}^T$ is arbitrary the terms in the square bracket must be zero, we can conclude

$$\int [\Omega]^T \{F_s\} \, dA + \int [\Omega]^T \{f\} \, dV - (\int ([D][\Omega])^T [G][D][\Omega] \, dV) \{q\} = 0 \tag{3.2.40}$$

We can write this in the following form:

$$[K]\{q\} = \{Q\} \tag{3.2.41}$$

where

$$[K] = \int [B]^T [G][B] \, dV \tag{3.2.42}$$

and where

$$\{Q\} = \int [\Omega]^T \{F_s\} \, dA + \int [\Omega]^T \{f\} \, dV \tag{3.2.43}$$

Since in a given problem everything in the integrals for $[K]$ and $\{Q\}$ are known, the stated integration can be carried out; thus, both are matrices of known coefficients. This makes Equation 3.2.41 a set of linear algebraic equations, which can be solved readily by well-known methods. In practice, $[K]$ is often called the *stiffness matrix*, $\{Q\}$ is called the *generalized force matrix*, and $\{q\}$ the *generalized displacement* or *generalized coordinate matrix*.

What we have just done is a presentation of the *Rayleigh-Ritz* method. With this method we can reduce the infinite degree of freedom system represented by the integral equation form of the statement of virtual work in terms of the unknown distributed functional value of $\{u\}$ to a finite degree of freedom system represented by a finite set of linear algebraic equations in terms of the unknown discrete quantities $\{q\}$. It is an approximation in most cases, although it can be shown that if enough of the proper admissible functions are used the answer converges to the true value.

It is easier to demonstrate the application of this method on a simple structure such as a beam than on a full three-dimensional body; therefore, before we show how the functions in $[\Omega]$ are selected and solutions are obtained, let us find the form of Equation 3.2.41 first for an axially loaded beam, next for a torsionally loaded beam, and then for one with transverse loads.

In a beam with only axial loads, the only displacement function carried is $u_0(x)$, and the $[G]$ and $[D]$ matrices reduce to the scalar quantities given in Equation 3.2.25. We see that

$$[B] = [D][\Omega] = \left[\frac{\partial}{\partial x}\right][\Omega] = [\Omega'] \qquad (3.2.44)$$

where primes are used to denote differentiation with respect to x. Note that $[\Omega']$ is a row matrix and $[\Omega']^T$ is a column matrix, since there is only one displacement component $u_0(x)$. It follows that the algebraic equations for the axially loaded beam analysis are of the form

$$[K]\{q\} = \{Q\} \qquad (3.2.45)$$

where

$$\begin{aligned}[K] &= \int [B]^T[G][B] \; dV \\ &= \iint [\Omega']^T E[\Omega'] \; dA \; dx = \int EA[\Omega']^T[\Omega'] \; dx \end{aligned} \qquad (3.2.46)$$

and

$$\{Q\} = \int [\Omega]^T f_x(x) \; dx + \Sigma \; [\Omega(x_i)]^T F_{xi} \qquad (3.2.47)$$

The matrix $[K]$ is seen to be a symmetrical square matrix with the components

$$K_{rs} = \int EA \; (\Omega_r)'(\Omega_s)' \; dx \qquad (3.2.48)$$

where r and s are indexes and $\{Q\}$ has components

$$Q_r = \int \Omega_r f_x(x) \; dx + \Sigma \; \Omega_r(x_i)F_{xi} \qquad (3.2.49)$$

The crucial step in any application of this method is in the selection of $[\Omega]$. The components of this matrix are supplied by the user, so to speak. Since $\{u\}$ and also $\{\delta u\}$ must be kinematically admissible, the Ω_r must be chosen carefully. The easiest way to ensure admissibility is to require that every Ω_r be admissible. Let us apply what we know to an example.

Example 3.2.2

Find the deflections and stresses by the Rayleigh-Ritz method for a uniform beam that is fixed at the left end, free at the right end, and has a uniform distributed axial load, as shown in figure (a). Compare the solutions to the exact answer found by solving the differential equation.

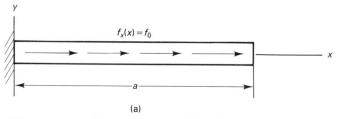

(a)

Solution: First, we must select some geometrically admissible functions. Since the only geometric constraint is

$$u_0(0) = 0 \tag{a}$$

the simple polynomials of the form

$$\Omega_r(x) = \left(\frac{x}{a}\right)^r, \qquad r = 1, 2, 3, \ldots \tag{b}$$

are admissible. Note that the nondimensional form is optional but desirable.

The next question is how many of these to choose. We have no clear guide, as yet, so let us choose just one term by letting $r = 1$. Then

$$u_0(x) = [\Omega]\{q\} = q_1 \frac{x}{a} \tag{c}$$

When we insert this in Equations 3.2.47, 3.2.48, and 3.2.49, we get

$$K_{11}q_1 = Q_1 \tag{d}$$

where

$$K_{11} = EA \int_0^a \frac{1}{a} \frac{1}{a} \, dx = \frac{EA}{a} \tag{e}$$

$$Q_1 = \int_0^a f_0 \frac{x}{a} \, dx = \frac{f_0 a}{2} \tag{f}$$

and therefore

$$q_1 = \frac{Q_1}{K_{11}} = \frac{f_0 a^2}{2EA} \tag{g}$$

and

$$u_0(x) = \frac{f_0 a^2}{2EA} \frac{x}{a} \tag{h}$$

If, instead, we let $r = 1, 2$, we have

$$u_0(x) = q_1 \left(\frac{x}{a}\right) + q_2 \left(\frac{x}{a}\right)^2 \tag{i}$$

or in matrix form

$$u_0(x) = [\Omega]\{q\} = \left[\frac{x}{a} \quad \frac{x^2}{a^2} \right] \begin{bmatrix} q_1 \\ q_2 \end{bmatrix} \tag{j}$$

The stiffness matrix elements are

$$K_{11} = \frac{EA}{a}, \qquad K_{22} = A \int_0^a \frac{2x}{a^2} \frac{2x}{a^2} \, dx = \frac{4EA}{3a}$$

$$K_{12} = K_{21} = EA \int_0^a \frac{2x}{a^2} \frac{1}{a} \, dx = \frac{EA}{a}$$

(k)

and the generalized forces are

$$Q_1 = \frac{f_0 a}{2}, \qquad Q_2 = f_0 \int_0^a \frac{x^2}{a^2} \, dx = \frac{f_0 a}{3}$$

(l)

The matrix Equation 3.2.45 is then

$$\frac{EA}{a} \begin{bmatrix} 1 & 1 \\ 1 & \frac{4}{3} \end{bmatrix} \begin{bmatrix} q_1 \\ q_2 \end{bmatrix} = f_0 a \begin{bmatrix} \frac{1}{2} \\ \frac{1}{3} \end{bmatrix}$$

(m)

from which we obtain

$$q_1 = \frac{f_0 a^2}{EA}, \qquad q_2 = -\frac{f_0 a^2}{2EA}$$

(n)

so that

$$u_0(x) = \frac{f_0 a^2}{EA} \left(\frac{x}{a} - \frac{x^2}{2a^2} \right)$$

(o)

Solving the same problem by the exact differential equation taken from Equation 2.2.15, we get

$$u_0(x) = \frac{f_0 a^2}{EA} \left(\frac{x}{a} - \frac{x^2}{2a^2} \right)$$

(p)

which is exactly the same answer as obtained by the two-term Rayleigh-Ritz solution. Normally, this Rayleigh-Ritz method will not give you the exact answer. It just happened in this case that the two admissible functions had the same polynomial form as the exact answer. It does give us some confidence that virtual work formulation is the equivalent of the differential equation formulation. It may be noted that had we added additional terms all the additional q_i's would have been found to be zero because the first two terms were sufficient to produce an exact answer.

The stress for the one-term solution is

$$\sigma_x = E \frac{\partial u_0}{\partial x} = \frac{f_0 a}{A}$$

(q)

and for two terms

$$\sigma_x = E \frac{\partial u_0}{\partial x} = \frac{f_0 a}{A} \left(1 - \frac{x}{a} \right)$$

(r)

We can compare the one-term solution with the two-term and exact solution by plotting the deflection curves, as shown in figure (b), and the stresses in figure (c). Note that the one-term solution is inadequate, but the two-term solution is exact.

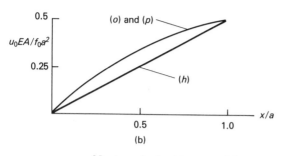
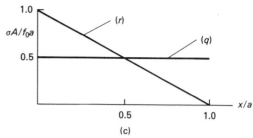

(b)   (c)

Next we look at beams with transverse loads in the x-y plane. In such a beam, the only displacement function carried is $v(x)$, and the $[G]$ and $[D]$ matrices for a beam are given in Equation 3.2.30. The form of the applied loading for a beam is shown in Equation 3.2.31. We see that

$$[B] = [D][\Omega] = \left[-y\frac{\partial^2}{\partial x^2}\right][\Omega] = -y[\Omega''] \qquad (3.2.50)$$

Note that $[\Omega'']$ is a row matrix and $[\Omega'']^T$ is a column matrix since there is only one displacement component $v(x)$. It follows that the algebraic equations for beam analysis are of the form

$$[K]\{q\} = \{Q\} \qquad (3.2.51)$$

where

$$
\begin{aligned}
[K] &= \int [B]^T[G][B] \, dV \\
&= \int [\Omega'']^T E[\Omega'']y^2 dA \, dx = \int EI_{zz}[\Omega'']^T[\Omega''] dx
\end{aligned}
\qquad (3.2.52)
$$

and

$$\{Q\} = \int [\Omega]^T f_y(x) \, dx + \Sigma [\Omega(x_i)]^T F_{yi} + \Sigma [\Omega'(x_j)]^T M_{zj} \qquad (3.2.53)$$

The matrix $[K]$ is seen to be a symmetrical square matrix with the elements

$$K_{rs} = \int EI_{zz} (\Omega_r)''(\Omega_s)'' \, dx \qquad (3.2.54)$$

where r and s are indexes and $\{Q\}$ has elements

$$Q_r = \int \Omega_r f_y(x) \, dx + \Sigma \Omega_r(x_i)F_{yi} + \Sigma \Omega'_r(x_j)M_{zj} \qquad (3.2.55)$$

If we add springs to the beam,

$$
\begin{aligned}
\delta\Delta U &= \Sigma \, \delta v_m(kv_m) + \Sigma \, \delta v_n'(\mu_n v_n') \\
&= \Sigma \, k_m\{\delta q\}^T[\Omega(x_m)]^T[\Omega(x_m)]\{q\} \\
&\quad + \Sigma \, \mu_n\{\delta q\}^T[\Omega'(x_n)]^T[\Omega'(x_n)]\{q\}
\end{aligned}
\qquad (3.2.56)
$$

and a term is added to each element of the stiffness matrix, as follows:

$$K_{rs} = \int EI_{zz}\Omega_r''\Omega_s'' \, dx + \Sigma \, k_m\Omega_r(x_m)\Omega_s(x_m) + \Sigma \, \mu_n\Omega_r'(x_n)\Omega_s'(x_n) \qquad (3.2.57)$$

Now let us apply what we know to some examples of beams with transverse shear loads.

Example 3.2.3

Find the deflections, strains, and stresses in a uniform cantilever beam with a uniform transverse load, as shown in figure (a), by the Rayleigh-Ritz method. Compare the answers with the exact solution obtained from the differential equation.

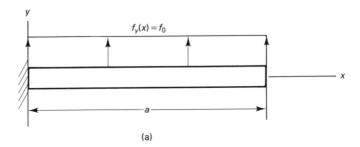

(a)

Solution: First, we must select some kinematically admissible functions for $[\Omega]$. Since the geometric constraints on the beam are

$$v(0) = 0, \qquad v'(0) = 0 \tag{a}$$

the most simple functions are polynomials of the form

$$\Omega_r(x) = \left(\frac{x}{a}\right)^{r+1}, \qquad r = 1, 2, 3, \ldots \tag{b}$$

Let us choose just one term by letting $r = 1$. Then

$$v(x) = [\Omega]\{q\} = \frac{x^2}{a^2} q_1 \tag{c}$$

When we insert this in Equations 3.2.51, 3.2.54, and 3.2.55, we get

$$K_{11}q_1 = Q_1 \tag{d}$$

where

$$K_{11} = \int EI_{zz} \frac{2}{a^2}\frac{2}{a^2} \, dx = \frac{4EI_{zz}}{a^3} \tag{e}$$

$$Q_1 = \int f_0 \frac{x^2}{a^2} \, dx = \frac{f_0 a}{3} \tag{f}$$

and therefore

$$q_1 = \frac{Q_1}{K_{11}} = \frac{f_0 a^4}{12EI_{zz}} \tag{g}$$

and

$$v(x) = q_1 \Omega_1 = \frac{f_0 a^4}{12EI_{zz}} \frac{x^2}{a^2} \tag{h}$$

If, instead, we let $r = 1, 2$, we have

$$v(x) = [\Omega]\{q\} = \begin{bmatrix} \dfrac{x^2}{a^2} & \dfrac{x^3}{a^3} \end{bmatrix} \begin{bmatrix} q_1 \\ q_2 \end{bmatrix} = q_1 \frac{x^2}{a^2} + q_2 \frac{x^3}{a^3} \tag{i}$$

The stiffness matrix elements are

$$K_{11} = \frac{4EI_{zz}}{a^3}, \qquad K_{22} = \int EI_{zz} \frac{6x}{a^3} \frac{6x}{a^3} dx = \frac{12EI_{zz}}{a^3}$$

$$K_{12} = K_{21} = \int EI_{zz} \frac{2}{a^2} \frac{6x}{a^3} dx = \frac{6EI_{zz}}{a^3}$$

(j)

and the generalized force elements are

$$Q_1 = \int f_0 \frac{x^2}{a^2} dx = \frac{f_0 a}{3}, \qquad Q_2 = \int f_0 \frac{x^3}{a^3} dx = \frac{f_0 a}{4}$$

(k)

The matrix Equation 3.2.52 is then

$$\frac{EI_{zz}}{a^3} \begin{bmatrix} 4 & 6 \\ 6 & 12 \end{bmatrix} \begin{bmatrix} q_1 \\ q_2 \end{bmatrix} = f_0 a \begin{bmatrix} \frac{1}{3} \\ \frac{1}{4} \end{bmatrix}$$

(l)

from which we obtain

$$q_1 = \frac{5 f_0 a^4}{24 EI_{zz}}, \qquad q_2 = -\frac{f_0 a^4}{12 EI_{zz}}$$

(m)

so that

$$v(x) = \frac{f_0 a^4}{EI_{zz}} \left(\frac{5}{24} \frac{x^2}{a^2} - \frac{1}{12} \frac{x^3}{a^3} \right)$$

(n)

Solving the same problem by the exact differential equation method from Chapter 2, we get

$$v(x) = \frac{f_0 a^4}{EI_{zz}} \left(\frac{1}{4} \frac{x^2}{a^2} - \frac{1}{6} \frac{x^3}{a^3} + \frac{1}{24} \frac{x^4}{a^4} \right)$$

(o)

The stresses for a one-term, two-term, and exact solution are, respectively,

$$\sigma_x = E\epsilon_x = -Eyv'' = -\frac{f_0 a^2}{I_{zz}} \frac{y}{6}$$

(p)

$$\sigma_x = -\frac{f_0 a^2}{I_{zz}} \left(\frac{5}{12} - \frac{x}{2a} \right) y$$

(q)

$$\sigma_x = -\frac{f_0 a^2}{I_{zz}} \left(\frac{1}{2} - \frac{x}{a} + \frac{x^2}{2a^2} \right) y$$

(r)

The two approximate solutions in equations (h) and (n) may be compared with the exact solution in equation (o) by plotting, as shown in figure (b), and similarly for the stresses in figure (c), where the maximum tensile stresses at $y = -c$ are plotted. It is seen that the one-term solution seriously underestimates the deflection, but that the two-term solution shows promise. We can speculate that three terms would be even better, and in the next section we shall show the rational basis for such a conclusion. In fact, adding a third term would have given us the exact answer in this special case. Once again, this would result from the fact that the exact answer and the chosen set of admissible functions happen to have the same polynomial form.

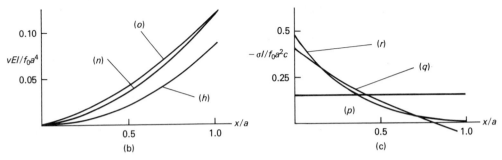

(b) (c)

Example 3.2.4

Consider the same cantilever beam with the same loading as in Example 3.2.3, but this time a lateral spring is attached to the right end, as shown in figure (a). Find the deflection by the Rayleigh-Ritz method.

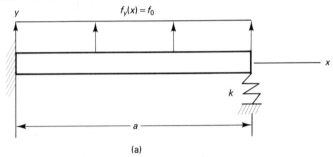

(a)

Solution: Let us use the same admissible functions used in the two-term solution in Example 3.2.3. First, we must modify the stiffness matrix to account for the spring. In this case,

$$K_{rs} = \int EI_{zz}\, \Omega_r{}''\Omega_s{}''\, dx + k\Omega_r(a)\Omega_s(a) \tag{a}$$

and since

$$\Omega_1(a) = 1, \qquad \Omega_2(a) = 1 \tag{b}$$

it follows that

$$K_{11} = \frac{4EI_{zz}}{a^3} + k(1)(1) = \frac{4EI_{zz}}{a^3} + k$$

$$K_{12} = K_{21} = \frac{6EI_{zz}}{a^3} + k(1)(1) = \frac{6EI_{zz}}{a^3} + k \tag{c}$$

$$K_{22} = \frac{12EI_{zz}}{a^3} + k(1)(1) = \frac{12EI_{zz}}{a^3} + k$$

The generalized forces are the same; therefore, the matrix equations are

$$\frac{EI_{zz}}{a^3}\begin{bmatrix} 4+\beta & 6+\beta \\ 6+\beta & 12+\beta \end{bmatrix}\begin{bmatrix} q_1 \\ q_2 \end{bmatrix} = f_0 a \begin{bmatrix} \dfrac{1}{3} \\[2mm] \dfrac{1}{4} \end{bmatrix} \tag{d}$$

where

$$\beta = \frac{ka^3}{EI_{zz}} \qquad \text{(e)}$$

These two algebraic equations may be solved for $\{q\}$ and hence $v(x)$.

At this point we may ask just how good is the answer we get from the Rayleigh-Ritz method, although it has been partially demonstrated by the convergence observed in using more terms in Example 3.2.4. In some cases, we even converged on the exact answer. In more complicated problems, we approach, but never reach, the exact answer. Before we try to show just how close we approach the exact answer, we shall look at another principle of mechanics, this one based on energy rather than work, which will give us a convenient way to answer that question. This new principle is important enough to have its own section.

3.3 STRAIN ENERGY. PRINCIPLE OF MINIMUM POTENTIAL ENERGY

We let the operator δ possess the same rules of operation as the differential operator d. Remember that it can only operate on displacements, since forces are held constant during a virtual displacement. We can write

$$\delta W_e - \delta U = \delta(W_e - U) = 0 \qquad (3.3.1)$$

where

$$W_e = \int \{u\}^T\{F_s\} \, dA + \int \{u\}^T\{f\} \, dV \qquad (3.3.2)$$

and

$$U = \frac{1}{2} \int \{\epsilon\}^T[G]\{\epsilon\} \, dV \qquad (3.3.3)$$

The quantity U is called the strain energy, and it gets its name from the fact that energy is stored in an elastic body when work is done on it. When a strained elastic body is allowed to slowly return to its unstrained state, that work is recovered. Consider a three-dimensional element such as that shown in Figure 1.2.8. Initially, the element is unloaded and unstrained, but let an end load be gradually applied in the x-direction only, as shown in that figure. The force, $F = \sigma_x \, dy \, dz$, would be zero initially and then grow to its final value. The deflection, $u = \epsilon_x \, dx$, would also be zero initially and would then grow to its final value. The load-deflection curve for this process is shown in Figure 3.3.1 for a linear elastic material. The work done on the body and hence the energy stored would be the area under the curve, or

$$\Delta U_x = \frac{(\sigma_x \, dy \, dz)(\epsilon_x \, dx)}{2} \qquad (3.3.4)$$

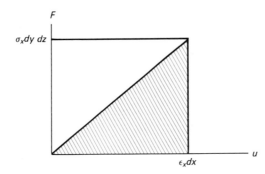

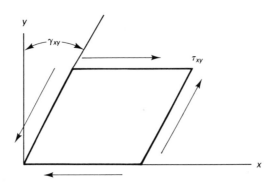

Figure 3.3.1 Load-Deflection Curve.

The additional contribution if we now add a normal end load in each of the y- and z-directions would be

$$\Delta U_y = \frac{(\sigma_y \, dx \, dz)(\epsilon_y \, dy)}{2}, \qquad \Delta U_z = \frac{(\sigma_z \, dy \, dx)(\epsilon_z \, dz)}{2} \qquad (3.3.5)$$

Figure 3.3.2 Shear Deflection of an Element.

Shear forces also do work. Look at Figure 3.3.2 for the effect of one component of shear stress. The force $\tau_{yx} \, dx \, dz$ displaces an amount $\gamma_{yx} \, dy$, and the strain energy stored is

$$\Delta U_{xy} = \frac{(\tau_{xy} \, dx \, dz)(\gamma_{xy} \, dy)}{2} \qquad (3.3.6)$$

and the contribution of the other two components is

$$\Delta U_{yz} = \frac{(\tau_{yz} \, dy \, dx)(\gamma_{yz} \, dz)}{2}, \qquad \Delta U_{zx} = \frac{(\tau_{zx} \, dz \, dy)(\gamma_{zx} \, dx)}{2} \qquad (3.3.7)$$

When all these components are added together and integrated over the whole of the body, the result is

$$U = \frac{1}{2} \int \{\epsilon\}^T \{\sigma\} \, dV \qquad (3.3.8)$$

which becomes Equation 3.3.3 when the stress–strain equations are used.

The term in parentheses, or more correctly the negative of the term in parentheses, in Equation 3.3.1 is called the *potential energy* of the solid body and is labeled as follows:

$$\Pi = U - W_e \qquad (3.3.9)$$

We shall show that $\delta\Pi = 0$ is equivalent to the following statement:

Of all possible kinematically admissible deflections, those that satisfy equilibrium make the potential energy a minimum.

This is called the *principle of minimum potential energy*.

We shall give a proof that of all admissible deflections the true deflection, which naturally satisfies equilibrium, results in the lowest value of the potential energy. This proof, given for a simple elastic beam in pure bending, is easily extended to the full three-dimensional problem of elasticity. Consider two kinematically admissible functions, one the true displacement $v(x)$ and another, $v(x) + \Delta v(x)$, which differs from the true by a small amount. Note that the strain energy for a beam in pure bending can be found to be

$$U = \frac{1}{2} \int EI_{zz}(v'')^2 \, dx \tag{3.3.10}$$

when Equations 3.2.30 are inserted into Equation 3.3.3.

Now denote the potential energy for the true displacement by Π and the other by Π_0. Then

$$\Pi_0 - \Pi = \frac{1}{2} \int EI_{zz}(v'' + \Delta v'')^2 \, dx - \int (v + \Delta v)f_y dx$$
$$-\frac{1}{2} \int EI_{zz}(v'')^2 \, dx - \int vf_y \, dx \tag{3.3.11}$$

which reduces to

$$\Pi_0 - \Pi = \int EI_{zz}(v'')(\Delta v'') \, dx - \int (\Delta v)f_y dx + \frac{1}{2} \int EI_{zz}(\Delta v'')^2 \, dx \tag{3.3.12}$$

Since Δv satisfies all the conditions of a virtual displacement, we can say that the virtual work is zero, or

$$\delta W = \int EI_{zz}(v'')(\Delta v'') \, dx - \int (\Delta v)f_y dx = 0 \tag{3.3.13}$$

and therefore

$$\Pi_0 - \Pi = \frac{1}{2} \int EI_{zz}(\Delta v'')^2 \, dx \tag{3.3.14}$$

Since this term is always positive (positive definite), it follows that

$$\Pi_0 - \Pi \geq 0 \tag{3.3.15}$$

and the principle of minimum potential energy for a beam is proved.

Once again we can use the calculus of variations to derive the equations of elasticity from this principle, or we can turn to direct methods, such as the Rayleigh-Ritz method. It will be useful to rederive the equations of the Rayleigh-Ritz method using minimum potential energy and then draw some general conclusions about the method based on energy considerations.

Again suppose it is possible to represent the distributed displacement field $\{u\}$ of a solid body by an expression of the form

$$\{u\} = [\Omega]\{q\} \qquad (3.3.16)$$

where $\{\Omega\}$ is a set of known admissible functions of the coordinates, and $\{q\}$ is a set of constants. Let us insert this into the expression for the potential energy for a general elastic body.

$$\Pi = \frac{1}{2}\int \{\epsilon\}^T\{\sigma\}\ dV - (\int \{u\}^T\{F_s\}\ dA + \int \{u\}^T\{f\}\ dV)$$

$$= \frac{1}{2}\int \{q\}^T[B]^T[G][B]\{q\}\ dV - (\int \{q\}^T[\Omega]^T\{F_s\}\ dA + \int \{q\}^T[\Omega]^T\{f\}\ dV) \qquad (3.3.17)$$

$$= \frac{1}{2}\{q\}^T[K]\{q\} - \{q\}^T\{Q\}$$

where

$$[K] = \int [B]^T[G][B]\ dV, \qquad \{Q\} = \int [\Omega]^T\{F_s\}\ dA + \int [\Omega]^T\{f\}\ dV] \qquad (3.3.18)$$

Since, for a given problem, everything within these integrals is either a constant or a known function of the coordinates, the indicated integrations can be carried out. The only unknowns are the constants $\{q\}$, and from the principle of minimum potential energy, the correct (or best) values for $\{q\}$ are those that make the potential energy a minimum. This minimum may be found by the simple act of finding the slope of Π for each q_i and setting it to zero, or

$$\frac{\partial \Pi}{\partial q_i} = \frac{\partial}{\partial q_i}\left(\frac{1}{2}\{q\}^T[K]\{q\} - \{q\}^T\{Q\}\right) = 0, \qquad i = 1, 2, 3, \ldots \qquad (3.3.19)$$

The differentiation of this matrix form is discussed in Apppendix A. When this is completed and the results collected, we get

$$[K]\{q\} = \{Q\} \qquad (3.3.20)$$

where

$$[K] = \int [B]^T[G][B]\ dV \qquad (3.3.21)$$

where

$$[B] = [D][\Omega] \qquad (3.3.22)$$

and where

$$\{Q\} = \int [\Omega]^T\{F_s\}\ dA + \int [\Omega]^T\{f\}\ dV \qquad (3.3.23)$$

This is exactly the same set of equations we derived from virtual work in Section 3.2.3. What is different is that we can talk about the accuracy of a given choice of functions and the convergence of the solutions when more admissible functions are added in terms of how close we reach the minimum of the potential energy. Generally, adding a term will reduce the value of the potential energy, thus improving the answer. At worst, an admissible but otherwise inappropriate term (for example, adding an antisymmetrical function when the solution is clearly symmetrical) will have little or no effect because the q_i for that term will prove to be very small or even zero. We are reminded that in the Rayleigh-Ritz method we are replacing an infinite degree of freedom system by one which has finite degrees of freedom. This tends to overestimate the stiffness of the system so that stress, strains, and deflection err on the side of a stiffer structure. The real value of the method is not in solving simple problems such as the one in Example 3.2.3, because the methods of Chapter 2 are entirely adequate. It is in solving more complicated problems, where exact answers are difficult or even impossible to obtain, that the real advantage lies. For example, beams with discontinuous distributed loadings, concentrated loads, multiple supports, or variable cross sections are difficult to solve by the differential equation method, but are little more difficult than the simple problems by the Rayleigh-Ritz method. Selecting efficient admissible functions may not be easy; in fact, it is something of an art form. Fortunately, many problems closely resemble problems that have been solved before, and the literature is full of examples that can serve as a guide.

The equations for specialized elastic structures, such as for beams given in Equations 3.2.50 through 3.2.57, are identical whether derived by virtual work or minimum potential energy, provided the same assumptions are carried in both. In the case of potential energy for beams, this means neglecting the strain energy of shear. As evidence that this can be neglected, compare the strain energy of bending, that is, due to normal stress, with the strain energy of shear for a typical beam. Take for a typical beam a cantilever with a rectangular cross section loaded by a single concentrated shear force at the right end, as shown in Figure 3.3.3.

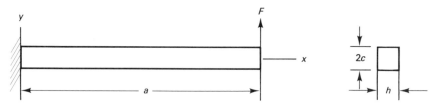

Figure 3.3.3 Cantilever Beam.

Noting that for this beam

$$\sigma_x = -\frac{M_z y}{I_{zz}}, \qquad M_z = F(a-x) \tag{3.3.24}$$

we see that the strain energy of bending can be written

$$U_b = \frac{1}{2} \int \epsilon_x \sigma_x dV = \frac{1}{2} \int_0^a \frac{(M_z)^2}{EI_{zz}} \, dx = \frac{F^2}{2EI_{zz}} \int_0^a (a-x)^2 \, dx = \frac{F^2 a^3}{6EI_{zz}} \tag{3.3.25}$$

Using the shear stress distribution in a beam with rectangular cross section from Equation 2.2.61, we have for the strain energy of shear

$$
\begin{aligned}
U_s &= \frac{1}{2} \int \gamma_{xy} \tau_{xy} \, dV = \frac{1}{2G} \int \left(\frac{S_y(c^2 - y^2)}{I_{zz}} \right)^2 dV \\
&= \left(\frac{F}{I_{zz}} \right)^2 \left(\frac{ah}{2G} \right) \int_0^a (c^2 - y^2)^2 \, dz = \left(\frac{F}{I_{zz}} \right)^2 \left(\frac{8c^5 ah}{15G} \right)
\end{aligned}
\tag{3.3.26}
$$

To assess the relative values of the strain energy of bending and shear, let us find their ratio:

$$\frac{U_b}{U_s} = \frac{F^2 a^3}{6EI_{zz}} \div \left(\frac{F}{I_{zz}}\right)^2 \left(\frac{8c^5 ah}{15G}\right) = \frac{5a^2 GI_{zz}}{8Ehc^5} \tag{3.3.27}$$

When we note that

$$G = \frac{E}{2(1 + v)}, \qquad I_{zz} = \frac{(2c)^3 h}{12} \tag{3.3.28}$$

we get

$$\frac{U_b}{U_s} = \frac{5}{24(1 + v)} \left(\frac{a}{c}\right)^2 \tag{3.3.29}$$

and since for many metals $v = 0.3$

$$\frac{U_b}{U_s} = 0.1602563 \left(\frac{a}{c}\right)^2 \tag{3.3.30}$$

For example, for a beam with a length to height ratio of 10, the value of the ratio of the strain energies would be

$$\frac{U_b}{U_s} = 0.1602563(20)^2 = 64.10252 \tag{3.3.31}$$

Thus, the strain energy of bending would be more than 64 times the strain energy of shear for this particular beam. This difference is often more than enough to justify neglecting the strain energy of shear in the application of potential energy to the solution of beam problems.

Two more examples will be given to show the relative ease in solving more complex problems using the Rayleigh-Ritz method.

Example 3.3.1

Find the deflection of the beam shown in figure (a) by the Rayleigh-Ritz method.

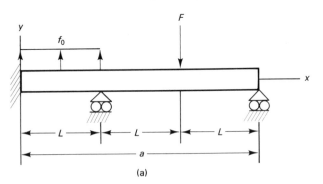

(a)

Solution: The geometric boundary conditions are

$$v(0) = 0, \qquad v'(0) = 0, \qquad v(L) = 0, \qquad v(3L) = 0 \tag{a}$$

For a one-term solution, an admissible function would be

Work and Energy Methods   Chap. 3

$$\Omega_1(x) = \left(\frac{x}{L}\right)^2 \left(1 - \frac{x}{L}\right)\left(3 - \frac{x}{L}\right) \tag{b}$$

You might want to sketch the shape of this function to see if it looks to be at all suitable. It follows that

$$K_{11} = EI_{zz}\int_0^{3L}\left(\frac{6}{L^2} - 24\frac{x}{L^3} + 12\frac{x^2}{L^4}\right)^2 dx = \frac{3132EI_{zz}}{5L^3}$$

$$Q_1 = \int_0^L f_0\left(\frac{x}{L}\right)^2\left(1 - \frac{x}{L}\right)\left(3 - \frac{x}{L}\right) dx - F(-4) = \frac{f_0 L}{5} + F \tag{c}$$

and

$$v(x) = q_1\left(\frac{x}{L}\right)^2\left(1 - \frac{x}{L}\right)\left(3 - \frac{x}{L}\right) \tag{d}$$

where

$$q_1 = \frac{Q_1}{K_{11}} = \frac{(f_0 L + 20F)L^3}{3132EI_{zz}} \tag{e}$$

We will not attempt to compare this solution with the exact solution obtained by solving the differential equation, because to solve exactly even a simple problem like this is rather tedious. We can improve the solution by using more admissible functions. That too can be a bit tedious. For now, let us accept the Rayleigh-Ritz method as a good way to get an approximate answer quickly if we use a small number of admissible functions.

Example 3.3.2

Set up the equations for the deflection of the tapered cantilever beam shown in figure (a). Let the beam have a linear variation in thickness so that

$$h = h_0\left(1 - \frac{x}{6L}\right), \qquad I_{zz} = I_0\left(1 - \frac{x}{6L}\right)^3 \tag{a}$$

where h_0 and I_0 are the thickness and the area moment of inertia, respectively, at $x = 0$. Use a two-term Rayleigh-Ritz solution.

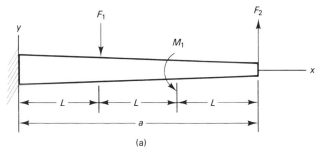

(a)

Solution: If we let

$$v(x) = q_1\left(\frac{x}{3L}\right)^2 + q_2\left(\frac{x}{3L}\right)^3 \tag{b}$$

then

$$K_{11} = \int_0^{3L} EI_0\left(1 - \frac{x}{6L}\right)^3\left(\frac{2}{9L^2}\right)\left(\frac{2}{9L^2}\right) dx$$

$$K_{12} = K_{21} = \int_0^{3L} EI_0 \left(1 - \frac{x}{6L}\right)^3 \left(\frac{2}{9L^2}\right)\left(\frac{2x}{9L^3}\right) dx \qquad \text{(c)}$$

$$K_{22} = \int_0^{3L} EI_0 \left(1 - \frac{x}{6L}\right)^3 \left(\frac{2x}{9L^3}\right)\left(\frac{2x}{9L^3}\right) dx$$

and

$$Q_1 = -\frac{F_1}{9} + \frac{4M_1}{9L} + F_2 \qquad \text{(d)}$$

$$Q_2 = -\frac{F_1}{27} + \frac{4M_1}{9L} + F_2$$

In matrix form, this gives us two simultaneous linear algebraic equations:

$$\begin{bmatrix} K_{11} & K_{12} \\ K_{21} & K_{22} \end{bmatrix} \begin{bmatrix} q_1 \\ q_2 \end{bmatrix} = \begin{bmatrix} Q_1 \\ Q_2 \end{bmatrix} \qquad \text{(e)}$$

which can be solved for $\{q\}$ and ultimately $v(x)$.

The Rayleigh-Ritz method normally does not produce an exact analytical solution; nevertheless, it has significant advantages over trying to solve the differential equations in all but the very simplest problems. It is a powerful method with a long and proud history. Among its major disadvantages are the difficulty in choosing good admissible functions and in carrying out the integrations required to find the elements of $[K]$ and $\{Q\}$. This latter step can be very tedious and fraught with error when the number of chosen admissible functions is large. Fortunately, a means of avoiding these disadvantages is with us. It will be introduced in the next section and expanded on in Chapter 4.

3.4 INTRODUCTION TO THE FINITE ELEMENT METHOD

About three decades ago, a method of structural analysis based on dividing the structure into simple parts, called elements, each of which could be analyzed by relatively straightforward methods, was formulated in practical form. These elements are then assembled into a complete complex structure for which a solution can be obtained by simple mathematical methods. This method grew out of a body of research that had been going on for some time, and, in time, it came to be known as the *finite element method*, or *FEM* for short. The history of the development of FEM is documented in the References 13 to 17 and in many other books and papers. Reference 15 has a list of 19 books on the subject, which includes many major works.

In this section, we are going to develop the general theory of FEM but apply it only for a specific case, the axially loaded bar. While some intuitive arguments may be given from time to time, there will not always be clear reasons for the steps taken. In the end, after this special case is carried all the way through, it will be easier to explain why those particular steps were taken. Then, in Chapter 4, the ideas will be fully developed for a wide class of structures.

3.4.1 General Theory of FEM

We can start out in much the same way as we did for the Rayleigh-Ritz method. Suppose it is possible to represent the distributed displacement field $\{u\}$ of any solid body by an expression of the form

$$\{u\} = [N]\{r\} \tag{3.4.1}$$

where $[N]$ is a set of known functions of the coordinates and $\{r\}$ is a set of constants. In this case, the r_i's have the physical meaning of discrete displacements at selected points in the structure called *nodes*. We have not shown that this is possible; nevertheless, if it is, and we shall show in a moment that it is, we can say

$$\{\delta u\} = [N]\{\delta r\} \tag{3.4.2}$$

and the expression for virtual work in Equation 3.2.24 may be written

$$
\begin{aligned}
\delta W &= \int \{\delta u\}^T\{F_s\}\, dA + \int \{\delta u\}^T\{f\}\, dV - \int ([D]\{\delta u\})^T[G][D]\{u\}\, dV \\
&= \int \{\delta r\}^T[N]^T\{F_s\}\, dA + \int \{\delta r\}^T[N]^T\{f\}\, dV \\
&\quad - \int ([D][N]\{\delta r\})^T[G][D][N]\{r\}\, dV \\
&= \{\delta r\}^T(\int [N]^T\{F_s\}\, dA + \int [N]^T\{f\}\, dV \\
&\quad - \int ([D][N])^T[G][D][N]\{r\}\, dV) = 0
\end{aligned}
\tag{3.4.3}
$$

This is exactly the same expression derived in Equations 3.2.37 and 3.2.38 for the Rayleigh-Ritz method. The only difference is in the interpretation and specification of $\{r\}$ and $[N]$, which we shall discuss in detail shortly. By precisely the same steps, if we let

$$[B] = [D][N] \tag{3.4.4}$$

we can conclude that

$$[K]\{r\} = \{R\} \tag{3.4.5}$$

where

$$[K] = \int [B]^T[G][B]\, dV \tag{3.4.6}$$

and where

$$\{R\} = \int [N]^T\{F_s\}\, dA + \int [N]^T\{f\}\, dV \tag{3.4.7}$$

Excuse the use of $[K]$ and $[B]$ to represent similar but different quantities in each method. They have much the same function in both the conventional Rayleigh-Ritz and the FEM methods, and there are just so many convenient symbols. Since we never mix the methods in the same problem, it should not be confusing.

In practice, just as in the Rayleigh-Ritz method, $[K]$ is often called the *stiffness matrix.* We shall soon see that the problem is so formulated that the elements of $\{r\}$ are discrete values of the *nodal displacements,* and the elements of $\{R\}$ are real and *equivalent nodal loads.*

The whole process depends on finding suitable $[N]$ and $\{r\}$ matrices. It is in the choice of these quantities that the power of FEM is determined. It is not easy to give a good general description of what happens next. It is much easier to demonstrate the application of this method on a simple structure, such as an axially loaded bar, and show how $\{r\}$ and the functions $[N]$ are selected and solutions are obtained. Then we can generalize to other cases.

3.4.2 Theory of FEM Applied to an Axially Loaded Bar

Let us find the form of Equation 3.4.5 for an axially loaded bar. In a bar with only axial loads, the only nonzero displacement function is $u(x)$, and the $[G]$ and $[D]$ matrices reduce to the scalar quantities given in Equation 3.2.25. We see that

$$\{u(x)\} = [N(x)]\{r\}, \qquad [B] = [D][N] = \left[\frac{d}{dx}\right][N] = [N'] \qquad (3.4.8)$$

It follows that the algebraic equations for the axially loaded bar analysis are of the form

$$[K]\{r\} = \{R\} \qquad (3.4.9)$$

where

$$\begin{aligned} [K] &= \int [B]^T[G][B] \ dV \\ &= \iint [N']^T E[N'] \ dA \ dx = \int EA[N']^T[N'] \ dx \end{aligned} \qquad (3.4.10)$$

and

$$\{R\} = \int [N]^T f_x(x) \ dx + \Sigma \ [N(x_i)]^T F_{xi} \qquad (3.4.11)$$

The matrix $[K]$ is seen to be a symmetrical square matrix with the elements

$$K_{rs} = \int EA(N_r)'(N_s)' \ dx \qquad (3.4.12)$$

where r and s are integer indexes and $\{R\}$ has elements

$$R_r = \int N_r f_x(x) \ dx + \Sigma \ N_r(x_i)F_{xi} \qquad (3.4.13)$$

These may be compared with Equations 3.2.45 through 3.2.49.

To demonstrate how these equations are applied in practice, let us consider an axially loaded bar restrained at the right end and loaded by several concentrated axial loads, as shown in Figure 3.4.1(a). Let us also show the support reaction, R_4, at the constraint along with the applied loads in a free-body diagram of the whole bar, as shown in Figure 3.4.1(b). Then let us divide the bar into segments, called *elements*, and number them and also number the points, called *nodes*, at the ends and where the segments join, as shown in Figure 3.4.1(c). Note that we have chosen to put nodes at the ends of the bar, at all interior points where concentrated loads are applied, and, in this case, at a point where nothing special can be noted. In this case, three elements and, consequently, four nodes are chosen for illustration, but this is quite arbitrary. Any number of elements and the proper accompanying number of nodes may be chosen, and later we shall discuss how many should be chosen.

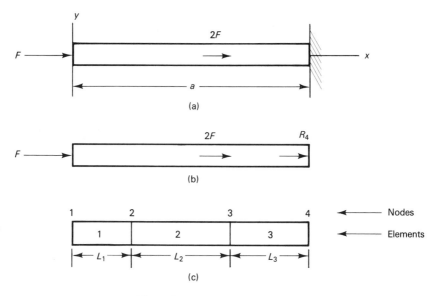

Figure 3.4.1 An Axially Loaded Bar.

For the moment, we ignore the specific constraints, in this case the fixed end boundary condition, and the specific applied loads, in this case discrete loads at nodes 1 and 3, and define, in a general way, $\{r\}$ to be the displacements at each node and $\{R\}$ to be the load at each node, including the support reaction as a load. Later we shall assign the specific values of the constraints and applied loads to these matrices, but for a while we shall carry them in symbolic form. Note that a symbolic value is assigned to the displacement of every node and to an applied load at every node, even though we know that some displacements are constrained to be zero and some loads are zero.

For this particular bar

$$\{r\} = \begin{bmatrix} r_1 \\ r_2 \\ r_3 \\ r_4 \end{bmatrix}, \qquad \{R\} = \begin{bmatrix} R_1 \\ R_2 \\ R_3 \\ R_4 \end{bmatrix} \qquad (3.4.14)$$

At this point, our goal is to find $[K]$, and to do this we must first find $[N]$. Finding $[N]$ is not done directly, but results from a two-part process involving the elements. Note that $[K]$, defined in Equation 3.4.10, requires an integration over the volume of the structure. This integration can be, and is, broken down into separate integrations over each element of the structure. This results in finding an *element stiffness matrix*, $[k]$, for each element, and these are then assembled into the structural or *global stiffness matrix*, $[K]$. Our first need then is to find the element stiffness matrix.

3.4.3 The Element Stiffness Matrix

Consider a single element, say, element j, with internal nodal forces V_i and V_{i+1} and nodal displacements as shown in Figure 3.4.2.

r_i, V_i ⟶ [j] ⟶ r_{i+1}, V_{i+1} **Figure 3.4.2**   Internal Forces on an Element.

These internal forces are, in fact, stress resultants, but they must be distinguished from the stress resultants, $\{S\}$, defined in Equation 2.2.3. First, they are defined only at the nodes and, second, they have a different sign convention. In FEM internal forces, as well as external forces, are positive if they act in the positive coordinate direction. Furthermore, internal moments are positive if they act according to the right-hand rule. You will remember that the stress resultants in Equations 2.2.3 have the same sign convention as stresses. The reason for the different convention will be realized in the next section when the assembly of elements into a global structure is discussed. To keep this difference before us, we shall use the notation $\{V\}$ to represent the three internal nodal forces and the three internal nodal moments that can act at any node in a beam.

If, for now, we imagine the element shown in Figure 3.4.2 as a complete structure and treat the internal forces shown as external applied loads, we can apply Equations 3.4.8 through 3.4.13 to this structure. To call attention to the fact that this is a single element, we shall adopt a distinguishing notation. First, we shall define an element displacement matrix of the form

$$\{u_e\}_i = [n]_i\{r_e\}_i \tag{3.4.15}$$

where $\{u_e\}_i$ is the distributed displacement over a single element numbered i, and $[n]_i$ is a set of functions to be determined. For example, for element 2 in Figure 3.4.1(b), $\{u_e\}_2$ would be the value of $\{u\}$ in the range $L_1 \leq x \leq (L_1 + L_2)$. A single element of an axially loaded bar has two nodes, and $\{r_e\}_i$ is a two-element matrix consisting of those two nodal displacements, as shown in Figure 3.4.2. For element number 2, for example,

$$\{r_e\}_2 = \begin{bmatrix} r_2 \\ r_3 \end{bmatrix} \tag{3.4.16}$$

Thus, $\{r_e\}_i$ is a subset of $\{r\}$, and in some sense $[n]_i$ will be a subset of $[N]$. Since for an element, as well as for the whole structure, $\{\sigma\} = [G]\{\epsilon\}$ and $\{\epsilon\} = [D]\{u\}$, we can say

$$\{\epsilon\}_i = [D][n]_i\{r_e\}_i = [b]_i\{r_e\}_i \tag{3.4.17}$$

where

$$[b]_i = [D][n]_i \tag{3.4.18}$$

and we can write

$$[k]_i\{r_e\}_i = \{V_e\}_i \tag{3.4.19}$$

where

$$[k]_i = \int [b]_i^T[G][b]_i \, dV = \iint [n']_i^T E[n']_i \, dA \, ds = \int_0^L EA[n']_i^T[n']_i \, ds \tag{3.4.20}$$

and where s is a local coordinate with origin at the left end of the element. The use of a local coordinate is added as a convenience and is not essential to the argument. This is called the *element stiffness matrix*, and the integration is carried out only over the volume of the element. Thus,

$$k_{rs} = \int_0^L EA(n_r)'(n_s)' \, ds \tag{3.4.21}$$

For a single element, say, the second element in Figure 3.4.1(b),

$$\{u_e(s)\}_2 = [n_1(s) \; n_2(s)]_2 \begin{bmatrix} r_2 \\ r_3 \end{bmatrix} \tag{3.4.22}$$

Since we have chosen each r_i to be a nodal displacement, by definition

$$r_i = u(x_i) \tag{3.4.23}$$

where x_i is the value of x at the ith node. Converting to the local coordinate s for each element, we have for element 2

$$u_e(0) = r_2, \qquad u_e(L_2) = r_3 \tag{3.4.24}$$

It follows that

$$\begin{aligned} n_1(0) &= 1, \qquad n_1(L_2) = 0 \\ n_2(0) &= 0 \qquad n_2(L_2) = 1 \end{aligned} \tag{3.4.25}$$

At this point we select a polynomial form for each $n_j(s)$. Why we do this will become clearer later. Note that we have two pieces of information about each $n_j(s)$. A polynomial with two undetermined coefficients could have those coefficients adjusted to satisfy those two conditions. The simplest polynomial with two coefficients is a linear function of the form

$$n_j(s) = a_j + b_j s \tag{3.4.26}$$

For this case, if we substitute Equation 3.4.26 into Equations 3.4.25 for $j = 1$ and 2 and evaluate the a_j's and b_j's, we obtain for the second element

$$n_1(s) = 1 - \frac{s}{L_2}, \qquad n_2(s) = \frac{s}{L_2} \tag{3.4.27}$$

The functions that make up each $[n]_j$ are called *shape functions*. We now have

$$\{u_e(s)\}_2 = \left[1 - \frac{s}{L_2} \quad \frac{s}{L_2} \right] \begin{bmatrix} r_2 \\ r_3 \end{bmatrix} = \left(1 - \frac{s}{L_2} \right) r_2 + \left(\frac{s}{L_2} \right) r_3 \tag{3.4.28}$$

It is seen that the distributed deflection for an element is a linear combination of the two shape functions. The plotted shape of the two functions is shown in Figure 3.4.3. Continuing,

$$[b]_2 = [D][n]_2 = \frac{d}{ds} \left[1 - \frac{s}{L_2} \quad \frac{s}{L_2} \right] = \left[-\frac{1}{L_2} \quad \frac{1}{L_2} \right] \tag{3.4.29}$$

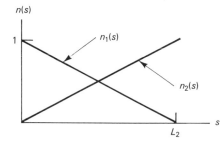

Figure 3.4.3 Shape Functions for an Axial Bar Element.

and

$$[k]_2 = \int_0^{L_2} [b_2]^T [G][b]_2 \, dV = \int_0^{L_2} \begin{bmatrix} -\dfrac{1}{L_2} \\ \dfrac{1}{L_2} \end{bmatrix} E_2 \begin{bmatrix} -\dfrac{1}{L_2} & \dfrac{1}{L_2} \end{bmatrix} A_2 \, ds \qquad (3.4.30)$$

$$= E_2 A_2 \int_0^{L_2} \begin{bmatrix} \left(\dfrac{1}{L_2}\right)^2 & -\left(\dfrac{1}{L_2}\right)^2 \\ -\left(\dfrac{1}{L_2}\right)^2 & \left(\dfrac{1}{L_2}\right)^2 \end{bmatrix} ds = \dfrac{E_2 A_2}{L_2} \begin{bmatrix} 1 & -1 \\ -1 & 1 \end{bmatrix}$$

We now have a $[k]$ matrix for element 2. Note that we assigned a subscript 2 to both the area, A, and Young's modulus, E. This emphasizes the fact that each element can differ in material properties and area, as well as length.

Had we done the same thing for element 1 or 3, we would have gotten exactly the same form for $[n]_i$ and $[k]_i$. Only the subscripted quantities might differ. Thus, the shape functions and the element stiffness matrix for each element can be written in the general form

$$[n]_i = \begin{bmatrix} 1 - \dfrac{s}{L_i} & \dfrac{s}{L_i} \end{bmatrix}, \qquad [k]_i = \dfrac{E_i A_i}{L_i} \begin{bmatrix} 1 & -1 \\ -1 & 1 \end{bmatrix} \qquad (3.4.31)$$

We may note that from Equation 3.4.19, given the element nodal displacements, we can find the nodal stress resultants:

$$\{V_e\}_i = [k]_i \{r_e\}_i \qquad (3.4.32)$$

Now $[K]$ is in some sense the sum of all the element $[k]_i$'s. The next step is to form the global stiffness matrix from the element stiffness matrices.

3.4.4 The Matrix Assembly Process

We note that $[k]_i$ is an integral over the volume of an element, and the sum of the integrals over all the elements is equal to the integral over the whole structure. Given all the element matrices, $[k]_i$, we must add them to find $[K]$. But summing matrices can be tricky, so we must proceed carefully. We first note a rule of matrix addition (see Appendix A) that two matrix equations of the form

$$\{X\} = [A]\{x\}, \qquad \{Y\} = [B]\{y\} \qquad (3.4.33)$$

can be added only if $\{y\}$ is exactly equal to $\{x\}$, and $[A]$ and $[B]$ have the same number of rows and columns, or the same *order*. We then obtain

$$\{Z\} = \{X\} + \{Y\} = ([A] + [B]) \{x\} \qquad (3.4.34)$$

The significance of this is found when trying to add up the element stiffness matrices, $[k]_i$, to obtain $[K]$. They simply do not add in their present form because the $\{r_e\}_i$ matrices, while each is a subset of $\{r\}$, are each different.

We take care of this by a simple trick. We can say that

$$[K]\{r\} = \Sigma \, [k]_i \{r\} \qquad (3.4.35)$$

where, for example, we embed the element stiffness matrix for the first element, $[k]_1$, in a global matrix format, which for the bar in Figure 3.4.1 is as follows:

$$[k]_1\{r\} = \begin{bmatrix} \dfrac{E_1A_1}{L_1} & -\dfrac{E_1A_1}{L_1} & 0 & 0 \\ -\dfrac{E_1A_1}{L_1} & \dfrac{E_1A_1}{L_1} & 0 & 0 \\ 0 & 0 & 0 & 0 \\ 0 & 0 & 0 & 0 \end{bmatrix} \begin{bmatrix} r_1 \\ r_2 \\ r_3 \\ r_4 \end{bmatrix} \tag{3.4.36}$$

By carrying out the indicated multiplication, we can confirm that exactly the same information is contained in Equation 3.4.36 as in the expression $[k]_1\{r_e\}_1$ using Equation 3.4.31.

In a similar way, for the second element we have

$$[k]_2\{r\} = \begin{bmatrix} 0 & 0 & 0 & 0 \\ 0 & \dfrac{E_2A_2}{L_2} & -\dfrac{E_2A_2}{L_2} & 0 \\ 0 & -\dfrac{E_2A_2}{L_2} & \dfrac{E_2A_2}{L_2} & 0 \\ 0 & 0 & 0 & 0 \end{bmatrix} \begin{bmatrix} r_1 \\ r_2 \\ r_3 \\ r_4 \end{bmatrix} \tag{3.4.37}$$

and, finally, for the third element

$$[k]_3\{r\} = \begin{bmatrix} 0 & 0 & 0 & 0 \\ 0 & 0 & 0 & 0 \\ 0 & 0 & \dfrac{E_3A_3}{L_3} & -\dfrac{E_3A_3}{L_3} \\ 0 & 0 & -\dfrac{E_3A_3}{L_3} & \dfrac{E_3A_3}{L_3} \end{bmatrix} \begin{bmatrix} r_1 \\ r_2 \\ r_3 \\ r_4 \end{bmatrix} \tag{3.4.38}$$

Now these may be added together to obtain

$$[K]\{r\} = ([k]_1 + [k]_2 + [k]_3)\{r\}$$

$$[K]\{r\} = \begin{bmatrix} \dfrac{E_1A}{L_1} & -\dfrac{E_1A_1}{L_1} & 0 & 0 \\ -\dfrac{E_1A_1}{L_1} & \dfrac{E_1A_1}{L_1} + \dfrac{E_2A_2}{L_2} & -\dfrac{E_2A_2}{L_2} & 0 \\ 0 & -\dfrac{E_2A_2}{L_2} & \dfrac{E_2A_2}{L_2} + \dfrac{E_3A_3}{L_3} & -\dfrac{E_3A_3}{L_3} \\ 0 & 0 & -\dfrac{E_3A_3}{L_3} & \dfrac{E_3A_3}{L_3} \end{bmatrix} \begin{bmatrix} r_1 \\ r_2 \\ r_3 \\ r_4 \end{bmatrix} \tag{3.4.39}$$

Thus, we have the $[K]$ matrix for a three-element axially loaded bar that has been assembled from the element stiffness matrices. The pattern for assembly of element stiffness matrices to form the global stiffness matrix for a bar with more elements can be seen from this example.

To complete the assembly of the set of equations, Equations 3.4.9, we need to establish the matrix $\{R\}$. In terms of the global nodal displacement matrix, we can say

$$[k]_i\{r\} = \{V\}_i \tag{3.4.40}$$

where $\{V\}_i$ is the form of $\{V_e\}_i$ when it is embedded in a matrix of the same order as $\{r\}$. We can now say

$$[K]\{r\} = \Sigma \ [k]_i\{r\} = \Sigma \ \{V\}_i = \{R\} \qquad (3.4.41)$$

where $\{R\}$ is a matrix of the applied nodal loads. A significant fact here is that the external force matrix, $\{R\}$, is equal to the sum of the internal nodal forces. We can use Equation 3.4.40 to find the internal stress resultants at each node after we have solved for $\{r\}$. For our case, in Figure 3.4.1 the matrix $\{R\}$ is seen to be

$$\{R\} = \begin{bmatrix} F \\ 0 \\ 2F \\ R_4 \end{bmatrix} \qquad (3.4.42)$$

where R_4 is the support reaction. Thus, the assembled set of equations is

$$\begin{bmatrix} \dfrac{E_1A_1}{L_1} & -\dfrac{E_1A_1}{L_1} & 0 & 0 \\[2mm] -\dfrac{E_1A_1}{L_1} & \dfrac{E_1A_1}{L_1} + \dfrac{E_2A_2}{L_2} & -\dfrac{E_2A_2}{L_2} & 0 \\[2mm] 0 & -\dfrac{E_2A_2}{L_2} & \dfrac{E_2A_2}{L_2} + \dfrac{E_3A_3}{L_3} & -\dfrac{E_3A_3}{L_3} \\[2mm] 0 & 0 & -\dfrac{E_3A_3}{L_3} & \dfrac{E_3A_3}{L_3} \end{bmatrix} \begin{bmatrix} r_1 \\ r_2 \\ r_3 \\ r_4 \end{bmatrix} = \begin{bmatrix} F \\ 0 \\ 2F \\ R_4 \end{bmatrix} \qquad (3.4.43)$$

This is the set of equations to be solved, but before we can solve them we must note that the right-hand side of these equations contains an unknown and, therefore, the equations are not yet in a form to be solved. The reason is that nothing about the boundary constraints has been introduced into the problem. Let us consider the boundary constraints.

3.4.5 Boundary Constraints and Support Reactions

What we have so far is a set of linear algebraic equations that, presumably, can be solved by familiar algebraic techniques for the unknowns in $\{r\}$. Not quite. An examination of the matrix $[K]$ will show that it is *singular*; that is, the determinant of $[K]$ is zero, or symbolically

$$|K| = 0 \qquad (3.4.44)$$

It is well known that a set of equations with a singular coefficient matrix does not have a unique solution.

There is a physical reason this is so. Since we have not applied the geometric constraints, as far as these equations are concerned the bar is free to move or, as we say, it has rigid body degrees of freedom. Adding the appropriate restraint against rigid body motion will modify the equations and produce a form of $[K]$ that is not singular. This may be done by noting that, for the problem shown in Figure 3.4.1, the constraint at the right end of the bar implies that $r_4 = 0$. If this is so, there are only three unknowns in Equation 3.4.43. With three unknowns, we need only three equations and, therefore, one of the equations is not needed. Furthermore, each element in the fourth column of $[K]$ is always multiplied by zero, that is, by r_4. Thus, the new, nonsingular form of the matrix $[K]$ is formed by crossing out the fourth row and the fourth column to obtain the final set of three equations. Let us illustrate this discussion with a numerical example.

Example 3.4.1

Use FEM to find deflection, strains, and stresses in the axially loaded bar shown in figure (a).

Solution: The bar has been divided into two elements, with nodes and elements shown numbered in figure (b).

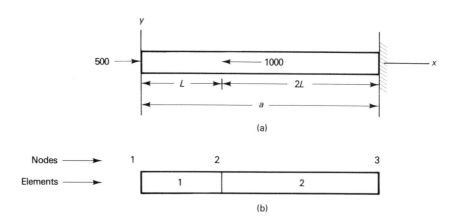

(a)

(b)

Using the element stiffness matrix given in Equation 3.4.31, we can assemble the set of equations as follows:

$$[K]\{r\} = ([k]_1 + [k]_2)\{r\} = \{R\} \tag{a}$$

or

$$
\begin{bmatrix}
\dfrac{E_1 A_1}{L_1} & -\dfrac{E_1 A_1}{L_1} & 0 & 0 \\[2mm]
-\dfrac{E_1 A_1}{L_1} & \dfrac{E_1 A_1}{L_1} + \dfrac{E_2 A_2}{L_2} & -\dfrac{E_2 A_2}{L_2} & 0 \\[2mm]
0 & -\dfrac{E_2 A_2}{L_2} & \dfrac{E_2 A_2}{L_2} & 0
\end{bmatrix}
\begin{bmatrix} r_1 \\ r_2 \\ r_3 \end{bmatrix} =
\begin{bmatrix} R_1 \\ R_2 \\ R_3 \end{bmatrix} \tag{b}
$$

Since for this case

$$E_1 = E_2 = E, \qquad A_1 = A_2 = A, \qquad L_1 = L = \frac{a}{3}$$

$$L_2 = 2L = \frac{2a}{3}, \qquad R_1 = 500, \qquad R_2 = -1000 \tag{c}$$

equation (b) reduces to

$$
\frac{3EA}{a}
\begin{bmatrix}
1 & -1 & 0 \\[2mm]
-1 & \dfrac{3}{2} & -\dfrac{1}{2} \\[2mm]
0 & -\dfrac{1}{2} & \dfrac{1}{2}
\end{bmatrix}
\begin{bmatrix} r_1 \\ r_2 \\ r_3 \end{bmatrix} =
\begin{bmatrix} 500 \\ -1000 \\ R_3 \end{bmatrix} \tag{d}
$$

The geometric constraint requires that $r_3 = 0$, so this set of equations may be reduced further by eliminating the last row and last column of the matrix $[K]$ to obtain

$$\frac{3EA}{a}\begin{bmatrix} 1 & -1 \\ -1 & \frac{3}{2} \end{bmatrix}\begin{bmatrix} r_1 \\ r_2 \end{bmatrix} = \begin{bmatrix} 500 \\ -1000 \end{bmatrix} \quad \text{(e)}$$

This set of equations can be solved by elementary algebra to obtain

$$r_1 = -\frac{500a}{3EA}, \qquad r_2 = -\frac{1000a}{3EA} \quad \text{(f)}$$

The element displacements are found from Equation 3.4.28 for the first element:

$$\{u_e(s)\}_1 = \begin{bmatrix} 1 - \frac{s}{L_1} & \frac{s}{L_1} \end{bmatrix}\begin{bmatrix} r_1 \\ r_2 \end{bmatrix} \quad \text{(g)}$$

$$= \begin{bmatrix} 1 - \frac{3s}{a} & \frac{3s}{a} \end{bmatrix}\begin{bmatrix} -\dfrac{500a}{3EA} \\ -\dfrac{1000a}{3EA} \end{bmatrix} = -\frac{500}{EA}\left(\frac{a}{3} + s\right)$$

For the second element,

$$\{u_e(s)\}_2 = \begin{bmatrix} 1 - \frac{s}{L_2} & \frac{s}{L_2} \end{bmatrix}\begin{bmatrix} r_2 \\ r_3 \end{bmatrix} \quad \text{(h)}$$

$$= \begin{bmatrix} 1 - \frac{3s}{2a} & \frac{3s}{2a} \end{bmatrix}\begin{bmatrix} -\dfrac{1000a}{3EA} \\ 0 \end{bmatrix} = -\frac{500}{EA}\left(\frac{2a}{3} - s\right)$$

Note that for both elements s is a local coordinate with an origin at the left end of the element. We can convert from the coordinate s to x, but it is not really necessary in this case.

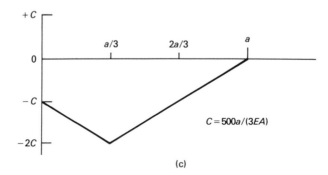

$$C = 500a/(3EA)$$

(c)

The deflection may be plotted as shown in figure (c). From the deflections we can obtain the strains in each element. Recognizing that the strain is given by the first derivative of the deflection, we have

$$\epsilon_{x1} = \frac{\partial}{\partial s}(u_e)_1 = -\frac{500}{EA}, \qquad \epsilon_{x2} = \frac{\partial}{\partial s}(u_e)_2 = \frac{500}{EA} \quad \text{(i)}$$

and from Hooke's law the stresses are

$$\sigma_{x1} = E\epsilon_{x1} = -\frac{500}{A}, \qquad \sigma_{x2} = E\epsilon_{x2} = \frac{500}{A} \quad \text{(j)}$$

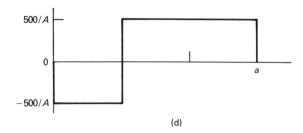

(d)

A plot of the stress is given in figure (d). Note that the strains and stresses are both constant over the full length of each element in this case. Had we solved this problem by the classical methods of Chapter 2, we would have gotten exactly the same answer. The reader can verify this.

Now we shall reexamine the full set of equations to show that there is additional information to be gleaned from them if we approach them properly. Actually, we threw away some valuable information when we crossed out the fourth row in Equations 3.4.43. The full set of equations governing this structure is now repeated (disregard for the moment the horizontal and vertical dashed lines between the third and fourth rows and columns of the matrices)

$$
\begin{bmatrix}
\dfrac{E_1A_1}{L_1} & -\dfrac{E_1A_1}{L_1} & 0 & \vdots & 0 \\[2mm]
-\dfrac{E_1A_1}{L_1} & \dfrac{E_1A_1}{L_1} + \dfrac{E_2A_2}{L_2} & -\dfrac{E_2A_2}{L_2} & \vdots & 0 \\[2mm]
0 & -\dfrac{E_2A_2}{L_2} & \dfrac{E_2A_2}{L_2} + \dfrac{E_3A_3}{L_3} & \vdots & -\dfrac{E_3A_3}{L_3} \\[1mm]
\hdashline \\[-3mm]
0 & 0 & -\dfrac{E_3A_3}{L_3} & \vdots & \dfrac{E_3A_3}{L_3}
\end{bmatrix}
\begin{bmatrix} r_1 \\[2mm] r_2 \\[2mm] r_3 \\[2mm] 0 \end{bmatrix}
=
\begin{bmatrix} F_1 \\[2mm] F_2 \\[2mm] F_3 \\[2mm] R_4 \end{bmatrix}
\qquad (3.4.45)
$$

where the known value of $r_4 = 0$ has been inserted. Looking just at the fourth equation, we can note that once $\{r\}$ has been found from the other three equations, we can insert those values and use the fourth equation to find R_4.

$$
0 \cdot r_1 + 0 \cdot r_2 - \frac{E_3A_3}{L_3} \cdot r_3 + \frac{E_3A_3}{L_3} \cdot 0 = R_4 \qquad (3.4.46)
$$

We can generalize this approach. This set of four equations contains four unknowns, but not all the unknowns are in the usual position. The three unknown nodal displacements are in the standard position in the matrix $\{r\}$, but a fourth is in the matrix $\{R\}$, where, normally, all values are known. We can deal with this situation with a process called *partitioning*. Now pay attention to the horizontal and vertical dashed lines in Equations 3.4.45. These lines indicate where the matrices will be partitioned. Symbolically, we have

$$
\begin{bmatrix} [K^{11}] & [K^{12}] \\ [K^{21}] & [K^{22}] \end{bmatrix}
\begin{bmatrix} \{r^1\} \\ \{r^2\} \end{bmatrix}
=
\begin{bmatrix} \{R^1\} \\ \{R^2\} \end{bmatrix}
\qquad (3.4.47)
$$

Equation 3.4.47 can be expanded by matrix multiplication to obtain two matrix equations of the form

$$[K^{11}]\{r^1\} + [K^{12}]\{r^2\} = \{R^1\}$$
$$[K^{21}]\{r^1\} + [K^{22}]\{r^2\} = \{R^2\}$$
(3.4.48)

where, in this case, $\{r^2\}$ and $\{R^1\}$ are known and $\{r^1\}$ and $\{R^2\}$ are unknowns. We can solve the first of these equations for displacements $\{r^1\}$:

$$[K^{11}]\{r^1\} = \{R^1\} - [K^{12}]\{r^2\} \longrightarrow \{r^1\} = [K^{11}]^{-1}(\{R^1\} - [K^{12}]\{r^2\}) \quad (3.4.49)$$

Then we can solve the second for the support reactions $\{R^2\}$:

$$\{R^2\} = [K^{21}]\{r^1\} + [K^{22}]\{r^2\} \tag{3.4.50}$$

For the problem posed in Figure 3.4.1, the quantities in Equations 3.4.49 and 3.4.50 are

$$[K^{11}] = \begin{bmatrix} \dfrac{E_1 A_1}{L_1} & -\dfrac{E_1 A_1}{L_1} & 0 \\[2mm] -\dfrac{E_1 A_1}{L_1} & \dfrac{E_1 A_1}{L_1} + \dfrac{E_2 A_2}{L_2} & -\dfrac{E_2 A_2}{L_2} \\[2mm] 0 & -\dfrac{E_2 A_2}{L_2} & \dfrac{E_2 A_2}{L_2} + \dfrac{E_3 A_3}{L_3} \end{bmatrix}$$

$$[K^{12}] = \begin{bmatrix} 0 \\ 0 \\ -\dfrac{E_3 A_3}{L_3} \end{bmatrix}, \quad [K^{21}] = \begin{bmatrix} 0 & 0 & -\dfrac{E_3 A_3}{L_3} \end{bmatrix}, \quad [K^{22}] = \begin{bmatrix} \dfrac{E_3 A_3}{L_3} \end{bmatrix} \quad (3.4.51)$$

$$\{r^1\} = \begin{bmatrix} r_1 \\ r_2 \\ r_3 \end{bmatrix}, \quad \{R^1\} = \begin{bmatrix} F_1 \\ F_2 \\ F_3 \end{bmatrix}, \quad \{r^2\} = [0], \quad \{R^2\} = [R_4]$$

Let us apply this approach to a numerical example.

Example 3.4.2

Find the support reaction for the bar in Example 3.4.1 using the method just described.

Solution: The full set of equations given in equation (d) in Example 3.4.1 is repeated here and is partitioned as shown.

$$\frac{3EA}{a} \begin{bmatrix} 1 & -1 & \vline & 0 \\ -1 & \frac{3}{2} & \vline & -\frac{1}{2} \\ \hline 0 & -\frac{1}{2} & \vline & \frac{1}{2} \end{bmatrix} \begin{bmatrix} r_1 \\ r_2 \\ 0 \end{bmatrix} = \begin{bmatrix} 500 \\ -1000 \\ R_3 \end{bmatrix} \tag{a}$$

The equation for finding the unknown deflections is found by expanding the partitioned matrix, which becomes

$$\frac{3EA}{a} \begin{bmatrix} 1 & -1 \\ -1 & \frac{3}{2} \end{bmatrix} \begin{bmatrix} r_1 \\ r_2 \end{bmatrix} = \begin{bmatrix} 500 \\ -1000 \end{bmatrix} \tag{b}$$

This equation is precisely the one we solved in Example 3.4.1 to obtain the displacements, which are given in equation (f).

$$
\begin{bmatrix} r_1 \\ r_2 \end{bmatrix} = \begin{bmatrix} -\dfrac{500a}{3EA} \\ -\dfrac{1000a}{3EA} \end{bmatrix}
\tag{c}
$$

The support reaction is found from

$$
R_3 = \frac{3EA}{a} \begin{bmatrix} 0 & -\dfrac{1}{2} \end{bmatrix} \begin{bmatrix} -\dfrac{500a}{3EA} \\ -\dfrac{1000a}{3EA} \end{bmatrix} = 500
\tag{d}
$$

The examples shown here are problems that can be solved very easily by hand calculation. They were used to show FEM in a simple setting so that we could concentrate on the method rather than the problems. Now we shall extend the method to include additional features. The first of these is how to deal with distributed loads.

3.4.6 Equivalent Nodal Loads

In FEM, distributed loads are converted into *equivalent nodal loads* for purposes of numerical analysis. The principle of virtual work establishes in Equations 3.4.11 and 3.4.13 the formula for converting distributed loads into equivalent nodal loads.

Consider the same bar as in Figure 3.4.1, but replace the loading with a distributed axial load $f_x(x)$. For the first element,

$$
\{R_e\}_1 = \int_0^{L_1} [n]_1^T f_x(s_1)\, ds_1 = \int_0^{L_1} \begin{bmatrix} 1 - \dfrac{s_1}{L_1} \\ \dfrac{s_1}{L_1} \end{bmatrix} f_x(s_1)\, ds_1
\tag{3.4.52}
$$

For the second element, we have

$$
\{R_e\}_2 = \int_0^{L_2} [n]_1^T f_x(s_2)\, ds_2 = \int_0^{L_2} \begin{bmatrix} 1 - \dfrac{s_2}{L_2} \\ \dfrac{s_2}{L_2} \end{bmatrix} f_x(s_2)\, ds_2
\tag{3.4.53}
$$

and similarly for the third element. Note that $f_x(x)$ has to be written in terms of local coordinates for each element. A numerical example follows.

Example 3.4.3

A tapered bar is restrained on both ends as shown in figure (a). The taper is linear from a value of cross-sectional area of A_0 at the left end to a value $A_0/2$ at the right end. It is loaded axially with a concentrated load, F, and a distributed load as shown in figure (b). Set up the equations for finding the deflections in the bar when $F = f_0 a/6$.

Solution: Let us divide the bar into four elements of equal length and number the nodes and elements as shown in figure (c). The simplest way of handling taper in FEM is to approximate the structure with elements that are of constant cross section. In this case, let us use the value of the cross-sectional area at the midpoint of each element. These are calculated at 1/8, 3/8, 5/8, and 7/8 of the length. Since the area is given by the formula

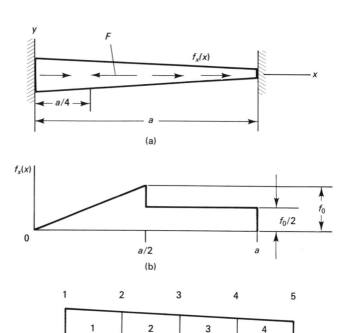

(a)

(b)

(c)

$$A = A_0\left(1 - \frac{x}{2a}\right) \qquad\qquad \text{(a)}$$

we find the element cross-sectional areas to be

$$A_1 = \frac{15}{16}A_0, \qquad A_2 = \frac{13}{16}A_0, \qquad A_3 = \frac{11}{16}A_0, \qquad A_4 = \frac{9}{16}A_0 \qquad \text{(b)}$$

This obviously introduces error into the solution, but that error can be lessened by using more and more elements.

The element stiffness matrix is given by Equation 3.4.31:

$$[k]_i = \frac{E_i A_i}{L_i}\begin{bmatrix} 1 & -1 \\ -1 & 1 \end{bmatrix} \qquad\qquad \text{(c)}$$

where A_i is as given previously, $E_i = E$, and $L_i = a/4$. When the stiffness matrices of all four elements are assembled into a global stiffness matrix, we have

$$[k] = \frac{A_0 E}{4a}\begin{bmatrix} 15 & -15 & 0 & 0 & 0 \\ -15 & 28 & -13 & 0 & 0 \\ 0 & -13 & 24 & -11 & 0 \\ 0 & 0 & -11 & 20 & -9 \\ 0 & 0 & 0 & -9 & 9 \end{bmatrix} \qquad\qquad \text{(d)}$$

The equivalent nodal loads from the distributed loading are

$$\{R_e\}_1 = \int_0^{a/4} [n]_1^T \frac{2f_0}{a} s \, ds = \int_0^{a/4} \begin{bmatrix} 1 - \dfrac{4s}{a} \\ \dfrac{4s}{a} \end{bmatrix} \frac{2f_0}{a} s \, ds = \begin{bmatrix} \dfrac{f_0 a}{48} \\ \dfrac{f_0 a}{24} \end{bmatrix}$$

$$\{R_e\}_2 = \int_0^{a/4} [n]_2^T \frac{2f_0}{a} \left(s + \frac{a}{4} \right) s \, ds = \int_0^{a/4} \begin{bmatrix} 1 - \dfrac{4s}{a} \\ \dfrac{4s}{a} \end{bmatrix} \frac{2f_0}{a} \left(s + \frac{a}{4} \right) s \, ds = \begin{bmatrix} \dfrac{f_0 a}{12} \\ \dfrac{5f_0 a}{48} \end{bmatrix} \qquad \text{(e)}$$

$$\{R_e\}_3 = \int_0^{a/4} [n]_3^T \frac{f_0}{2} \, ds = \int_0^{a/4} \begin{bmatrix} 1 - \dfrac{4s}{a} \\ \dfrac{4s}{a} \end{bmatrix} \frac{f_0}{2} \, ds = \begin{bmatrix} \dfrac{f_0 a}{16} \\ \dfrac{f_0 a}{16} \end{bmatrix}$$

$$\{R_e\}_4 = \int_0^{a/4} [n]_4^T \frac{f_0}{2} \, ds = \int_0^{a/4} \begin{bmatrix} 1 - \dfrac{4s}{a} \\ \dfrac{4s}{a} \end{bmatrix} \frac{f_0}{2} \, ds = \begin{bmatrix} \dfrac{f_0 a}{16} \\ \dfrac{f_0 a}{16} \end{bmatrix}$$

When these values plus that of the concentrated force are assembled into a global applied load matrix, along with the reactions R_1 and R_5 at nodes 1 and 5, we have

$$\{R\} = \begin{bmatrix} R_1 + \dfrac{f_0 a}{48} \\[2mm] \dfrac{f_0 a}{8} - F \\[2mm] \dfrac{f_0 a}{6} \\[2mm] \dfrac{f_0 a}{8} \\[2mm] R_5 + \dfrac{f_0 a}{16} \end{bmatrix} \qquad \text{(f)}$$

Recognizing that $r_1 = r_5 = 0$, and $F = f_0 a/6$, we have the following set of equations:

$$\frac{A_0 E}{4a} \begin{bmatrix} 15 & -15 & 0 & 0 & 0 \\ -15 & 28 & -13 & 0 & 0 \\ 0 & -13 & 24 & -11 & 0 \\ 0 & 0 & -11 & 20 & -9 \\ 0 & 0 & 0 & -9 & 9 \end{bmatrix} \begin{bmatrix} 0 \\ r_2 \\ r_3 \\ r_4 \\ 0 \end{bmatrix} = \begin{bmatrix} R_1 + \dfrac{f_0 a}{48} \\[2mm] \dfrac{-f_0 a}{24} \\[2mm] \dfrac{f_0 a}{6} \\[2mm] \dfrac{f_0 a}{8} \\[2mm] R_5 + \dfrac{f_0 a}{16} \end{bmatrix} \qquad \text{(g)}$$

This matrix is not in the standard form for partitioning. To partition, we need to collect the unknown and the known r_i's and place them in adjacent rows. Let us move the fifth equation into the second row. When we do this, we must also rearrange the columns of the matrix $[K]$ so that the correct coefficients are multiplying the appropriate r_i's. The result is

$$\frac{A_0 E}{4a} \begin{bmatrix} 15 & 0 & -15 & 0 & 0 \\ 0 & 9 & 0 & 0 & -9 \\ -15 & 0 & 28 & -13 & 0 \\ 0 & 0 & -13 & 24 & -11 \\ 0 & -9 & 0 & -11 & 20 \end{bmatrix} \begin{bmatrix} 0 \\ 0 \\ r_2 \\ r_3 \\ r_4 \end{bmatrix} = \begin{bmatrix} R_1 + \dfrac{f_0 a}{48} \\ R_5 + \dfrac{f_0 a}{16} \\ \dfrac{-f_0 a}{24} \\ \dfrac{f_0 a}{6} \\ \dfrac{f_0 a}{8} \end{bmatrix} \qquad \text{(h)}$$

where the partition lines are shown.

To find the unknown deflections, solve the following set of equations

$$\begin{bmatrix} 28 & -13 & 0 \\ -13 & 24 & -11 \\ 0 & -11 & 20 \end{bmatrix} \begin{bmatrix} r_2 \\ r_3 \\ r_4 \end{bmatrix} = \frac{f_0 a^2}{6A_0 E} \begin{bmatrix} -1 \\ 4 \\ 3 \end{bmatrix} \qquad \text{(i)}$$

Then, once r_2, r_3, and r_4 are known, we can find R_1 and R_5 from

$$\frac{A_0 E}{4a} \begin{bmatrix} -15 & 0 & 0 \\ 0 & 0 & -9 \end{bmatrix} \begin{bmatrix} r_2 \\ r_3 \\ r_4 \end{bmatrix} = \begin{bmatrix} R_1 + \dfrac{f_0 a}{48} \\ R_5 + \dfrac{f_0 a}{16} \end{bmatrix} \qquad \text{(j)}$$

3.4.7 Equation Solvers

So far, we have solved the linear algebraic equations generated by the finite element method and, earlier, by the Rayleigh-Ritz method by elementary means. These elementary methods work fine when only a few equations need be solved. In both methods, however, the complexity of the structure and the requirements for accuracy often demand sets of equations of the order of 10, 100, or even 1000. Such sets of equations require special methods of solution. These special methods are usually a variation of a method called *Gauss elimination*. We shall offer here an elementary discussion of Gauss elimination and one of its special forms sometimes called the *Gauss-Doolittle factorization*.

Gauss elimination is based on the fact that by successive manipulation of equations you can replace equations of the form

$$[A]\{x\} = \{B\} \qquad (3.4.54)$$

with an equivalent set of the form

$$[A]\{x\} = \{B\} \qquad (3.4.55)$$

where

$$[A] = \begin{bmatrix} A_{11} & 0 & 0 & 0 & \cdots \\ A_{21} & A_{22} & 0 & 0 & \cdots \\ A_{31} & A_{32} & A_{33} & 0 & \cdots \\ \cdot & \cdot & \cdot & \cdot & \cdots \\ \cdots & \cdots & \cdots & \cdots & \cdots \end{bmatrix} \qquad (3.4.56)$$

This is called *triangularization* of the matrix $[A]$. Note that the triangularization could occur in the upper right as well as the lower left, as shown. Once triangularization is complete, the newly formed equations can be solved one by one. An example will show this clearly.

Example 3.4.4

Solve the following set of linear algebraic equations by Gauss elimination.

$$\begin{bmatrix} 5 & -4 & 1 & 0 \\ -4 & 6 & -4 & 1 \\ 1 & -4 & 6 & -4 \\ 0 & 1 & -4 & 5 \end{bmatrix} \begin{bmatrix} x_1 \\ x_2 \\ x_3 \\ x_4 \end{bmatrix} = \begin{bmatrix} 0 \\ 1 \\ 0 \\ 0 \end{bmatrix} \tag{a}$$

Solution: Multiply the first equation by $-1/5$ and add it to the third equation, as follows:

$$\begin{array}{rcl} -\dfrac{5}{5}x_1 + \dfrac{4}{5}x_2 - \dfrac{1}{5}x_3 + \dfrac{0}{5}x_4 &=& -\dfrac{0}{5} \\[1mm] x_1 - 4x_2 + 6x_3 - 4x_4 &=& 0 \\[1mm] \hline 0x_1 - \dfrac{16}{5}x_2 + \dfrac{29}{5}x_3 - 4x_4 &=& 0 \end{array} \tag{b}$$

This new equation replaces the third equation in the original set to obtain a new set:

$$\begin{bmatrix} 5 & -4 & 1 & 0 \\ -4 & 6 & -4 & 1 \\ 0 & -\dfrac{16}{5} & \dfrac{29}{5} & -4 \\ 0 & 1 & -4 & 5 \end{bmatrix} \begin{bmatrix} x_1 \\ x_2 \\ x_3 \\ x_4 \end{bmatrix} = \begin{bmatrix} 0 \\ 1 \\ 0 \\ 0 \end{bmatrix} \tag{c}$$

Now multiply the first equation by $4/5$ and add to the second equation:

$$\begin{array}{rcl} 4x_1 - \dfrac{16}{5}x_2 + \dfrac{4}{5}x_3 - 0x_4 &=& 0 \\[1mm] -4x_1 + 6x_2 - 4x_3 + x_4 &=& 1 \\[1mm] \hline 0x_1 + \dfrac{14}{5}x_2 - \dfrac{16}{5}x_3 + x_4 &=& 1 \end{array} \tag{d}$$

This new equation replaces the second equation to form the following new set:

$$\begin{bmatrix} 5 & -4 & 1 & 0 \\ 0 & \dfrac{14}{5} & -\dfrac{16}{5} & 1 \\ 0 & -\dfrac{16}{5} & \dfrac{29}{5} & -4 \\ 0 & 1 & -4 & 5 \end{bmatrix} \begin{bmatrix} x_1 \\ x_2 \\ x_3 \\ x_4 \end{bmatrix} = \begin{bmatrix} 0 \\ 1 \\ 0 \\ 0 \end{bmatrix} \tag{e}$$

Now multiply the second equation by successive numbers and add to rows 3 and 4 to obtain new equations for those rows. The result is

$$\begin{bmatrix} 5 & -4 & 1 & 0 \\ 0 & \dfrac{14}{5} & -\dfrac{16}{5} & 1 \\ 0 & 0 & \dfrac{15}{7} & -\dfrac{20}{7} \\ 0 & 0 & -\dfrac{20}{7} & \dfrac{65}{14} \end{bmatrix} \begin{bmatrix} x_1 \\ x_2 \\ x_3 \\ x_4 \end{bmatrix} = \begin{bmatrix} 0 \\ 1 \\ \dfrac{8}{7} \\ -\dfrac{5}{14} \end{bmatrix} \tag{f}$$

One more time. Multiply the third equation by a number so that when it is added to the fourth we get a new equation with a zero in the third column. The final result is

$$
\begin{bmatrix}
5 & -4 & 1 & 0 \\
0 & \dfrac{14}{5} & -\dfrac{16}{5} & 1 \\
0 & 0 & \dfrac{15}{7} & -\dfrac{20}{7} \\
0 & 0 & 0 & \dfrac{5}{6}
\end{bmatrix}
\begin{bmatrix}
x_1 \\ x_2 \\ x_3 \\ x_4
\end{bmatrix}
=
\begin{bmatrix}
0 \\ 1 \\ \dfrac{8}{7} \\ \dfrac{7}{6}
\end{bmatrix}
\tag{g}
$$

Now the equations are solved in order from the bottom up.

$$
\frac{5}{6}x_4 = \frac{7}{6} \longrightarrow x_4 = \frac{7}{5}
$$

$$
\frac{15}{7}x_3 - \frac{20}{7}\frac{7}{5} = \frac{8}{7} \longrightarrow x_3 = \frac{12}{5}
$$

$$
\frac{14}{5}x_2 - \frac{16}{5}\frac{12}{5} + \frac{7}{5} = 1 \longrightarrow x_2 = \frac{81}{35}
$$
(h)

$$
5x_1 - 4\frac{81}{35} + \frac{12}{5} = 0 \longrightarrow x_1 = \frac{48}{35}
$$

In practice, this method of solution is programmed and performed on a digital computer. One of the most efficient implementations is based on an extension of Gauss elimination called the Gause-Doolittle factorization. In this formulation, the matrix $[K]$ is decomposed into two new matrices, as follows

$$
[K] = [L] [D] [L]^T \tag{3.4.57}
$$

where $[L]$ and $[D]$ are of the form

$$
[L] =
\begin{bmatrix}
1 & 0 & 0 & \dots \\
L_{21} & 1 & 0 & \dots \\
L_{31} & L_{32} & 1 & \dots \\
\cdot & \cdot & \cdot & \dots \\
\dots & \dots & \dots & \dots
\end{bmatrix},
\qquad
[D] =
\begin{bmatrix}
D_{11} & 0 & 0 & \dots \\
0 & D_{22} & 0 & \dots \\
0 & 0 & D_{33} & \dots \\
\cdot & \cdot & \cdot & \dots \\
\dots & \dots & \dots & \dots
\end{bmatrix}
\tag{3.4.58}
$$

Note that D and L do not have the same meaning as in other notation in this text. Note also that decomposing $[K]$ is a mechanical numerical process that has been programmed.

The solution process continues by noting that

$$
[K]\{r\} = [L] [D] [L]^T\{r\} = \{R\} \tag{3.4.59}
$$

and defining a new unknown

$$
\{r*\} = [D] [L]^T\{r\} \tag{3.4.60}
$$

so that

$$
[L] \{r*\} = \{R\} \tag{3.4.61}
$$

Equation 3.4.61 is in triangular form and can be easily solved for $\{r*\}$. With $\{r\}$ known, Equation 3.4.60, which also is in triangular form, can also be easily solved for $\{r\}$. This

Work and Energy Methods   Chap. 3

solution method is implemented in BASIC and is presented in Appendix H. This code is used in the next example to find the solution.

Example 3.4.5

Complete the analysis of Example 3.4.3 by solving equations (i) and (j) from that example for the following specific values of the various parameters.

$$a = 1000 \text{ mm} \qquad\qquad A_0 = 625 \text{ mm}^2$$
$$E = 68{,}950.0 \text{ MPa (N/mm}^2), \qquad f_0 = 250 \text{ N/mm} \tag{a}$$

Solution: Equation (i) becomes

$$\begin{bmatrix} 28 & -13 & 0 \\ -13 & 24 & -11 \\ 0 & -11 & 20 \end{bmatrix} \begin{bmatrix} r_1 \\ r_2 \\ r_3 \end{bmatrix} = \begin{bmatrix} -0.96688 \\ 3.86754 \\ 2.90065 \end{bmatrix} \tag{b}$$

From the computer program in Appendix H, we obtain

$$\{r\} = \begin{bmatrix} 0.16086 \\ 0.43084 \\ 0.37649 \end{bmatrix} \tag{c}$$

Using equation (j) to find the support reactions,

$$10{,}773.4 \begin{bmatrix} -15 & 0 & 0 \\ 0 & 0 & -9 \end{bmatrix} \begin{bmatrix} 0.16086 \\ 0.43084 \\ 0.37649 \end{bmatrix} = \begin{bmatrix} R_1 + 5208.3 \\ R_2 + 5208.3 \end{bmatrix} \tag{d}$$

from which

$$R_1 = 37{,}203.5, \qquad R_5 = -52{,}729.8 \tag{e}$$

Given these nodal displacements, we can find the distributed displacements from Equation 3.4.15 to be

$$\{u_e\}_1 = \left(1 - \frac{4s}{a}\right) r_1 + \frac{4s}{a} r_2$$

$$= \frac{4s}{1000} 0.16086 = 0.64344s \cdot 10^{-3}$$

$$\{u_e\}_2 = \left(1 - \frac{4s}{a}\right) r_2 + \frac{4s}{a} r_3$$

$$= \left(1 - \frac{4s}{1000}\right) 0.16086 + \frac{4s}{1000} 0.42084$$

$$= 0.16086 + 1.0399s \cdot 10^{-3}$$

$$\{u_e\}_3 = \left(1 - \frac{4s}{a}\right) r_3 + \frac{4s}{a} r_4 \tag{f}$$

$$= \left(1 - \frac{4s}{1000}\right) 0.42084 + \frac{4s}{1000} 0.37649$$

$$= 0.42084 - 0.17740s \cdot 10^{-3}$$

$$\{u_e\}_4 = \left(1 - \frac{4s}{a}\right)r_4 + \frac{4s}{a}r_5$$

$$= \left(1 - \frac{4s}{1000}\right)0.37649 = 0.37649 - 1.5060s \cdot 10^{-3}$$

where s is measured from the left end of each element. This deflection is plotted in figure (a).

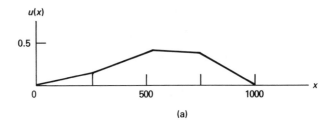

(a)

The strains in each element are given by $\{\epsilon_e\} = [D]\{n\}$ and are

$$\epsilon_1 = 0.64344 \cdot 10^{-3}, \qquad \epsilon_2 = 1.0399 \cdot 10^{-3}$$

$$\epsilon_3 = -0.17740 \cdot 10^{-3}, \qquad \epsilon_4 = -1.5060 \cdot 10^{-3}$$

(g)

The stresses are given by $\{\sigma_e\} = [E]\{\epsilon_e\}$ and are

$$\sigma_1 = 44.365 \text{ N/mm}^2, \qquad \sigma_2 = 71.702$$

$$\sigma_3 = -12.232, \qquad \sigma_4 = -103.834$$

(h)

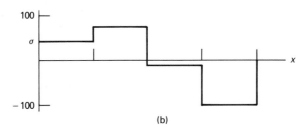

(b)

A plot of the stresses is given in figure (b). Note that the tapered beam is approximated by a four-stepped beam with uniform cross section in each element. This, coupled with the choice of shape function, produces a constant strain and a constant stress within each element. The true stress and strain would show a smoother variation, which could be approached by using more elements.

We may also note that $\{V_e\} = [k]\{r_e\}$, which for element 1 leads to

$$\begin{bmatrix} V_1 \\ V_2 \end{bmatrix} = 10{,}773.4 \begin{bmatrix} 15 & -15 \\ -15 & 15 \end{bmatrix} \begin{bmatrix} 0.00000 \\ 0.16086 \end{bmatrix} = \begin{bmatrix} -25{,}995.0 \\ 25{,}995.0 \end{bmatrix}$$

(i)

and for element 2

$$\begin{bmatrix} V_2 \\ V_3 \end{bmatrix} = 10{,}773.4 \begin{bmatrix} 13 & -13 \\ -13 & 13 \end{bmatrix} \begin{bmatrix} 0.16086 \\ 0.42084 \end{bmatrix} = \begin{bmatrix} -36{,}411.3 \\ 36{,}411.3 \end{bmatrix}$$

(j)

and for element 3

$$\begin{bmatrix} V_3 \\ V_4 \end{bmatrix} = 10{,}773.4 \begin{bmatrix} 11 & -11 \\ -11 & 11 \end{bmatrix} \begin{bmatrix} 0.42084 \\ 0.37649 \end{bmatrix} = \begin{bmatrix} 5255.8 \\ -5255.8 \end{bmatrix}$$

(k)

and for element 4

$$\begin{bmatrix} V_4 \\ V_5 \end{bmatrix} = 10,773.4 \begin{bmatrix} 9 & -9 & 0.37649 \\ -9 & 9 & 0.00000 \end{bmatrix} = \begin{bmatrix} 36,504.0 \\ -36,504.0 \end{bmatrix} \qquad (1)$$

We can check these figures by multiplying the stresses in equation (h) by the cross-sectional areas.

3.4.8 Pin-Jointed Trusses. Coordinate Transformations

A classical structural idealization is the *pin-jointed truss* made from an assemblage of axially loaded bar elements. In trusses, loads can only be applied at the joints to preserve the condition that only axial forces are carried in the members. Examples of plane trusses are shown in Figure 3.4.4.

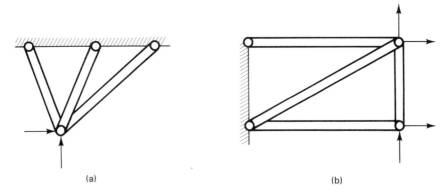

(a)   (b)

Figure 3.4.4 Examples of Simple Plane Trusses.

The FEM approach is ideal for such problems; however, the preceding equations require a simple modification before they can be used. The element stiffness matrix derived in Section 3.4.3 is aligned with the x-axis, but the truss elements may be aligned in any direction. A transformation of the stiffness matrix to a general coordinate orientation is needed before the global stiffness matrix of the truss structure can be assembled. We shall do this first for a plane element and then generalize to three dimensions.

The problem before us is, given the stiffness matrix for an axial element as shown in local coordinates in Figure 3.4.5(a), find the stiffness matrix for an element oriented in global coordinates as shown in Figure 3.4.5(b). Shown in the figure are the nodal displacements. There is a set of corresponding nodal forces, $\{V_e\}$, as well, which are not shown in the figure.

We shall use the matrix form of the transformation of vector components from one coordinate system to another. There is no change in a vector quantity from the translation of rectangular coordinate systems, so consider a vector, $\{u\}$, with two coordinate systems having the same origin, but with one rotated with respect to the other, as shown in Figure 3.4.6. For a vector $\{u\}$, we have

$$\{u\}_n = \begin{bmatrix} u \\ v \end{bmatrix}, \qquad \{u_\alpha\}_n = \begin{bmatrix} u_\alpha \\ v_\alpha \end{bmatrix} \qquad (3.4.62)$$

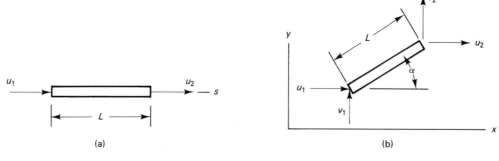

(a)

(b)

Figure 3.4.5 An Axial Element in Two Coordinate Systems.

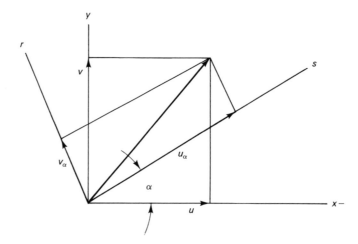

Figure 3.4.6 Coordinate Transformation of a Vector.

where $\{u\}_n$ repesents the vector components in the x-y, or global, coordinate system, and $\{u_\alpha\}_n$ represents the components in the r-s, or element, coordinate system, which is rotated an amount α with respect to the first. We may see that

$$\{u_\alpha\}_n = \begin{bmatrix} u_\alpha \\ v_\alpha \end{bmatrix} = \begin{bmatrix} \cos\alpha & \sin\alpha \\ -\sin\alpha & \cos\alpha \end{bmatrix} \begin{bmatrix} u \\ v \end{bmatrix} = [t]\{u\}_n \qquad (3.4.63)$$

This useful relation can be applied to the problem at hand. There is a displacement vector at each node of our axial element. We can write for the two-node element

$$\{r_{e\alpha}\} = \begin{bmatrix} \{u_\alpha\}_1 \\ \{u_\alpha\}_2 \end{bmatrix} = \begin{bmatrix} [t]_1 & [0] \\ [0] & [t]_2 \end{bmatrix} \begin{bmatrix} \{u\}_1 \\ \{u\}_2 \end{bmatrix} = [T]\{r_e\} \qquad (3.4.64)$$

or

$$\begin{bmatrix} u_{\alpha 1} \\ 0 \\ u_{\alpha 2} \\ 0 \end{bmatrix} = \begin{bmatrix} \cos\alpha & \sin\alpha & 0 & 0 \\ -\sin\alpha & \cos\alpha & 0 & 0 \\ 0 & 0 & \cos\alpha & \sin\alpha \\ 0 & 0 & -\sin\alpha & \cos\alpha \end{bmatrix} \begin{bmatrix} u_1 \\ v_1 \\ u_2 \\ v_2 \end{bmatrix} \qquad (3.4.65)$$

Once again we must plead guilty to the use of the same symbols, in this case t and T, to mean different things; however, the context in which the symbols are used should help to keep things clear. To the extent possible, we wish to retain notation that is widely used in other texts.

Now $[T]$ has the interesting property that $[T]^{-1} = [T]^T$, characteristic of what is called an *orthogonal matrix*. You can verify that $[T]^T[T] = [1]$. Thus, we see that

$$\{r_e\} = [T]^T\{r_{e\alpha}\}, \qquad \{r_{e\alpha}\} = [T]\{r_e\} \tag{3.4.66}$$

and since nodal forces are vectors

$$\{V_e\} = [T]^T\{V_{e\alpha}\}, \qquad \{V_{e\alpha}\} = [T]\{V_e\} \tag{3.4.67}$$

This may be applied to the stiffness matrix. From the preceding equations and Equation 3.4.19,

$$[k_\alpha]\{r_{e\alpha}\} = \{V_{e\alpha}\}$$

$$[k_\alpha][T]\{r_e\} = [T]\{V_e\} \tag{3.4.68}$$

$$[T]^T[k_\alpha][T]\{r_e\} = \{V_e\} = [k]\{r_e\}$$

For these operations to be defined, we have to cast $[k_\alpha]$ in the proper form of a 4×4 matrix. Equation 3.4.31 becomes

$$[k_\alpha] = \frac{EA}{L}\begin{bmatrix} 1 & 0 & -1 & 0 \\ 0 & 0 & 0 & 0 \\ -1 & 0 & 1 & 0 \\ 0 & 0 & 0 & 0 \end{bmatrix} \tag{3.4.69}$$

An example will show this best of all.

Example 3.4.6

Consider the three-bar truss shown in figure (a). Set up and solve the FEM equations for the nodal deflections. It is made of steel with E = 206,840 MPa and $\nu = 0.3$. The dimensions are $a = 1000$ mm and the areas for each element are $A_1 = 400$ mm^2, $A_2 = 800$ mm^2, and $A_3 = 400$ mm^2. The applied load $F = 90,000$ N.

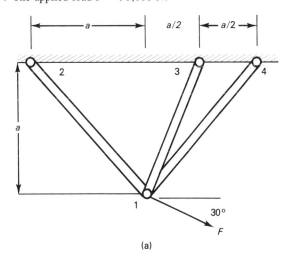

(a)

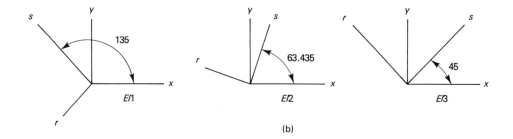

(b)

Solution: The nodes are numbered in the figure. This is a plane problem so at each node there are two displacement components. Nodes 1 and 2 go with element 1, nodes 1 and 3 with element 2, and nodes 1 and 4 with element 3. The appropriate rotations between local, or element, coordinates and global coordinates are shown in figure (b) for each element. Note that for all elements the origin of the local and global coordinates is at node 1. Given that

$$\sin 135° = 0.70711, \qquad \cos 135° = -0.70711$$

$$\sin 63.435° = 0.89442, \qquad \cos 63.435° = 0.44721 \qquad \text{(a)}$$

$$\sin 45° = 0.70711, \qquad \cos 45° = 0.70711$$

for element 1, from Equation 3.4.65,

$$[T] = \begin{bmatrix} -0.70711 & 0.70711 & 0 & 0 \\ -0.70711 & -0.70711 & 0 & 0 \\ 0 & 0 & -0.70711 & 0.70711 \\ 0 & 0 & -0.70711 & -0.70711 \end{bmatrix} \qquad \text{(b)}$$

and similarly for elements 2 and 3. Using this and Equation 3.4.69, $[T]^T[k_\alpha][T] = [k]$, we have for element 1

$$[k]_1 = 29{,}251.9 \begin{bmatrix} 1.0 & -1.0 & -1.0 & 1.0 \\ -1.0 & 1.0 & 1.0 & -1.0 \\ -1.0 & 1.0 & 1.0 & -1.0 \\ 1.0 & -1.0 & -1.0 & 1.0 \end{bmatrix} \qquad \text{(c)}$$

and for element 2

$$[k]_2 = 148{,}003.0 \begin{bmatrix} 0.2 & 0.4 & -0.2 & -0.4 \\ 0.4 & 0.8 & -0.4 & -0.8 \\ -0.2 & -0.4 & 0.2 & 0.4 \\ -0.4 & -0.8 & 0.4 & 0.8 \end{bmatrix} \qquad \text{(d)}$$

and for element 3

$$[k]_3 = 29{,}251.9 \begin{bmatrix} 1.0 & 1.0 & -1.0 & -1.0 \\ 1.0 & 1.0 & -1.0 & -1.0 \\ -1.0 & -1.0 & 1.0 & 1.0 \\ -1.0 & -1.0 & 1.0 & 1.0 \end{bmatrix} \qquad \text{(e)}$$

These may now be assembled to form the matrix $[K]$. We can also form the matrix $\{R\}$ from the components of the force at node 1. We note that at nodes 2, 3, and 4 we have unknown support reactions. Thus, we have

$$[K] = \begin{bmatrix} 88,104 & 59,201 & -29,252 & 29,252 & -29,601 & -59,201 & -29,252 & -29,252 \\ 59,201 & 176,906 & 29,252 & -29,252 & -59,201 & -118,402 & -29,252 & -29,252 \\ -29,252 & 29,252 & 29,252 & -29,252 & 0 & 0 & 0 & 0 \\ 29,252 & -29,252 & -29,252 & 29,252 & 0 & 0 & 0 & 0 \\ -29,601 & -59,201 & 0 & 0 & 29,601 & 59,201 & 0 & 0 \\ -59,201 & 118,402 & 0 & 0 & 59,201 & 118,402 & 0 & 0 \\ -29,252 & 29,252 & 0 & 0 & 0 & 0 & 29,252 & 29,252 \\ -29,252 & -29,252 & 0 & 0 & 0 & 0 & 29,252 & 29,252 \end{bmatrix} \quad \text{(f)}$$

and

$$\{R\} = \begin{bmatrix} 866.03 \\ -500.0 \\ R_{2x} \\ R_{2y} \\ R_{3x} \\ R_{3y} \\ R_{4x} \\ R_{4y} \end{bmatrix} \quad \text{(g)}$$

After imposing the boundary conditions of no displacements at nodes 2, 3, and 4, we have

$$\begin{bmatrix} 88,104 & 59,201 \\ 59,201 & 176,906 \end{bmatrix} \begin{bmatrix} r_1 \\ r_2 \end{bmatrix} = \begin{bmatrix} 866.03 \\ -500.0 \end{bmatrix} \quad \text{(h)}$$

from which

$$r_1 = 0.01513 \text{ mm}, \qquad r_2 = 0.00789 \text{ mm} \quad \text{(i)}$$

The support reactions, the internal forces, and the stresses may now be found by straightforward means.

3.5 OTHER WORK AND ENERGY METHODS

In the literature you will find a number of methods based on work and energy that have been brought to bear on problems of structural analysis. Many of these methods are powerful and still useful, while others have fallen into disuse in recent years. Generally, those methods based on displacement formulation, such as the methods of virtual work and potential energy used in this and subsequent chapters, have prevailed, while those based on stress formulation have tended to be used more sparingly. Some of the other methods are outlined briefly in Appendix I.

3.6 CONCLUSIONS AND SUMMARY

We have used powerful methods of mechanics based on the principle of virtual work to formulate the equations of linear elastic solid bodies in the form of integral equations. Two closely related methods for reducing these integral equations to sets of linear algebraic equations, the Rayleigh-Ritz method and the finite element method, were introduced. These are, as formulated here, in fact, one method, since FEM based on shape functions is a special case of the Rayleigh-Ritz method.

The Rayleigh-Ritz method has a long history of solving problems successfully in an analytical but approximate way that removes many of the barriers to solutions found in the classical differential equation approach. However, many limitations remain. Many of these have been removed by FEM. There is a price to pay in the form of greater numerical dependence on solutions. Fortunately, the digital computer has made the numerical calculation part easy. And new ways of assimilating and presenting numerical results have overcome much of the disadvantages of numerical over analytical formulations.

Referring back to the symbolic representation of the principle of virtual work in Figure 3.2.1, we shall summarize the virtual work expression for the general elastic solid:

$$\delta W = \int \{\delta u\}^T \{F_s\}\, dA + \int \{\delta u\}^T \{f\}\, dV - \int ([D]\{\delta u\})^T [G]\, [D]\{u\}\, dV = 0 \qquad (3.2.24)$$

where $[D]$ and $[G]$ are obtained from Equations 1.2.20 and 1.2.31. For slender bars with purely axial loads, this reduces to

$$\delta W = \delta W_e - \delta U = \int \delta u_0 f_x(x)\, dx + \Sigma\, \delta u_{0i} F_{xi} - \int \{\delta\epsilon\}^T \{\sigma\}\, dV$$

$$= \int \delta_0 f_x(x)\, dx + \Sigma\, \delta u_{0i} F_{xi} - \int \frac{\partial \delta u_0}{\partial x}\, E\, \frac{\partial u_0}{\partial x}\, A\, dx = 0 \qquad (3.2.27)$$

For purely torsionally loaded bars,

$$\delta W = \delta W_e - \delta U = \int \delta\beta t_x(x)\, dx + \Sigma\, \delta\beta_i T_{xi} - \int \{\delta\epsilon\}^T \{\sigma\}\, dV$$

$$= \int \delta\beta t_x(x)\, dx + \Sigma\, \delta\beta_i T_{xi} - \int \frac{\partial \delta\beta}{\partial x}\, G\, \frac{\partial \beta}{\partial x}\, (\int r^2\, dA)\, dx \qquad (3.2.29)$$

$$= \int \delta\beta t_x(x)\, dx + \Sigma\, \delta\beta_i T_{xi} - \int \frac{\partial \delta\beta}{\partial x}\, GJ\, \frac{\partial \beta}{\partial x}\, dx$$

For bars loaded with pure moments about the z-axis and loaded transversely in the y-direction,

$$\delta W = \int \delta v f_y(x)\, dx + \Sigma\, F_{yi}\, \delta v_i + \Sigma\, M_{zj}(\delta v_j)' - \int \{\delta\epsilon\}^T \{\sigma\}\, dV$$

$$= \int \delta v f_y(x)\, dx + \Sigma\, F_{yi}\, \delta v_i + \Sigma\, M_{zj}(\delta v_j)' - \int \frac{\partial^2 \delta v}{\partial x^2}\, E\, \frac{\partial^2 v}{\partial x^2}\, (\int y^2\, dA)\, dx \qquad (3.2.32)$$

$$= \int \delta v f_y(x)\, dx + \Sigma\, F_{yi}\, \delta v_i + \Sigma\, M_{zj}(\delta v_j)' - \int \frac{\partial^2 \delta v}{\partial x^2}\, EI_{zz}\, \frac{\partial^2 v}{\partial x^2}\, dx$$

In Equation 3.2.32, it is assumed that $I_{yz} = 0$ and that a similar equation is obtained for applied loads in the z-direction.

To use virtual work to solve practical problems, we often turn to the Rayleigh-Ritz method, which is represented symbolically in Figure 3.2.2. Starting with a set of admissible functions of the form

$$\{u\} = [\Omega]\{q\} \qquad (3.2.35)$$

we obtain

$$[K]\{q\} = \{Q\} \qquad (3.2.41)$$

where

$$[K] = \int [B]^T[G][B] \ dV \tag{3.2.42}$$

with

$$[B] = [D][\Omega] \tag{3.2.39}$$

and where

$$\{Q\} = \int [\Omega]^T\{F_s\} \ dA + \int [\Omega]^T\{f\} \ dV \tag{3.2.43}$$

For beams with purely axial loads, this reduces to

$$[K] = \iint [\Omega']^T EA[\Omega'] \ dx \tag{3.2.46}$$

and

$$\{Q\} = \int [\Omega]^T f_x(x) \ dx + \Sigma \ [\Omega(x_i)]^T F_{xi} \tag{3.2.47}$$

and for transversely loaded beams

$$[K] = \iint [\Omega'']^T EI_{zz}[\Omega''] \ dx \tag{3.2.52}$$

and

$$\{Q\} = \int [\Omega]^T f_y(x) \ dx + \Sigma \ [\Omega(x_i)]^T F_{yi} + \Sigma \ [\Omega'(x_j)]^T M_{zj} \tag{3.2.53}$$

An alternative derivation of the very same equations is provided by the principle of minimum potential energy utilizing a defined external work term

$$W_e = \int \{u\}^T\{F_s\} \ dA + \int \{u\}^T\{f\} \ dV \tag{3.3.2}$$

and the strain energy

$$U = \frac{1}{2} \int \{\epsilon\}^T\{\sigma\} \ dV \tag{3.3.3}$$

The finite element method (FEM) is a special case of the Rayleigh-Ritz method, starting with a set of functions of the form

$$\{u\} = [N]\{r\} \tag{3.4.1}$$

where the $\{r\}$ are nodal displacements and $[N]$ are called shape functions. Using the principle of virtual work, we obtain

$$[K]\{r\} = \{R\} \tag{3.4.5}$$

where

$$[K] = \int [B]^T[G][B] \ dV \tag{3.4.6}$$

now with

$$[B] = [D][N] \tag{3.4.4}$$

and where

$$\{R\} = \int [N]^T\{F_s\} \ dA + \int [N]^T\{f\} \ dV \tag{3.4.7}$$

In contrast to the conventional Rayleigh-Ritz method, these equations are assembled from the element equations, starting with

$$\{u_e\}_i = [n]_i\{r_e\}_i \tag{3.4.15}$$

and

$$[b]_i = [D][n]_i \tag{3.4.18}$$

from which we obtain

$$[k]_i\{r_e\}_i = \{V_e\}_i \tag{3.4.19}$$

where

$$[k]_i = \int [b]_i^T [G] [b]_i \, dV \tag{3.4.20}$$

For a purely axially loaded bar,

$$[k]_i = \int [n']_i^T EA[n']_i \, dx \tag{3.4.20}$$

where

$$[n]_i = \left[1 - \frac{s}{L_i} \quad \frac{s}{L_i} \right] \tag{3.4.31}$$

The global matrices are formed from the element matrices by modifying them so that

$$[K]\{r\} = \Sigma \, [k]_i\{r\} = \Sigma \, \{V\}_i = \{R\} \tag{3.4.41}$$

When boundary constraints are added, this set of equations is solved for the displacements and the support reactions by partitioning

$$\begin{bmatrix} [K^{11}] & [K^{12}] \\ [K^{21}] & [K^{22}] \end{bmatrix} \begin{bmatrix} \{r^1\} \\ \{r^2\} \end{bmatrix} = \begin{bmatrix} \{R^1\} \\ \{R^2\} \end{bmatrix} \tag{3.4.47}$$

and using a programmed form of Gauss elimination to solve the linear algebraic equations.

PROBLEMS

1. Consider an axially loaded bar with constraints and loading as shown.
   (a) Write down the virtual work for this particular structure.
   (b) Which of the following would or would not be acceptable admissible functions for a Rayleigh-Ritz solution?

   i. $\Omega = \sin \dfrac{\pi x}{a}$

   ii. $\Omega = \cos \dfrac{\pi x}{a}$

   iii. $\Omega = \dfrac{x}{a}$

   iv. $\Omega = \dfrac{x}{a}\left(1 - \dfrac{x}{a}\right)$

Work and Energy Methods   Chap. 3

(c) Complete a one-term Rayleigh-Ritz solution to the problem.
(d) Suggest suitable functions for a two-term solution.

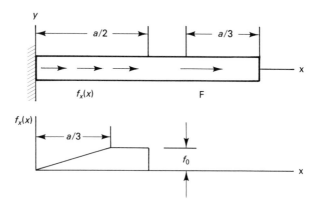

2. A simply supported beam is loaded as shown.
   (a) Write down the virtual work for this particular structure.
   (b) Which of the following would or would not be acceptable admissible functions for a Rayleigh-Ritz solution?

   i. $\Omega = \dfrac{x}{a}\left(1 - \dfrac{x}{a}\right)$

   ii. $\Omega = \dfrac{x^2}{a^2}$

   iii. $\Omega = 3\dfrac{x^2}{a^2} - 5\dfrac{x^3}{a^3} + 2\dfrac{x^4}{a^4}$

   (c) Solve the problem using one term in the Rayleigh-Ritz method for each admissible case.
   (d) Solve the differential equation for the exact deflection of the beam and compare with the one-term solutions by plotting them. What is there about the function in (iii) that suggests that it is not a very good one-term choice even though it is admissible?
   (e) Find a two-term solution using at least one of the terms in part (b) and plot the answer with the others.

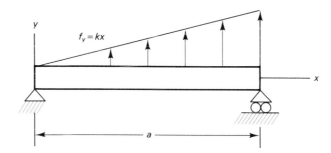

3. For the beam shown, find a two-term Rayleigh-Ritz solution using the same functions as in Problem 2. Suggest a third term for the case where the spring constant is very large.

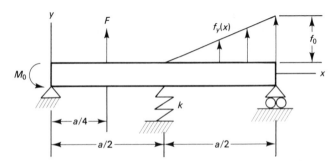

4. Extend virtual work and potential energy to the torsional problem. Consider the standard, uniform, cantilever shaft with a solid, circular cross section and a single concentrated moment at the free end.

   (a) Write down the expression for the virtual work in terms of the displacement β.

   (b) Write down the expression for the potential energy in terms of the displacement β.

   (c) Define K_{rs} and Q_r.

   (d) Find the solution by the Rayleigh-Ritz method based on a single assumed function $\Omega_1 = x/a$ and compare to the exact solution.

   (e) How much will adding a second term, say, $\Omega_2 = x^2/a^2$, improve the answer?

5. Consider the same shaft of Problem 4 but with both ends fixed. A distributed torque with units of moment per unit length, $t_x(x) = t_0 x/a$, is applied.

   (a) Write down the expression for the virtual work in terms of the displacement β.

   (b) Write down the expression for the potential energy in terms of the displacement β.

   (c) Define K_{rs} and Q_r.

   (d) What are the conditions for a function to be admissible in this case?

   (e) Find the solution by the Rayleigh-Ritz method based on a single assumed function

$$\Omega = \frac{x}{a}\left(1 - \frac{x}{a}\right)$$

   and compare to the exact solution.

   (f) What is the maximum stress in the shaft and where is it located?

6. An axially loaded bar has the dimensions and constraints shown. The cross section of the left half is $2A$ and the right half is A. A uniform distributed load acts over the second quarter of the bar, and there is a concentrated load at the right end.

   (a) Set up the FEM equations using three elements.

   (b) Show how the constraints are applied.

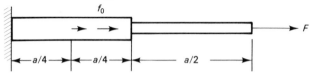

7. Divide the structure in Problem 1 into an appropriate number of elements and find the equivalent nodal loads.

8. Consider Problem 6 for the specific case of a steel bar where

$$a = 500 \text{ mm} \qquad A = 25 \text{ mm}^2$$

$$E = 206,840.0 \text{ MPa} \qquad f_0 = 400 \text{ N/mm} \qquad F = 860 \text{ N}$$

   (a) Find the nodal deflections and the distributed deflections.

   (b) Find the strains and stresses in the bar.

9. Consider the truss in Example 3.4.6. Instead of a load, a deflection is imposed on the horizontal component of value $r_1 = 0.025$ mm. What are the deflections and stresses in the truss?

10. Consider the pin-jointed truss shown. The members are made of steel and have a cross-sectional area of 200 mm². The two diagonal members are not joined where they cross.

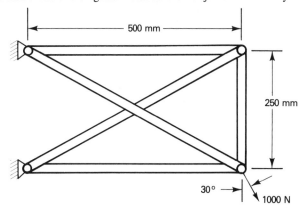

Set up the FEM equations, apply constraints, and partition the equations into the set for finding the displacements and the set for finding the support reactions.

11. Solve the equations in Problem 10 using some form of Gauss elimination.

12. The diagonal members in Problem 10 are divided in two and pinned at the midpoint where they cross. Set up and partition the FEM equations.

APPLICATIONS OF THE FINITE ELEMENT METHOD

4

4.1 INTRODUCTION

The axially loaded slender bar was especially useful for introducing the finite element method because of the simple form of its element stiffness matrix, applied loading, geometric constraints, and assembly process. Now we shall turn our attention to structures and structural elements that are increasingly more complex to represent in FEM, but are not necessarily more difficult to solve. First, we shall extend the axially loaded member to three dimensions, and then we shall adapt the general equations of the finite element method given in Section 3.4.1 to a variety of practical problems, including torsion in shafts, bending and shear, and then combined loadings in beams and frames, two-dimensional stress in plane bodies, and three-dimensional stress in general solid bodies, and then bending and shear in plates.

Perhaps, the greatest advantage of FEM has been in the solution of three-dimensional solid bodies and structures. Simple beam and plate theory had reached a high level of sophistication, but three-dimensional problems had grown in importance without a corresponding growth in the ability of classical methods to deal with them. In a sense, FEM came along just in time to handle the more complex problems.

The FEM problems introduced in this chapter are, for the most part, best solved by FEM computer codes because of the complexity of the element matrices and the large number of degrees of freedom required. Both the complexity and size call for automated assembly, automated equation solving, and automated handling of the data for both input and output. Examples are introduced to emphasize these points, and a number of solutions are provided. References 13 through 17, already cited, provided additional information on all the FEM topics and References 4 through 11 on the work and energy topics considered here.

4.2 GENERAL EQUATIONS OF FEM

When the symbolic representation of the equations in Figure 3.2.2 are adapted to the notation of the finite element method, we have for the global structure the representation in Figure 4.2.1. The associated equations are summarized in Equations 3.4.1 through 3.4.7.

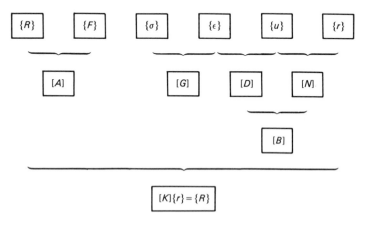

Figure 4.2.1 Symbolic Representation of the Global FEM Form of the Equations of Solid Mechanics.

As we have learned, these equations are assembled from the element equations that have the symbolic form shown in Figure 4.2.2. The equations associated with this representation are given in Equations 3.4.15 through 3.4.20; however, a new term has been inserted between the matrices $\{u\}$ and $\{r\}$, leading to two new equations:

$$\{u\} = [\emptyset]\{a\}, \qquad \{r\} = [h]\{a\}, \longrightarrow \{a\} = [h]^{-1}\{r\} \tag{4.2.1}$$

which leads to

$$[n] = [\emptyset][h]^{-1} \tag{4.2.2}$$

We bypassed these steps with the axially loaded bar but will use them for the more complex beam bending element and all subsequent elements. We shall also discover that in some cases it is convenient to develop shape functions that relate nodal displacements directly to the constrained displacements $\{u_c\}$; then we modify the strain–displacement equations by defining $[D_c]$ so that

$$\{\epsilon\} = [D_c]\{u_c\} \tag{4.2.3}$$

The symbolic representation in Figure 4.2.2 holds if we replace $\{u\}$ with $\{u_c\}$ and $[D]$ with $[D_c]$. We shall be doing this for beam and plate bending elements.

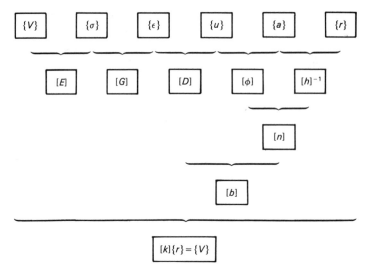

Figure 4.2.2 Symbolic Representation of the Element FEM Form of the Equations of Solid Mechanics.

4.3 AXIALLY LOADED BAR, CONTINUED

We have spent so much time with the axially loaded bar in Chapter 3 and now spend more time with it in this chapter because it is the simplest expression of nearly all the features of FEM. In Section 3.4, we carried the development of the axially loaded bar through its use in plane trusses. Now we shall extend the development to three-dimensional trusses, or *space trusses*.

The expression in three dimensions of a vector in two coordinate systems, one rotated with respect to the other, can best be expressed by direction cosines. Such a representation is discussed in Appendix B for stress transformation. If we consider a vector $\{u\}$ with components in the x, y, z coordinate system and the same vector $\{u'\}$ with components in the x', y', z' coordinate system, we get

$$\{u'\} = [t]\{u\} \tag{4.3.1}$$

or

$$\begin{bmatrix} u' \\ v' \\ w' \end{bmatrix} = \begin{bmatrix} l_1 & m_1 & n_1 \\ l_2 & m_2 & n_2 \\ l_3 & m_3 & n_3 \end{bmatrix} \begin{bmatrix} u \\ v \\ w \end{bmatrix} \tag{4.3.2}$$

where, for example, $l_1 = \cos xx'$, where xx' signifies the angle between the x and x' and, similarly, $m_3 = \cos yz'$, where yz' signifies the angle between the y and z' axes. When this is applied to the axially loaded line element, we can extend the two-dimensional development in Equations 3.4.64 through 3.4.69 to three dimensions. Note that primes do not denote differention.

First, we have

$$\{r_e\}' = \begin{bmatrix} [t]_1 & [0] \\ [0] & [t]_2 \end{bmatrix} = \begin{bmatrix} \{u\}_1 \\ \{u\}_2 \end{bmatrix} = [T]\{r_e\} \tag{4.3.3}$$

where $\{r_e\}$ and $\{r_e\}'$ now have six components and $[T]$ is a 6×6 matrix. Since the nodal forces are vectors,

$$\{V_e\}' = [T] \{V_e\} \tag{4.3.4}$$

The coordinate transformation matrix, $[T]$, continues to have the property that $[T]^{-1} = [T]^T$, and it follows that

$$[k] = [T]^T [k]' \, [T] \tag{4.3.5}$$

where

$$[k]' = \frac{EA}{L} \begin{bmatrix} 1 & 0 & 0 & -1 & 0 & 0 \\ 0 & 0 & 0 & 0 & 0 & 0 \\ 0 & 0 & 0 & 0 & 0 & 0 \\ -1 & 0 & 0 & 1 & 0 & 0 \\ 0 & 0 & 0 & 0 & 0 & 0 \\ 0 & 0 & 0 & 0 & 0 & 0 \end{bmatrix} \tag{4.3.6}$$

We have now reached that point in FEM where it becomes burdensome to write out the actual form of the element matrices and assemble them by hand. In actual practice, we rarely see the matrix $[k]$ for a particular element and certainly we do not print out the matrix $[K]$ for the global structure, as we have done, for example, in Example 3.4.6. Instead we rely on computer programs that perform the appropriate coordinate transformations, assemble the matrices, and solve the resulting set of equations.

It is not our purpose in this text to teach the programing of FEM codes. Information about these codes is given in several of the FEM references, for example, References 14 through 17. Many computer codes for FEM analysis are available, and some are published in various references, for example, Reference 17. Some of these are home-grown products that generally apply to a selected class of problems, while others, which are sold or leased as commercial products, are often large, sophisticated, comprehensive, and expensive. We shall assume that such codes have been written and illustrate their use.

The task of a structural analyst using a typical FEM code is reduced to the following seven steps:

1. Divide the structure into suitable elements and number the nodes and elements.
2. Record the coordinates of each node.
3. Identify the nodes that go with each element.
4. Specify the material properties of each element.
5. Specify the geometric properties of each element.
6. Specify the geometric constraints.
7. Specify the equivalent nodal loads.

The assembly and solution of the equations then proceeds automatically. There remains only the interpretation of the results.

In the next example, we show a problem that is sufficiently complicated that it would be difficult to write down the assembled equations and be even more difficult to solve them by the usual hand-calculation methods. What we shall do is identify all the information that must be fed to a typical FEM computer code. We shall also show the results that were obtained when this information was actually used in one of the major commercial codes called MSC/NASTRAN. MSC stands for the MacNeal Schwendler Corporation, which developed and leases the software.

Example 4.3.1

The pin-jointed space truss with dimensions shown in figure (a) is constrained at its base, joints *ABCD*, and is to be analyzed for two sets of loads. The first set of loads is essentially a transverse set acting in the *y-z* plane and applied as shown in figure (b), and the second, shown in figure (c), produces a torsional load. All joint loads are 1000 N (newtons). After the initial analyses, the member connecting joints *B* and *E* is to be removed to simulate a damaged part, and the structure is reanalyzed to determine the effect on maximum stress and deflection in the remaining parts. The members of the truss are circular tubes made of steel for which $E =$ 206,840 MPa (megapascals) and $\upsilon = 0.3$.

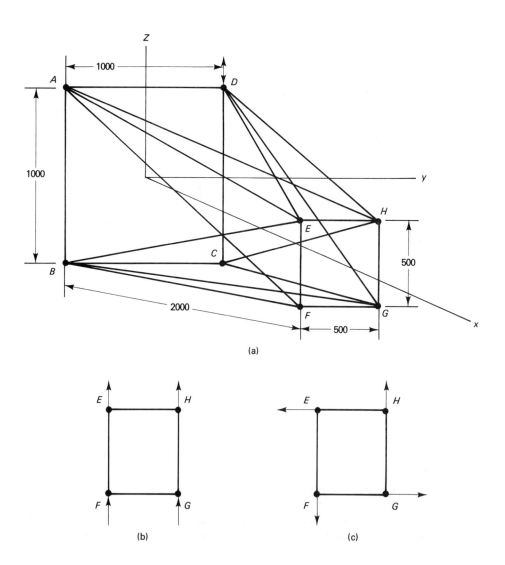

(a)

(b)   (c)

Solution: Let us go through the seven steps outlined previously.

1. Divide the structure into elements and number nodes and elements: For a pin-jointed truss the joints are the nodes. Nodal and element numbers are shown on figure (d).

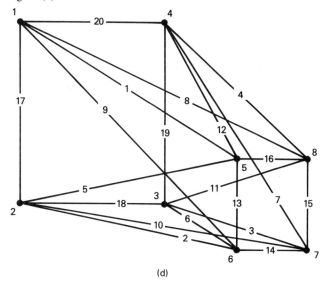

(d)

2. Record the coordinates of each node:

Nodal Number	Coordinates (x, y, z)		
1	0.0	-500.0	500.0
2	0.0	-500.0	-500.0
3	0.0	500.0	-500.0
4	0.0	500.0	500.0
5	2000.0	-250.0	250.0
6	2000.0	250.0	250.0
7	2000.0	250.0	-250.0
8	2000.0	250.0	250.0

3. Identify the nodes that go with each element:

Element Number	Nodes	Element Number	Nodes
1	1,5	11	3,8
2	2,6	12	4,5
3	3,7	13	5,6
4	4,8	14	6,7
5	2,5	15	7,8
6	3,6	16	8,5
7	4,7	17	1,2
8	1,8	18	2,3
9	1,6	19	3,4
10	2,7	20	4,1

4. Specify the material properties in each element: As stated previously,

$$E = 206,840.0 \text{ MPa}, \qquad \upsilon = 0.3$$

5. Specify the geometric properties of each element: The only geometric property needed is the cross-sectional area of each rod. We have used two sizes. For elements 1 to 4 and 13 to 20, we have used a tube of radius 15 mm and a wall thickness of 5 mm and have calculated

$$A = 2\pi rh = 471.24 \text{ mm}^2$$

For the rest of the elements we have taken a 10-mm radius and and a 5-mm wall thickness and have calculated

$$A = 2\pi rh = 314.16 \text{ mm}^2$$

6. Specify the geometric constraints: Nodes 1 to 4 are fixed to zero displacement in all coordinate directions.

7. Specify equivalent nodal loads: All loads in a pin-jointed truss must be applied at the nodes. In this case, there are two separate sets of loads. The first, which we shall call the shear load, is

Node Number	Load Component (x, y, z)		
5	0.0	0.0	1000.0
6	0.0	0.0	1000.0
7	0.0	0.0	1000.0
8	0.0	0.0	1000.0

and the second, which we shall call the torsional load, is

Node Number	Load Component (x, y, z)		
5	0.0	-1000.0	0.0
6	0.0	0.0	-1000.0
7	0.0	1000.0	0.0
8	0.0	0.0	1000.0

This problem has been solved with MSC/NASTRAN. The input file required is listed and briefly described in Appendix J. The results are described next. First, the nodal displacements (abbreviated to four significant figures) for the shear load case were found to be

Nodal Number	Displacements (x, y, z)		
1	0.0	0.0	0.0
2	0.0	0.0	0.0
3	0.0	0.0	0.0
4	0.0	0.0	0.0
5	$-4.551E-3$	$-2.737E-3$	$6.253E-1$
6	$4.551E-3$	$2.737E-3$	$6.253E-1$
7	$4.551E-3$	$-2.737E-3$	$6.253E-1$
8	$-4.551E-3$	$2.737E-3$	$6.253E-1$

and for the torsional case

Nodal Number	Displacements (x, y, z)		
1	0.0	0.0	0.0
2	0.0	0.0	0.0
3	0.0	0.0	0.0
4	0.0	0.0	0.0
5	$-5.532E-5$	$-2.334E-1$	$-2.258E-1$
6	$-5.532E-5$	$2.258E-1$	$-2.334E-1$
7	$-5.532E-5$	$2.334E-1$	$2.258E-1$
8	$-5.532E-5$	$-2.258E-1$	$2.334E-1$

where, for example, the notation $E-5$ means 10^{-5}. These displacements are shown in figure (e) for the shear case and in figure (f) for the torsional case. These plots were generated by the NASTRAN plot routines. They show both the deflected and undeflected structure and are normalized to show the maximum deflection as a fraction of a characteristic length of the structure; that is, they are not to scale.

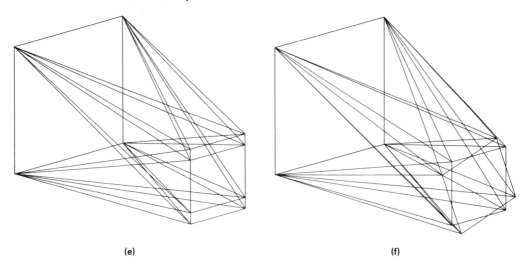

(e)   (f)

The stresses, expressed in N/mm^2, for the shear load are

Element No.	Axial Stress	Element No.	Axial Stress
1	-2.777	11	6.845
2	2.777	12	-2.436
3	2.777	13	0.0
4	-2.777	14	-0.755
5	6.845	15	0.0
6	2.436	16	0.755
7	-6.845	17	0.0
8	-2.436	18	0.0
9	-6.845	19	0.0
10	2.436	20	0.0

and for the torsional load are

Element No.	Axial Stress	Element No.	Axial Stress
1	$-3.375E-2$	11	3.450
2	$-3.375E-2$	12	3.450
3	$-3.375E-2$	13	1.053
4	$-3.375E-2$	14	1.053
5	-3.396	15	1.053
6	-3.396	16	1.053
7	-3.396	17	0.0
8	-3.396	18	0.0
9	3.450	19	0.0
10	3.450	20	0.0

Finally, the forces at the constraints are for the shear load

Nodal Number	Constraint Forces (x, y, z)		
1	4000.0	677.9	-1000.0
2	-4000.0	-677.9	-1000.0
3	-4000.0	677.9	-1000.0
4	4000.0	-677.9	-1000.0

and for the torsional load

Nodal Number	Constraint Forces (x, y, z)		
1	0.0	248.0	252.0
2	0.0	-252.0	248.0
3	0.0	-248.0	-252.0
4	0.0	252.0	-248.0

The second part of this problem is to reexamine the structure when element 5 is removed. This would simulate structural behavior when a member is damaged, for example. It also demonstrates one of the very great strengths of the FEM. All that need be done is to remove the line in the input file that identifies that element. Just delete the line and run the problem again. All the other input information is unchanged. This has been done and the following results obtained.

First, the nodal displacements for the shear load case were found to be

Nodal Number	Displacements (x, y, z)		
1	0.0	0.0	0.0
2	0.0	0.0	0.0
3	0.0	0.0	0.0
4	0.0	0.0	0.0

continued

Nodal Number	Displacements (x, y, z)		
5	1.701E−1	1.189E−1	1.309
6	1.027E−2	1.244E−1	1.293
7	3.686E−3	1.160E−1	6.248E−1
8	−5.415E−3	1.215E−1	6.258E−1

and for the torsional case

Nodal Number	Displacements (x, y, z)		
1	0.0	0.0	0.0
2	0.0	0.0	0.0
3	0.0	0.0	0.0
4	0.0	0.0	0.0
5	−8.670E−2	−2.938E−1	−5.648E−1
6	−2.894E−3	1.654E−1	−5.648E−1
7	3.734E−4	1.745E−1	2.260E−1
8	3.734E−4	−2.847E−1	2.331E−1

These displacements are shown in figure (g) for the shear case and in figure (h) for the torsional case. These plots also were generated by the NASTRAN plot routines.

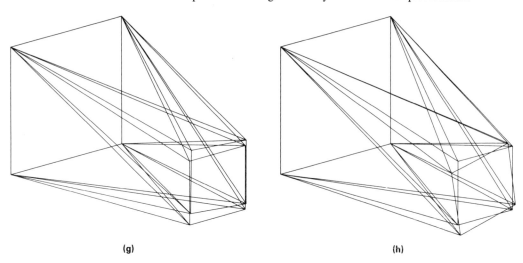

(g)   (h)

The stresses for the shear load are

Element No.	Axial Stress	Element No.	Axial Stress
1	0.714	11	6.383
2	6.268	12	−1.135
3	2.249	13	2.122
4	−3.304	14	−1.158
			continued

Element No.	Axial Stress	Element No.	Axial Stress
		15	0.143
6	3.736	16	0.352
7	−7.308	17	0.0
8	−1.135	18	0.0
9	−13.691	19	0.0
10	3.736	20	0.0

and for the torsional load are

Element No.	Axial Stress	Element No.	Axial Stress
1	−1.766	11	3.679
2	−1.766	12	2.805
3	0.228	13	0.0
4	0.228	14	1.253
		15	0.982
6	−4.041	16	1.253
7	−3.166	17	0.0
8	−4.041	18	0.0
9	6.845	19	0.0
10	2.805	20	0.0

Finally, the forces at the constraints are for the shear load

Nodal Number	Constraint Forces (x, y, z)		
1	4000.0	582.9	−1500.0
2	−4000.0	−772.8	−500.0
3	−4000.0	772.8	−966.2
4	4000.0	−582.9	−1033.8

and for the torsional load

Nodal Number	Constraint Forces (x, y, z)		
1	0.0	295.1	500.0
2	0.0	−204.9	0.0
3	0.0	−295.1	−268.7
4	0.0	204.9	−231.3

As might be expected, the removal of member 5 results in a large increase in the stress level in member 9 for both loading cases, with some redistribution of loads among the other members. With regard to displacement, nodes 5 and 6 show significant increases in both cases.

4.4 BEAM APPLICATIONS

We shall now do for the beam in torsion and bending what was done for the axially loaded bar. In general, a beam can deflect in three dimensions under axial and transverse forces in three directions and torsion and bending moments about three axes; however, we have learned that the axial and torsional deformations are uncoupled from those of bending and shear. Furthermore, we know that the bending and shear in the two component directions can be uncoupled by choosing to work with principal axes of inertia of the cross section. Therefore, we shall start with simple beams in torsion, then with bending and shear in one plane, and then show how all the components of deformation can be expressed in the full three-dimensional case.

4.4.1 Beams in Torsion

With the help of the torsional equations studied in Sections 1.3.3 and 2.2.2, we can quickly adapt the process used for the axially loaded bar to derive the FEM equations for a rod in pure torsion. Consider such a rod as shown in Figure 4.4.1. In such a bar, the displacement can be represented by a single quantity, $\beta(x)$, and the nodal displacements, $\{r_e\}_i$, are the rotations about the x-axis at each element node. If we define the element shape functions in terms of the constrained displacement $\{u_c\} = \beta(s)$ and the nodal displacements, we have

$$\{\beta_e\}_i = [n]_i\{r_e\}_i \qquad (4.4.1)$$

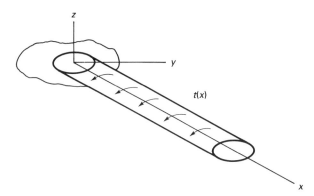

Figure 4.4.1 Rod with Distributed Torsional Load.

We must then modify the strain–displacement equations to work with the constrained displacement

$$\{\epsilon\}_i = [D_c]_i\{u_c\}_i = [D_c]_i[n]_i\{r_e\}_i = [b]_i\{r_e\}_i \qquad (4.4.2)$$

Finally, the stiffness matrix satisfies the equation

$$[k]_i\{r_e\}_i = \{V_e\}_i \qquad (4.4.3)$$

where

$$[k]_i = \int [b]_i^T[G]\{b]_i \, dV \qquad (4.4.4)$$

Let us first consider a rod with a circular cross section. Using equations (i) and (j) from Example 1.3.6 along with Hooke's law, we have

$$\gamma = \frac{\tau}{G} = \frac{Tr}{GJ} = r\theta = r\frac{d\beta}{ds} \qquad (4.4.5)$$

and we get

$$[D_c]_i = r\frac{d}{ds} \qquad (4.4.6)$$

To find $[n]_i$, we note that $\{r_e\}_i$ has just two components; therefore, $[n]_i$ has two components, and each is of the form

$$n_i(s) = a_i + b_i s \qquad (4.4.7)$$

By the same arguments used for the axial line element in Section 3.4.3, we conclude that

$$[n]_i = \left[1 - \frac{s}{L_i} \quad \frac{s}{L_i} \right] \qquad (4.4.8)$$

Thus, we get

$$[b]_i = \frac{r}{L_i}[-1 \ 1] \qquad (4.4.9)$$

and

$$[k]_i = \int [b]_i^T[G][b]_i \ dV = \int_0^{L_i} \int_0^{2\pi} \int_0^R \left(\frac{r}{L_i}\right)^2 \begin{bmatrix} -1 \\ 1 \end{bmatrix} G[-1 \ 1] r \ dr \ d\varphi \ ds$$

$$= \frac{G_i J_i}{L_i} \begin{bmatrix} 1 & -1 \\ -1 & 1 \end{bmatrix} \qquad (4.4.10)$$

where for the circular cross section $J = \pi R^4/2$. When a different cross section is used, we can insert the proper torsional constant for that cross section, as found by the methods in Section 1.3.3.

The assembly of element stiffness matrices to form the global stiffness matrix and the application of the method to structures in torsion is so like that for the axial elements that we shall not pause here for an example.

4.4.2 Simple Beams in Bending and Shear

Now let us consider a beam in bending and shear uncoupled from axial or torsional loads or deflections. What we are seeking is the FEM equivalent of the study of beams in Sections 2.2.3 and 2.2.4. Let us restrict ourselves to a simple beam in the special case where $I_{yz} = 0$ and there are no z-components of loads or deflections. Once we master this in FEM, we shall extend it to general three-dimensional beam theory.

In this case, $\{u_c\}_i$ reduces to single term, $v(s)$. If we denote the shape functions to be

$$\{v_e\}_i = [n]_i\{r_e\}_i \qquad (4.4.11)$$

then

$$[D_c] = -y\frac{d^2}{ds^2} \tag{4.4.12}$$

and

$$[b]_i = [D_c][n]_i = -y\frac{d^2}{ds^2}[n]_i \tag{4.4.13}$$

and, finally,

$$[k]_i = \int [b]_i^T[G][b]_i \, dV = \iint \left(-y\frac{d^2}{ds^2}[n]_i\right)^T E\left(-y\frac{d^2}{ds^2}[n']_i\right) dA \, ds$$
$$= \int_0^L EI_{zz}[n'']_i^T[n'']_i \, ds \tag{4.4.14}$$

where, once again, primes denote differentiation with respect to x.

When $\{R_e\}_i$ is adapted to the transversely loaded beam, we get

$$\{R_e\}_i = \int [n]_i^T f_y(s) \, ds \tag{4.4.15}$$

where $f_y(s)$ includes both body forces and lateral applied loads and has the units of force per unit length.

Now we must find the shape functions. A simple beam element is shown in Figure 4.4.2(a). The corresponding line element is shown in Figure 4.4.2(b) for element 1. It has a node at each end and two degrees of freedom at each node. The two degrees of freedom allow for a lateral displacement v_i in the y-direction and a rotational displacement θ_i at each node. Thus, the nodal displacements for a beam element, say, element 1, are

$$\{r_e\}_1 = \begin{bmatrix} r_1 \\ r_2 \\ r_3 \\ r_4 \end{bmatrix} = \begin{bmatrix} v_1 \\ \theta_1 \\ v_2 \\ \theta_2 \end{bmatrix} \tag{4.4.16}$$

We note that both the rotations are positive in the counterclockwise directions in the x-y plane and that

$$\theta_1 = \left(\frac{dv}{ds}\right)_1, \qquad \theta_2 = \left(\frac{dv}{ds}\right)_2 \tag{4.4.17}$$

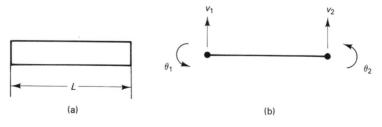

(a)   (b)

Figure 4.4.2 Beam Line Element.

The element shape functions may be found by the method used for the axial element. It is more convenient to introduce a two-step process as indicated in Figure 4.2.2 and Equations 4.2.1 through 4.2.3. First, because a simple beam element has four degrees of freedom, the simplest polynomial that can represent its shape is a cubic polynomial. So let

$$v(s) = a_1 + a_2 s + a_3 s^2 + a_4 s^3 \qquad (4.4.18)$$

$$= [1 \ s \ s^2 \ s^3] \begin{bmatrix} a_1 \\ a_2 \\ a_3 \\ a_4 \end{bmatrix} = [\emptyset]\{a\}$$

We now take the second step to form the equation

$$\{r_e\}_i = [h]\{a\} \qquad (4.4.19)$$

which for element 1 in expanded notation is

$$\begin{bmatrix} v_1 \\ \theta_1 \\ v_2 \\ \theta_2 \end{bmatrix} = \begin{bmatrix} 1 & 0 & 0 & 0 \\ 0 & 1 & 0 & 0 \\ 1 & L_1 & L_1^2 & L_1^3 \\ 0 & 1 & 2L_1 & 3L_1^2 \end{bmatrix} \begin{bmatrix} a_1 \\ a_2 \\ a_3 \\ a_4 \end{bmatrix} \qquad (4.4.20)$$

We can invert $[h]$ to obtain the following equation

$$\{a\} = [h]^{-1}\{r_e\}_1 \qquad (4.4.21)$$

where

$$[h]^{-1} = \frac{1}{L_1^3} \begin{bmatrix} L_1^3 & 0 & 0 & 0 \\ 0 & L_1^3 & 0 & 0 \\ -3L_1 & -2L_1^2 & 3L_1 & -L_1^2 \\ 2 & L_1 & -2 & L_1 \end{bmatrix} \qquad (4.4.22)$$

From Equations 4.4.18 and 4.4.20, we have

$$[n]_i = [\emptyset][h]^{-1} = \frac{1}{L_1^3} [L_1^3 - 3L_1 s^2 + 2s^3, \ L_1^3 s - 2L_1^2 s^2 + L_1 s^3, \qquad (4.4.23)$$
$$3L_1 s^2 - 2s^3, \ -L_1^2 s^2 + L_1 s^3]$$

where commas have been added to clarify each term in this row matrix.

It is seen that the distributed deflection of a beam element is a linear combination of four terms of a cubic polynomial. We can generalize this by replacing the subscript 1 with i and noting that the shape functions for all elements have the same form. It is interesting to see them plotted in Figure 4.4.3.

Note that n_1 has the value 1.0 at $x = 0$ and the value 0.0 at $x = L_i$, while n_3 has the value 0.0 at $x = 0$ and the value 1.0 at $x = L_i$, and that the first derivatives of both are zero at both nodes. Also note that the first derivative of n_2 has the value 1.0 at $x = 0$ and the value 0.0 at $x = L_i$, while the first derivative of n_4 has the value 0.0 at $x = 0$ and the value 1.0 at $x = L_i$, and that both have the value of zero at both nodes. We shall see similar characteristics in shape functions in other elements, and we will learn that these characteristics play an important role in the simplicity of the method.

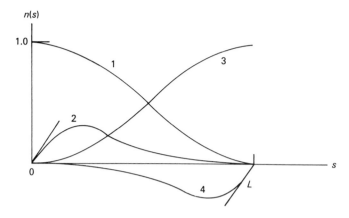

Figure 4.4.3 Simple Beam Element Shape Functions.

For the contribution of a single beam element to the $[K]$ matrix, we have, from Equations 4.4.14 and 4.4.23,

$$[k]_i = \int EI_{zz}[n'']_i^T [n'']_i \, ds$$

$$= \frac{(EI_{zz})_i}{L_i^3} \begin{bmatrix} 12 & 6L_i & -12 & 6L_i \\ 6L_i & 4L_i^2 & -6L_i & 2L_i^2 \\ -12 & -6L_i & 12 & -6Li \\ 6L_i & 2L_i^2 & -6L_i & 4L_i^2 \end{bmatrix} \tag{4.4.24}$$

As described in Section 3.4.4, the assembly process requires $[k]_i$ to be embedded in a matrix the size of the global $[K]$ matrix and summed, as follows:

$$[K]\{r\} = \Sigma \, [k]_i\{r\} \tag{4.4.25}$$

The structural equivalent nodal loads are also assembled from the element nodal loads. Concentrated forces and moments are added directly to $\{R\}$ at the appropriate nodes. All this is best illustrated by an example.

Example 4.4.1

Find the displacements and the stresses in the beam shown in figure (a).

Solution: Let us divide the beam into two elements and number the nodes and elements as shown in figure (b). There are six degrees of freedom in the global structure and hence there will be six elements in $\{r\}$, and $[K]$ will have six rows and six columns. Before we assemble

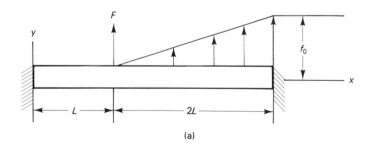

(a)

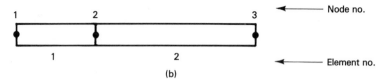

(b)

the matrix $[K]$, let us find the equivalent nodal loads from the distributed load. In terms of a local coordinate s, with origin at the left end of element 2, the distributed load is

$$f_y(s) = f_0 \frac{s}{L_2} = f_0 \frac{s}{2L} \tag{a}$$

From Equation 4.4.15, for element 2,

$$\{R_e\}_2 = \int [n]_2^T (f_y)_2 \, ds$$

$$= \int_0^{L_2} \frac{1}{L_2^3} \begin{bmatrix} L_2^3 - 3L_2 s^2 + 2s^3 \\ L_2^3 s - 2L_2^2 s^2 + L_2 s^3 \\ 3L_2 s^2 - 2s^3 \\ -L^2 s^2 + L_2 s^3 \end{bmatrix} f_0 \frac{s}{L_2} \, ds \tag{b}$$

$$= \begin{bmatrix} 0.30000 f_0 L \\ 0.13333 f_0 L^2 \\ 0.70000 f_0 L \\ -0.20000 f_0 L^2 \end{bmatrix}$$

Now we can assemble the matrix equation

$$[K]\{r\} = \{R\} \tag{c}$$

for this particular structure. For element 1, we have

$$[k]_1 \{r_e\}_1 = \frac{EI_{zz}}{L^3} \begin{bmatrix} 12 & 6L & -12 & 6L \\ 6L & 4L^2 & -6L & 2L^2 \\ -12 & -6L & 12 & -6L \\ 6L & 2L^2 & -6L & 4L^2 \end{bmatrix} \begin{bmatrix} v_1 \\ \theta_1 \\ v_2 \\ \theta_2 \end{bmatrix} \tag{d}$$

and for element 2

$$[k]_2 \{r_e\}_2 = \frac{EI_{zz}}{L^3} \begin{bmatrix} 1.5 & 1.5L & -1.5 & 1.5L \\ 1.5L & 2L^2 & -1.5L & L^2 \\ -1.5 & -1.5L & 1.5 & -1.5L \\ 1.5L & L^2 & -1.5L & 2L^2 \end{bmatrix} \begin{bmatrix} v_2 \\ \theta_2 \\ v_3 \\ \theta_3 \end{bmatrix} \tag{e}$$

When these are embedded in a global matrix form and assembled, we have

$$\frac{EI_{zz}}{L^3} \begin{bmatrix} 12 & 6L & -12 & 6L & 0 & 0 \\ 6L & 4L^2 & -6L & 2L^2 & 0 & 0 \\ -12 & -6L & 13.5 & -4.5L & -1.5 & 1.5L \\ 6L & 2L^2 & -4.5L & 6L^2 & -1.5L & L^2 \\ 0 & 0 & -1.5 & -1.5L & 1.5 & -1.5L \\ 0 & 0 & 1.5L & L^2 & -1.5L & 2L^2 \end{bmatrix} \begin{bmatrix} v_1 \\ \theta_1 \\ v_2 \\ \theta_2 \\ v_3 \\ \theta_3 \end{bmatrix} = \begin{bmatrix} R_1 \\ M_1 \\ F + 0.3 f_0 L \\ 0.13333 f_0 L^2 \\ R_3 + 0.7 f_0 L \\ M_3 - 0.8 f_0 L^2 \end{bmatrix} \tag{f}$$

There are several things to note. The element stiffness matrices and the assembled global stiffness matrix $[K]$ are expressed in terms of L, which is the length of element 1 and one-half the length of element 2. The support reactions are shown explicitly in the matrix $\{R\}$

Applications of the Finite Element Method   Chap. 4

as R_1, M_1, R_3, and M_3. The equations so assembled cannot be solved until we specify the geometric boundary conditions, or constraints, because $|K| = 0$.

The boundary conditions are

$$v_1 = \theta_1 = v_3 = \theta_3 \tag{g}$$

Thus, the partitioned equations used for solving for the unknown displacements are

$$\frac{EI_{zz}}{L^3} \begin{bmatrix} 13.5 & -4.5L \\ -4.5L & 6L^2 \end{bmatrix} \begin{bmatrix} v_2 \\ \theta_2 \end{bmatrix} = \begin{bmatrix} F + 0.3f_0L \\ 0.13333f_0L^2 \end{bmatrix} \tag{h}$$

from which we obtain

$$v_2 = \frac{L^3}{EI_{zz}} (0.0987654F + 0.0395062f_0L)$$

$$\theta_2 = \frac{L^3}{EI_{zz}} (0.0740741F + 0.0518519f_0L) \tag{i}$$

The support reactions can be found by substituting equations (i) into (f). They can also be found by noting the relationship between the support reactions and the stress resultants at the two ends of the beam. The internal stress resultants may be obtained from the element stiffness matrix for each element. For element 1, we have

$$\{V_e\}_1 = [k]_1\{r_e\}_1 = \frac{EI_{zz}}{L^3} \begin{bmatrix} 12 & 6L & -12 & 6L \\ 6L & 4L^2 & -6L & 2L^2 \\ -12 & -6L & 12 & -6L \\ 6L & 2L^2 & -6L & 4L^2 \end{bmatrix} \begin{bmatrix} 0 \\ 0 \\ v_2 \\ \theta_2 \end{bmatrix} \tag{j}$$

from which we get

$$(V_{e1})_1 = \frac{EI_{zz}}{L^3} (-12v_2 + 6L\theta_2) = -0.74074F - 0.16296f_0L = R_1$$

$$(V_{e2})_1 = \frac{EI_{zz}}{L^3} (-6Lv_2 + 2L^2\theta_2) = -0.14815F - 0.13333f_0L^2 = M_1$$

$$(V_{e3})_1 = \frac{EI_{zz}}{L^3} (12v_2 - 6L\theta_2) = 0.74074F + 0.16296f_0L \tag{k}$$

$$(V_{e4})_1 = \frac{EI_{zz}}{L^3} (-6Lv_2 + 4L^2\theta_2) = -0.29630F - 0.02963f_0L^2$$

where we have noted that $(V_{e1})_1 = R_1$ and $(V_{e2})_1 = M_1$.

Similarly, for element 2,

$$\{v_e\}_2 = [k]_2\{r_e\}_2 = \frac{EI_{zz}}{L^3} \begin{bmatrix} 1.5 & 1.5L & -1.5 & 1.5L \\ 1.5L & 2L^2 & -1.5L & L^2 \\ -1.5 & -1.5L & 1.5 & -1.5L \\ 1.5L & L^2 & 1.5L & 2L^2 \end{bmatrix} \begin{bmatrix} v_2 \\ \theta_2 \\ 0 \\ 0 \end{bmatrix} \tag{l}$$

from which we get

$$(V_{e1})_2 = \frac{EI_{zz}}{L^3} (1.5v_2 + 1.5L\theta_2) = 0.25926F + 0.13654f_0L$$

$$(V_{e2})_2 = \frac{EI_{zz}}{L^3} (1.5Lv_2 + 2L^2\theta_2) = -0.29630F + 0.16296f_0L^2 \tag{m}$$

$$(V_{e3})_2 = \frac{EI_{zz}}{L^3}(-1.5v_2 - 1.5L\theta_2) = -0.25926F - 0.13654f_0L$$

$$(V_{e4})_2 = \frac{EI_{zz}}{L^3}(1.5Lv_2 + L^2\theta_2) = 0.22222F + 0.11111f_0L^2$$

where we note that $(V_{e3})_2 = R_3 + 0.7f_0L$ and $(V_{e4})_2 = M_3 - 0.8f_0L^z$ from which we can solve for R_3 and M_3.

The bending stresses at the nodes are found from the usual beam bending stress formula:

$$\sigma_x = -\frac{M_z y}{I_{zz}} \tag{n}$$

where each element moment is inserted in turn. Given suitable cross-sectional shapes, the shearing stresses at the nodes can also be found from the shearing forces, using the formulas found in Chapter 2. The deflections and stresses at points intermediate to nodes can also be found by working backward through the equations used in the derivation of FEM. Distributed displacements are given by Equation 4.4.11,

$$\{v_e\}_i = [n]_i\{r_e\}_i \tag{o}$$

and distributed strains and stresses by

$$\{\epsilon_e\}_i = [D_c]\{v_e\}_i, \qquad \{\sigma_e\}_i = [G]\{\epsilon_e\}_i \tag{p}$$

If we add springs to the beam, following the same steps as in Equations 3.2.56 and 3.2.57, we add terms to the matrix $[K]$ so that the elements of the stiffness matrix become

$$K_{rs} = \int EI_{zz}(N_r)''(N_s)''\,dx + \Sigma\,k_m N_r(x_m)N_s(x_m) + \Sigma\,\mu_n N_r'(x_n)N_s'(x_n) \tag{4.4.26}$$

where k_m and μ_n are the linear and rotational spring constants, respectively, and x_m and x_n are the locations of those springs. Because of the choice of shape functions, this reduces to a very simple form when the spring is located at a node. Each shape function has the characteristics at the nodes shown in Figure 4.4.3 and as described for the figure. The net effect of the shape functions and their first derivatives having the value of either zero or one at each node is that the spring constants add directly to the appropriate K_{rs} term. Perhaps, an example will do more than words to clear this up.

Example 4.4.2

Find the FEM equations for the beam shown in figure (a).

Solution: The beam differs from the one in Example 4.4.1 only in that at the right end there is a linear and rotational spring in place of the former fixed support; thus, much of what we obtained in the previous example can be used. We shall divide the beam into elements and number them and the nodes exactly as before, as shown in figure (b). The equivalent nodal loads found in equation (b) are not changed. In fact, the assembled equations are changed only by the addition of spring terms at the appropriate nodes in the $[K]$ matrix and the removal of the support reactions from node 3 in the $\{R\}$ matrix, as we have done in equation (a).

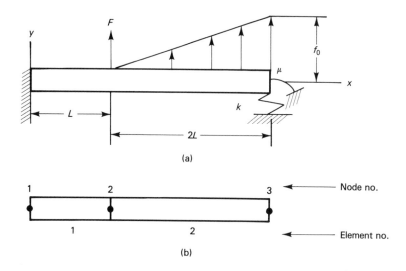

(a)

(b)

$$\frac{EI_{zz}}{L^3}\begin{bmatrix} 12 & 6L & -12 & 6L & 0 & 0 \\ 6L & 4L^2 & -6L & 2L^2 & 0 & 0 \\ -12 & -6L & 13.5 & -4.5L & -1.5 & 1.5L \\ 6L & 2L^2 & -4.5L & 6L^2 & -1.5L & L^2 \\ 0 & 0 & -1.5 & -1.5L^2 & 1.5+k & -1.5L \\ 0 & 0 & 1.5L & L^2 & -1.5L & 2L^2+\mu \end{bmatrix}\begin{bmatrix} v_1 \\ \theta_1 \\ v_2 \\ \theta_2 \\ v_3 \\ \theta_3 \end{bmatrix} = \begin{bmatrix} R_1 \\ M_1 \\ F+0.3f_0L \\ 0.13333f_0L^2 \\ 0.7f_0L \\ -0.8f_0L^2 \end{bmatrix} \quad \text{(a)}$$

where

$$k = \frac{kL^3}{EI_{zz}}, \qquad \mu = \frac{\mu L^3}{EI_{zz}} \qquad \text{(b)}$$

The only other thing that changes is that the geometric constraints are now just two in number:

$$w_1 = \theta_1 = 0 \qquad \text{(c)}$$

The solution proceeds exactly as in 4.2.1, but it is considerably more difficult to solve by analytical methods because four equations and four unknowns remain in the partitioned equations.

4.4.3 Beams with Combined Loads

Now we shall extend what we have learned about beams to the general case where we have bending in two planes and axial and torsional loads and deflections as well. A finite element of such a beam is still a line element that has two nodes, but at each node there are six degrees of freedom. This is depicted in Figure 4.4.4.

Note that the positive directions of the rotations are by the right-hand rule. This means that rotations about the y axis are defined as

$$\varphi_1 = -\left(\frac{dw}{dx}\right)_1, \qquad \varphi_2 = -\left(\frac{dw}{dx}\right)_2 \qquad (4.4.27)$$

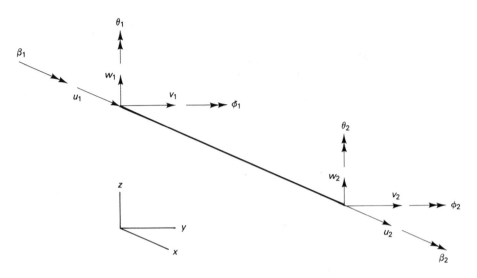

Figure 4.4.4. General Beam Element Degrees of Freedom.

This definition, which differs from that for rotations about the z-axis given in Equation 4.4.17, results in the following element stiffness matrix for deflections in the x-z plane:

$$[k]_i = \int EI_{yy} [n'']_i^T [n'']_i \, ds = \frac{(EI_{yy})_i}{L_i^3} \begin{bmatrix} 12 & -6L_i & -12 & -6L_i \\ -6L_i & 4L_i^2 & 6L_i & 2L_i^2 \\ -12 & 6L_i & 12 & 6L_i \\ -6L_i & 2L_i^2 & 6L_i & 4L_i^2 \end{bmatrix} \qquad (4.4.28)$$

Now we can combine the four uncoupled stiffness matrices from Equations 3.4.31, 4.4.10, 4.4.24, and 4.4.28 for axial loading, torsional loading, and bending in two planes. It is conventional to order the nodal displacements as follows:

$$\{r_e\} = \begin{bmatrix} u_1 \\ v_1 \\ w_1 \\ \beta_1 \\ \varphi_1 \\ \theta_1 \\ u_2 \\ v_2 \\ w_2 \\ \beta_2 \\ \varphi_2 \\ \theta_2 \end{bmatrix} \qquad (4.4.29)$$

We shall drop the explicit listing of the subscript i and assume that it is implied. And we shall simplify the notation by letting

$$I_{zz} = I_1, \qquad I_{yy} = I_2 \qquad (4.4.30)$$

Thus, for the general beam element,

$$[k]_i = \tag{4.4.31}$$

$$
\begin{bmatrix}
\dfrac{EA}{L} & 0 & 0 & 0 & 0 & 0 & -\dfrac{EA}{L} & 0 & 0 & 0 & 0 & 0 \\[2mm]
0 & \dfrac{12EI_1}{L^3} & 0 & 0 & 0 & \dfrac{6EI_1}{L^2} & 0 & -\dfrac{12EI_1}{L^3} & 0 & 0 & 0 & \dfrac{6EI_1}{L^2} \\[2mm]
0 & 0 & \dfrac{12EI_2}{L^3} & 0 & -\dfrac{6EI_2}{L^2} & 0 & 0 & 0 & -\dfrac{12EI_2}{L^3} & 0 & -\dfrac{6EI_2}{L^2} & 0 \\[2mm]
0 & 0 & 0 & \dfrac{GJ}{L} & 0 & 0 & 0 & 0 & 0 & -\dfrac{GJ}{L} & 0 & 0 \\[2mm]
0 & 0 & -\dfrac{6EI_2}{L^2} & 0 & \dfrac{4EI_2}{L} & 0 & 0 & 0 & \dfrac{6EI_2}{L^2} & 0 & \dfrac{2EI_2}{L} & 0 \\[2mm]
0 & \dfrac{6EI_1}{L^2} & 0 & 0 & 0 & \dfrac{4EI_1}{L} & 0 & -\dfrac{6EI_1}{L^2} & 0 & 0 & 0 & \dfrac{2EI_1}{L} \\[2mm]
-\dfrac{EA}{L} & 0 & 0 & 0 & 0 & 0 & \dfrac{EA}{L} & 0 & 0 & 0 & 0 & 0 \\[2mm]
0 & -\dfrac{12EI_1}{L^3} & 0 & 0 & 0 & -\dfrac{6EI_1}{L^2} & 0 & \dfrac{12EI_1}{L^3} & 0 & 0 & 0 & -\dfrac{6EI_1}{L^2} \\[2mm]
0 & 0 & -\dfrac{12EI_2}{L^3} & 0 & \dfrac{6EI_2}{L^2} & 0 & 0 & 0 & \dfrac{12EI_2}{L^3} & 0 & \dfrac{6EI_2}{L^2} & 0 \\[2mm]
0 & 0 & 0 & -\dfrac{GJ}{L} & 0 & 0 & 0 & 0 & 0 & \dfrac{GJ}{L} & 0 & 0 \\[2mm]
0 & 0 & -\dfrac{6EI_2}{L^2} & 0 & \dfrac{2EI_2}{L} & 0 & 0 & 0 & \dfrac{6EI_2}{L^2} & 0 & \dfrac{4EI_2}{L} & 0 \\[2mm]
0 & \dfrac{6EI_1}{L^2} & 0 & 0 & 0 & \dfrac{2EI_1}{L} & 0 & -\dfrac{6EI_1}{L^2} & 0 & 0 & 0 & \dfrac{4EI_1}{L}
\end{bmatrix}
$$

This form of the beam element stiffness matrix is usually provided in computer codes for FEM. As noted before, the formal assembly and explicit recording of the global stiffness matrix are seldom performed except for illustrative purposes, such as was done in Examples 4.4.1 and 4.4.2. Clearly, if very many elements are chosen, the matrix becomes quite large, and the act of assembling the element matrices and writing down the complete $[K]$ becomes a major chore. Furthermore, the solving of the assembled equations is, in itself, a formidable task.

Once an element stiffness matrix is obtained, it is available for the solution of a large number of problems. Note that the element stiffness matrix for a general beam element depends only on the material properties, E, and v, and on certain geometrical properties, L, A, I_{yy}, I_{zz}, and J, *of the element*. The global geometry, the geometric constraints, and the loading of a particular structure do not enter into the formulation of the element. Thus, the characteristics of these elements can be stored in a library in the computer to be made available for any structural problem that can be represented by a set of line elements. The analyst provides information according to the same seven steps outlined in Section 4.3 for all static structural problems, regardless of the type of element or type of structure. We shall illustrate this for a beam in the next example.

Example 4.4.3

Find the deflections and stresses of the cantilever tapered beam shown in figure (a) using a FEM computer code. The beam has a rectangular cross section as shown in figure (b). The value of $b(x)$ at the free end is one-half $b(x)$ at the fixed end; and c is a uniform value. The beam is loaded with a moment, $M_x = M$, at the free end; a concentrated force, F_3, in the z-direction at one-half the length and F_5 at a; and a uniform distributed load, f_0, in the y-direction. The beam is made of aluminum.

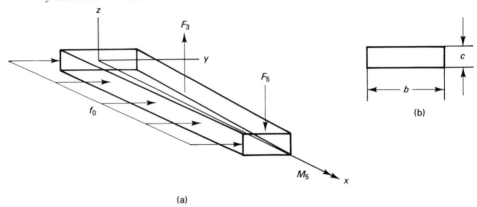

(a)

(b)

Solution: FEM computer codes require all information in numerical form. Let us select the following values for the various parameters.

$$a = 4.0 \text{ m} = 4000.0 \text{ mm}$$

$$b(0.0) = 150.0 \text{ mm}$$

$$c(0.0) = 100.0 \text{ mm}$$

Now let us set up the problem by identifying the seven items of input data.

1. Divide the structure into suitable elements and number the nodes and elements: Let us use four elements of equal length as shown in figure (c).

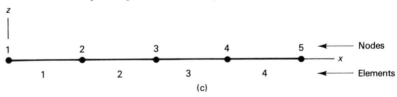

(c)

2. Record the coordinates of each node:

Node	x	y	z
1	0.0	0.0	0.0
2	1000.0	0.0	0.0
3	2000.0	0.0	0.0
4	3000.0	0.0	0.0
5	4000.0	0.0	0.0

3. Identify the nodes that go with each element:

Element	Nodes (from left to right)	
1	1	2
2	2	3
3	3	4
4	4	5

4. Specify the material properties of each element: The material for all elements is aluminum, for which

$$E = 68950.0 \text{ MPa (megapascals)}$$

$$G = 26520.0 \text{ MPa}$$

$$\upsilon = 0.3$$

5. Specify the geometric properties of each element: This is a tapered beam. We shall approximate it with elements of constant cross section using the geometric properties at the midpoint of each element. The length of the element is given in the nodal coordinate data. The remaining quantities to be specified are area, area moments of inertia about two axes, torsional constant, and points where normal stresses are to be output. Thus,

Element	A	I_1	I_2	J
1	14062.5	23.17E+6	11.72E+6	51.9E+6
2	12187.5	15.09E+6	10.16E+6	30.4E+6
3	10312.5	9.14E+6	8.59E+6	15.4E+6
4	0.84375	5.01E+6	7.03E+6	13.1E+6

The torsional constant was obtained by interpolating data from a table on page 277 of Reference 6. We have chosen the four corners of the cross section for stress output. The coordinates at the nodes are given next in order y, z.

Element	Upper Right	Upper Left	Lower Left	Lower Right
1	70.3125, 50.0	−70.3125, 50.0	−70.3125, −50.0	70.3125, −50.0
2	60.9375, 50.0	−60.9375, 50.0	−60.9375, −50.0	60.9375, −50.0
3	51.5625, 50.0	−51.5625, 50.0	−51.5625, −50.0	51.5625, −50.0
4	42.1875, 50.0	−42.1875, 50.0	−42.1875, −50.0	42.1875, −50.0

6. Specify geometric constraints: The beam is a cantilever with restraint at node 1 in all six degrees of freedom. None of the other nodes are restrained in any degree of freedom.

7. Specify equivalent nodal loads: These values of the loads were chosen:

$$f_0 = 0.5 \text{ N/mm}, \quad F_3 = 5000.0 \text{ N}, \quad F_5 = -3000.0 \text{ N}, \quad M_5 = 200,000.0 \text{ N-mm}$$

Node	R_x	R_y	R_z	M_x	M_y	M_z
1	0.0	250.0	0.0	0.0	0.0	41650.0
2	0.0	500.0	0.0	0.0	0.0	0.0
3	0.0	500.0	5000.0	0.0	0.0	0.0
4	0.0	500.0	0.0	0.0	0.0	0.0
5	0.0	250.0	−3000.0	200,000.0	0.0	−41650.0

The input file for MSC/NASTRAN has been prepared and is listed in Appendix J for this problem. It has been executed, and output data on stresses and deflections are presented next.

Node	Displacements (x, y, z)			Rotations (x, y, z)		
1	0.0	0.0	0.0	0.0	0.0	0.0
2	0.0	1.056	−1.650	1.453E-4	3.712E-3	1.930E-3
3	0.0	3.847	−9.693	3.934E-4	1.085E-2	3.452E-3
4	0.0	7.861	−23.764	8.831E-4	1.845E-2	4.377E-3
5	0.0	12.420	−44.275	1.459E-3	2.154E-2	4.619E-3

The constraint forces at the root are

Node	Resultant Forces (x, y, z)			Resultant Moments (x, y, z)		
1	0.0	−2.0E+3	−2.0E+3	−2.0E+5	−2.0E+5	−4.0E+6

The stresses are harder to specify. First, we note that they are reported at the four corners of the cross section for each element as noted in item 5. Second, they are reported at the location of the nodes at each end of the element. Because we have chosen to replace the taper of the column by a set of elements that are uniform, we have in fact solved a stepped beam, as shown in figure (d).

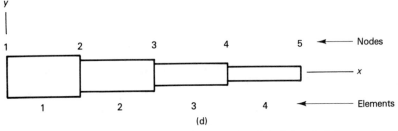

(d)

Thus, the stresses reported at the right end of element 1 and the left end of element 2 are both the stresses at node 2, but they are reported for different values of y and z because the two elements have different cross sections and, therefore, different coordinates for the corners. The true value should lie somewhere in between. We can approach the true value by using many more elements so that the differences in cross section between elements becomes very small. In the following table, $++$ means the corner of the cross section at the plus y and plus z coordinate and $+-$ means the corner of the cross section at the plus y and minus z coordinate, and so on.

Element	Node	$++$	$+-$	$--$	$-+$
1	1	−3.480	20.545	3.480	−20.545
1	2	10.633	23.766	−10.633	−23.766
2	2	10.767	28.603	−10.767	−28.603
2	3	25.657	33.398	−25.657	−33.398
3	3	29.518	40.331	−29.518	−40.331
3	4	16.218	18.638	−16.218	−18.638
4	4	19.583	23.092	−19.583	−23.092
4	5	0.035	−0.035	−0.035	0.035

Note that the stresses are not quite zero at the free end, as they should be, but the error is small. The axial stresses are given as zero for all elements at all nodes. Note that the transverse and torsional shearing stresses are not given. These depend on properties of the cross section that are not included in the description of the beam in the program. The shearing and torsion stress resultants are given, however, and using the methods of Chapters 1 and 2, the shearing stresses can be obtained. These stress resultants are

Element	Node	M_x	M_y	M_z	S_x	S_y	S_z
1	1	2.0E+5	−2.0E+6	3.958E+6	0.0	1.750E+3	2.0E+3
1	2	2.0E+5	−4.0E+6	2.208E+6	0.0	1.750E+3	2.0E+3
2	2	2.0E+5	−4.0E+6	2.208E+6	0.0	1.250E+3	2.0E+3
2	3	2.0E+5	−6.0E+6	9.583E+5	0.0	1.250E+3	2.0E+3
3	3	2.0E+5	−6.0E+6	9.583E+5	0.0	7.500E+2	−3.0E+3
3	4	2.0E+5	−3.0E+6	2.083E+5	0.0	7.500E+2	−3.0E+3
4	4	2.0E+5	−3.0E+6	2.083E+5	0.0	2.500E+2	−3.0E+3
4	5	2.0E+5	0.0	−4.165E+4	0.0	2.500E+2	−3.0E+3

The best use of FEM for beams is in still more complicated problems. Among these are those involving assemblies of beams called frames. In the next section, we shall examine some examples of frames.

4.4.4 Frames

In Section 2.2.4, near the end, we briefly mentioned that better methods for solving frame problems were coming later. The time has now arrived. No new theory is needed, so we shall proceed directly to examples. In frame problems, as in truss problems, the elements are oriented in global coordinates; thus, the element stiffness matrix for the beam that is oriented in local coordinates must undergo coordinate transformation before the elements are assembled.

Example 4.4.4

A steel frame has dimensions in millimeters and loads as shown in figure (a). It has a square cross section with dimensions 24 mm × 24 mm. $P = 6000$ N.

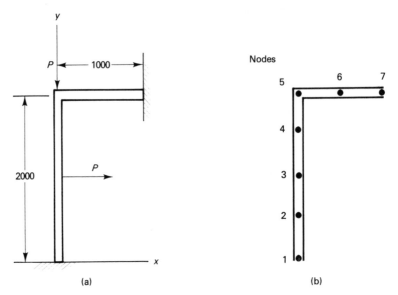

(a)   (b)

Solution: The following seven steps are taken to prepare the FEM model for analysis:

1. Divide the structure into suitable elements and number the nodes and elements: We shall use six elements of equal length as shown in figure (b).

2. Record the coordinates of each node:

Node	x	y	z
1	0.0	0.0	0.0
2	0.0	500.0	0.0
3	0.0	1000.0	0.0
4	0.0	1500.0	0.0
5	0.0	2000.0	0.0
6	500.0	2000.0	0.0
7	1000.0	2000.0	0.0

3. Identify the nodes that go with each element:

Element	Nodes
1	1, 2
2	2, 3
3	3, 4
4	4, 5
5	5, 6
6	6, 7

4. Specify the material properties for all elements: Let us use a steel for which

$$E = 206{,}800.0 \text{ MPa}, \qquad G = 79{,}550.0 \text{ MPa}, \qquad \upsilon = 0.3$$

5. Specify the geometric properties of each cross section: For a solid square cross section of 24 mm on a side,

$$A = 576 \text{ mm}^2, \qquad I_1 = I_2 = 27{,}648 \text{ mm}^4$$

If we choose the four corners of the cross section for stress output, the coordinates are ± 12 at each corner.

6. Specify geometric constraints: All degrees of freedom are constrained at nodes 1 and 7. The frame is also constrained to deflect only in the x-y plane.

7. Specify equivalent nodal loads:

Node	R_x	R_y	R_z
3	6000.0	0.0	0.0
5	0.0	−6000.0	0.0

This problem has been executed on MSC/NASTRAN, and the input file is listed and discussed briefly in Appendix J. The output data on stress and deflections have been obtained and are summarized as follows.

Node	Displacements (x, y, z)			Rotations (x, y, z)		
1	0.0	0.0	0.0	0.0	0.0	0.0
2	77.91	−0.09	0.0	0.0	0.0	−0.2378
3	164.05	−0.19	0.0	0.0	0.0	−0.0329
4	102.61	−0.28	0.0	0.0	0.0	0.2213
5	0.06	−0.38	0.0	0.0	0.0	0.1315
6	0.03	16.25	0.0	0.0	0.0	−0.0323
7	0.0	0.0	0.0	0.0	0.0	0.0

The constraint forces at the root are

Node	Resultant Forces (x, y, z)			Resultant Moments (x, y, z)		
1	−3.376E+3	7.495E+3	0.0	0.0	0.0	1.751E+6
7	−2.624E+3	−1.495E+3	0.0	0.0	0.0	4.971E+5

The stresses are given next. In the table $++$ means the corner of the cross section at the plus y and plus z coordinate, $+-$ means the corner of the cross section at the plus y and minus z coordinate, and so on.

Element	Node	+ +	+ -	- -	- +	Axial
1	1	−759.93	759.93	759.93	−759.93	
1	2	−27.24	27.24	27.24	−27.24	−13.01
2	2	−27.24	27.24	27.24	−27.24	−13.01
2	3	705.44	−705.44	−705.44	705.44	−13.01
3	3	705.44	−705.44	−705.44	705.44	−13.01
3	4	136.05	−136.05	−136.05	136.05	−13.01
4	4	136.05	−136.05	−136.05	136.05	−13.01
4	5	−433.35	433.35	433.35	−433.35	−13.01
5	5	433.35	−433.35	−433.35	433.35	−4.56
5	6	108.81	−108.81	−108.81	108.81	−4.56
6	6	108.81	−108.81	−108.81	108.81	−4.56
6	7	−215.74	215.74	215.74	−215.74	−4.56

Note that the axial and bending stresses must be added to provide the total normal stress at the corners of the cross section.

Again we wish to emphasize that the greater value of FEM is for large complex problems for which hand methods are just too laborious. The next example, while not out of the question for hand analysis, is large enough to illustrate this point.

Example 4.4.5

The space truss of Example 4.2.1 can be turned into a space frame by replacing the pinned joints with fixed joints, for example, by welding them together. The structure in this problem will have the same dimensions, material properties, and loading of Example 4.2.1. Only the nature of the joints has changed.

Solution: The seven steps are the same with only minor changes. Let us divide the structure into elements and number the nodes and elements (1) as before. The same coordinates are recorded (2); the same nodes go with each element (3): the material properties are the same (4). The geometric properties are the same, but now we must calculate the moments of inertia and the torsional constants for use on the property cards. For elements 1 to 4 and 13 to 20, they are

$$I_1 = I_2 = 53,014.4, \qquad J = 106,028.8$$

and for the rest of the elements

$$I_1 = I_2 = 15,708.0, \qquad J = 31,416.0$$

The geometric constraints are the same (6) and the nodal loads are the same (7).

All that has to be done is to modify the input file for the truss problem in Example 4.2.1 to define the elements as beam elements and add the beam properties to the file. The input file for MSC/NASTRAN is listed in Appendix J. The displacement results are presented next. It is interesting to note that the displacements are not much different from those for the space truss, which gives validity to the time-honored practice of modeling such structures as trusses even when the joints are not pinned. This use of a truss model depends on the loads being applied only at the joints. The frame model would permit loads to be applied at places other than the joints.

First, the nodal displacements (abbreviated to four significant figures) for the shear load case were found to be

Nodal Number	Displacements (x, y, z)		
1	0.0	0.0	0.0
2	0.0	0.0	0.0
3	0.0	0.0	0.0
4	0.0	0.0	0.0
5	−4.993E−3	−2.727E−3	6.229E−1
6	4.993E−3	2.727E−3	6.229E−1
7	4.993E−3	−2.727E−3	6.229E−1
8	−4.993E−3	2.727E−3	6.228E−1

The largest displacements in the z-direction are just 0.038% smaller than for the pin-jointed truss. The other displacements are much smaller in both cases and the differences are not significant. There are rotations at each node in the frame model, but these are very small. For the torsional case,

Nodal Number	Displacements (x, y, z)		
1	0.0	0.0	0.0
2	0.0	0.0	0.0
3	0.0	0.0	0.0
4	0.0	0.0	0.0
5	−5.537E−5	−2.303E−1	−2.226E−1
6	−5.537E−5	2.226E−1	−2.303E−1
7	−5.537E−5	2.303E−1	2.226E−1
8	−5.537E−5	−2.226E−1	2.303E−1

In this case the largest deflections change only by about 1%. For both cases, the displacement plots generated by the NASTRAN plot routines are indistinguishable between the truss and frame cases.

The axial stresses, expressed in N/mm^2, for the shear load are

Element No.	Axial Stress	Element No.	Axial Stress
1	−2.781	11	6.806
2	2.781	12	−2.440
3	2.781	13	0.0
4	−2.781	14	−0.752
5	6.806	15	0.0
6	2.440	16	0.752
7	−6.806	17	0.0
8	−2.440	18	0.0
9	−6.806	19	0.0
10	2.440	20	0.0

and for the torsional load are

Element No.	Axial Stress	Element No.	Axial Stress
1	−3.375E−2	11	3.403
2	−3.375E−2	12	3.403
3	−3.375E−2	13	1.053
4	−3.375E−2	14	1.053
5	−3.349	15	1.053
6	−3.349	16	1.053
7	−3.349	17	0.0
8	−3.349	18	0.0
9	3.403	19	0.0
10	3.403	20	0.0

In both cases the change in the largest axial load is 1% ± 0.5%. To these axial stresses must be added the stresses of bending in the frame model. For example, let us look at element 11, which has a large axial stress of 6.806 for the case of shear loading. To this must be added the normal stresses of bending at the four corners of the cross section. At node 3, these values are

$$-1.988E-3 \qquad -4.561E-1 \qquad 1.988E-3 \qquad 4.561E-1$$

These are small, changing only at the third significant figure. Similar results are obtained for the torsional loading case.

In many frames, the members are much more rigid and the loads are applied off the joints. In such cases the pin-jointed truss approximation to the frame may not apply. Then it is necessary to model the structure as a frame.

4.5 PLANE STRESS APPLICATIONS

The plane stress problem introduced briefly in Section 1.3.1 has a particularly elegant solution in finite element analysis. It is elegant because of its simplicity and its power when compared to classical methods. It is normal practice to divide the structure into either triangular elements, as shown in Figure 4.5.1(a), or quadrilateral elements, as shown in Figure 4.5.1(b).

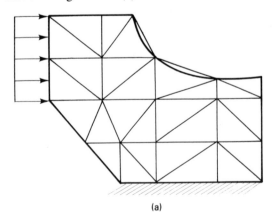

(a)

Applications of the Finite Element Method   Chap. 4

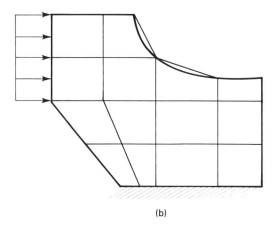

(b)

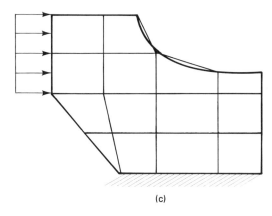

(c)

Figure 4.5.1 Plane Stress Elements.

It is possible to mix element types. Figure 4.5.1(c) is a case of mixed rectangular and triangular elements. No claim is made that in the figures these are the best or even a good distribution of elements. They merely illustrate that the structure can be represented by an assembly of elements. In these cases, the curved edge is approximated by the straight edge segments of adjacent elements.

4.5.1 Constant Strain Triangle

First, we shall consider that the plane stress structure is formed from a group of triangular elements. The element stiffness matrix is derived in terms of global coordinates by taking a general triangular element, as shown in Figure 4.5.2.

There are three nodes and two degrees of freedom at each node; thus, the element stiffness matrix will be of order 6×6. The first step is to decide on the polynomial form of the shape functions. The simplest form is as follows:

$$u_e(x, y) = a_1 + a_2 x + a_3 y$$
$$v_e(x, y) = a_4 + a_4 x + a_6 y$$

(4.5.1)

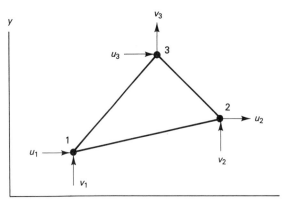

Figure 4.5.2 General Triangular Element for Plane Stress.

Written in matrix form

$$\{u_e\} = [\emptyset]\{a\} = \begin{bmatrix} 1 & x & y & 0 & 0 & 0 \\ 0 & 0 & 0 & 1 & x & y \end{bmatrix} \begin{bmatrix} a_1 \\ a_2 \\ a_3 \\ a_4 \\ a_5 \\ a_6 \end{bmatrix} \tag{4.5.2}$$

From this, we can construct the matrix equation for the plane stress triangle:

$$\{r_e\} = [h]\{a\} \tag{4.5.3}$$

which in expanded notation is

$$\begin{bmatrix} u_1 \\ v_1 \\ u_2 \\ v_2 \\ u_3 \\ v_3 \end{bmatrix} = \begin{bmatrix} 1 & x_1 & y_1 & 0 & 0 & 0 \\ 0 & 0 & 0 & 1 & x_1 & y_1 \\ 1 & x_2 & y_2 & 0 & 0 & 0 \\ 0 & 0 & 0 & 1 & x_2 & y_2 \\ 1 & x_3 & y_3 & 0 & 0 & 0 \\ 0 & 0 & 0 & 1 & x_3 & y_3 \end{bmatrix} \begin{bmatrix} a_1 \\ a_2 \\ a_3 \\ a_4 \\ a_5 \\ a_6 \end{bmatrix} \tag{4.5.4}$$

The inversion of $[h]$ is documented in several references (for example, Reference 15). We shall carry on with symbolic representation of the equations and not attempt to write out all the details. Once $[h]$ is inverted, we can obtain

$$[n] = [\emptyset][h]^{-1}$$

$$[b] = [D][n] \tag{4.5.5}$$

$$[k] = \iint [b]^T[G][b] \, dx \, dy$$

where, in this case (see Equations 1.3.10 and 1.3.11),

$$[G] = \begin{bmatrix} \dfrac{E}{1-v^2} & \dfrac{E}{1-v^2} & 0 \\ \dfrac{E}{1-v^2} & \dfrac{E}{1-v^2} & 0 \\ 0 & 0 & G \end{bmatrix}, \qquad [D] = \begin{bmatrix} \dfrac{\partial}{\partial x} & 0 \\ 0 & \dfrac{\partial}{\partial y} \\ \dfrac{\partial}{\partial y} & \dfrac{\partial}{\partial x} \end{bmatrix} \tag{4.5.6}$$

The matrix $[k]_i$ is written out in component form in References 15 and 16, among others. For an element of a global structure, the nodes take on various nodal numbers. The assembly process is the same as for line and beam elements considered earlier. The element stiffness matrix $[k]$ must be embedded in a larger matrix the size of the global matrix $[K]$ and summed as follows:

$$[K]\{r\} = \Sigma \, [k]_i\{r\} \qquad (4.5.7)$$

In practice, the matrix $[k]_i$ would be stored in computer memory and called up by the FEM program as needed by the portion of the program that assembles the elements. The applied loads are converted to element equivalent nodal loads by the integral

$$\{R_e\}_i = \iint [n]_i\{F_s\} \, dx \, dy \qquad (4.5.8)$$

and are assembled to form the global load matrix $\{R\}$. Constraints must now be imposed on $\{r\}$ and constraint reactions added to $\{R\}$. The set of equations is now complete.

The triangular element just described is called a *constant strain triangle* because of the linear polynomial used in defining the shape functions $[n]$. Strain is obtained from the first derivatives of the shape functions; thus, it may be seen that all strains and, therefore, stresses in this case are constants within any given element.

An important feature of this element is that the edges remain straight after deformation; therefore, the elements are *compatible* or *conformable*. This means that the displacements of two edges connecting the common nodes of two adjacent elements are always identical. That is, there are no gaps or overlaps of edges. Such compatibility has always been a requirement in the theory of solid mechanics, but it is sometimes elusive in finite element equations. Fortunately, it is always satisfied when these elements are used.

The element has been found to be satisfactory for practical analysis but, generally, because of the approximation of constant strain in each element, many elements must be used before the solution closely approximates the true answer. Therefore, steps have been taken to find improved elements. Some of the steps taken are described in the next section.

4.5.2 Rectangular Element

The rectangular element shown in Figure 4.5.3(a) has four nodes with two degrees of freedom at each node for a total of eight degrees of freedom. This permits us to use a higher-order polynomial, which allows variation in the strain across the element and, therefore, achieves greater accuracy. It is convenient to convert to a set of nondimensional coordinates, as shown in Figure 4.5.3(b) and defined by

$$\xi = \frac{x}{a}, \qquad \eta = \frac{y}{b} \qquad (4.5.9)$$

The simplest polynomial form for the deflection is

$$u_e(\xi, \eta) = a_1 + a_2\xi + a_3\eta + a\xi\eta$$
$$v_e(\xi, \eta) = a_5 + a_6\xi + a_7\eta + a_8\xi\eta \qquad (4.5.10)$$

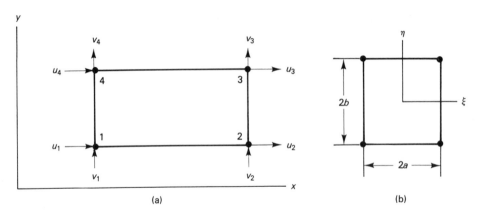

Figure 4.5.3 Rectangular Plane Stress Element.

Written in matrix form, we have

$$\{u_e\} = [\emptyset]\{a\} = \begin{bmatrix} 1 & \xi & \eta & \xi\eta & 0 & 0 & 0 & 0 \\ 0 & 0 & 0 & 0 & 1 & \xi & \eta & \xi\eta \end{bmatrix} \begin{bmatrix} a_1 \\ a_2 \\ a_3 \\ a_4 \\ a_5 \\ a_6 \\ a_7 \\ a_8 \end{bmatrix} \tag{4.5.11}$$

and from it we can construct the matrix $[h]$ and invert it. Because of our choice of coordinates, $[h]$ is relatively simple (for example, see Reference 15).

$$[h] = \begin{bmatrix} 1 & -1 & -1 & 1 & 0 & 0 & 0 & 0 \\ 0 & 0 & 0 & 0 & 1 & -1 & -1 & 1 \\ 1 & 1 & -1 & -1 & 0 & 0 & 0 & 0 \\ 0 & 0 & 0 & 0 & 1 & 1 & -1 & -1 \\ 1 & 1 & 1 & 1 & 0 & 0 & 0 & 0 \\ 0 & 0 & 0 & 0 & 1 & 1 & 1 & 1 \\ 1 & -1 & 1 & -1 & 0 & 0 & 0 & 0 \\ 0 & 0 & 0 & 0 & 1 & -1 & 1 & -1 \end{bmatrix} \tag{4.5.12}$$

and its inverse is

$$[h]^{-1} = \frac{1}{4} \begin{bmatrix} 1 & 0 & 1 & 0 & 1 & 0 & 1 & 0 \\ -1 & 0 & 1 & 0 & 1 & 0 & -1 & 0 \\ -1 & 0 & -1 & 0 & 1 & 0 & 1 & 0 \\ 1 & 0 & -1 & 0 & 1 & 0 & -1 & 0 \\ 0 & 1 & 0 & 1 & 0 & 1 & 0 & 1 \\ 0 & -1 & 0 & 1 & 0 & 1 & 0 & -1 \\ 0 & -1 & 0 & -1 & 0 & 1 & 0 & 1 \\ 0 & 1 & 0 & -1 & 0 & 1 & 0 & -1 \end{bmatrix} \tag{4.5.13}$$

The shape functions are

$$[n] = [\emptyset][h]^{-1} = \frac{1}{4}\begin{bmatrix} N_1 & 0 & N_2 & 0 & N_3 & 0 & N_4 & 0 \\ 0 & N_1 & 0 & N_2 & 0 & N_3 & 0 & N_4 \end{bmatrix} \qquad (4.5.14)$$

where

$$N_1 = (1 - \xi)(1 - \eta), \qquad N_2 = (1 + \xi)(1 - \eta)$$
$$N_3 = (1 + \xi)(1 + \eta), \qquad N_4 = (1 - \xi)(1 + \eta) \qquad (4.5.15)$$

These shape functions can be used to find the stiffness matrix. This is sometimes called the *bilinear* element because of the form of the shape functions. Fortunately, when this element is used, the edges between any two nodes remain straight after deformation, thus ensuring compatibility.

The problem with this element is that rectangularity is confining as a shape for modeling, especially at the boundaries. So while it would be an improvement over the triangular element internally, it simply is not used because of modeling restrictions. For modeling purposes, a general quadrilateral element, such as some of those shown in Figure 4.5.1, would be much better than a rectangular element; however, the simple method of forming the polynomial that we have used here does not work for the quadrilateral because it leads to nonconforming elements. Fortunately, there is a way to overcome this, called the *isoparametric* method of element derivation. This formulation is so important that we shall give it a section of its own. There the reader will see why we went into this much detail for a rectangular element that is rarely, if ever, used in practical analysis.

4.6 ISOPARAMETRIC ELEMENTS

The isoparametric formulation makes possible nonrectangular quadrilateral elements for plane stress and plates, elements with curved edges and mid-edge nodes in one, two, and three dimensions, and, generally, elements defined by higher-order polynomials that are not possible without the isoparametric formulation.

An element is isoparametric if the same shape function used to derive the element stiffness matrix defines the global coordinates in terms of the nodal coordinates. To explain this, we need to define the matrix of nodal coordinates $\{g\}$ as a column matrix of the coordinates of the nodes on any given element and the matrix of global coordinates as the column matrix $\{c\}$. Then, if the shape functions of the element are $[n]$, the element is isoparametric if

$$\{c\} = [n]\{g\} \qquad (4.6.1)$$

The presentation of the isoparametric formulation is best done in a set of dimensionless coordinates called *natural coordinates*. These are best explained in the context of a specific element formulation. Actually, some of the elements we have already considered are isoparametric. Let us take a look back at the line element and then go forward with new derivations.

4.6.1 Line Elements

Consider a simple axial element oriented along the x-axis, as shown in Figure 4.6.1(a). The very same element is represented in natural coordinates along the ξ-axis with ends

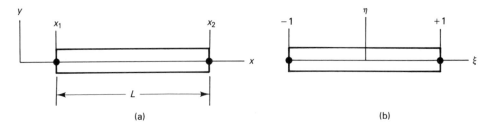

Figure 4.6.1 Two-Node Line Element in Regular and Natural Coordinates.

defined at $\xi = \pm 1$, regardless of the length of the bar, as shown in Figure 4.6.1(b). Let us develop the shape functions in natural coordinates using the method of Equations 4.2.1 and 4.2.2. We note that there is one degree of freedom at each node; thus

$$\{u_e\} = [\emptyset]\{a\} = [1 \ \xi] \begin{bmatrix} a_1 \\ a_2 \end{bmatrix} \tag{4.6.2}$$

and

$$\{r_e\} = [h]\{a\} = \begin{bmatrix} 1 & -1 \\ 1 & 1 \end{bmatrix} \begin{bmatrix} a_1 \\ a_2 \end{bmatrix} \tag{4.6.3}$$

from which

$$[n] = [\emptyset][h]^{-1} = [1 \ \xi] \begin{bmatrix} \dfrac{1}{2} & \dfrac{1}{2} \\ -\dfrac{1}{2} & \dfrac{1}{2} \end{bmatrix} = \begin{bmatrix} \dfrac{1-\xi}{2} & \dfrac{1+\xi}{2} \end{bmatrix} \tag{4.6.4}$$

From the definition of isoparametric in Equation 4.6.1, we have

$$\{c\} = [n]\{g\} = x = \begin{bmatrix} \dfrac{1-\xi}{2} & \dfrac{1+\xi}{2} \end{bmatrix} \begin{bmatrix} x_1 \\ x_2 \end{bmatrix} \tag{4.6.5}$$

which we can check against the values in Figure 4.6.1. Sure enough, when $\xi = -1$, Equation 4.6.4 gives us x_1 and when $\xi = +1$, we get x_2.

Continuing with the derivation of the element stiffness matrix, we note that

$$\{b\} = [D][n] = \frac{d}{dx} \begin{bmatrix} \dfrac{1-\xi}{2} & \dfrac{1+\xi}{2} \end{bmatrix} \tag{4.6.6}$$

To carry out this integration, we must use the rule

$$\frac{d}{dx} = \frac{d\xi}{dx} \frac{d}{d\xi} \tag{4.6.7}$$

From Equation 4.6.5, after noting that $x_2 - x_1 = L$, we have

$$\frac{dx}{d\xi} = \begin{bmatrix} -\dfrac{1}{2} & \dfrac{1}{2} \end{bmatrix} \begin{bmatrix} x_1 \\ x_2 \end{bmatrix} = \frac{L}{2} \tag{4.6.8}$$

Thus

$$\frac{d\xi}{dx} = \frac{2}{L} \tag{4.6.9}$$

The scale factor between two coordinate systems is called the Jacobian and is denoted by J (not to be confused with the torsional constant), and in this case

$$dx = J\, d\xi \longrightarrow J = \frac{dx}{d\xi} = \frac{L}{2} \tag{4.6.10}$$

Thus, from Equation 4.6.6,

$$[b] = \frac{1}{J}\frac{d}{d\xi}\,[n] = \left[-\frac{1}{L}\quad\frac{1}{L}\right] \tag{4.6.11}$$

and

$$[k] = \int [b]^T[G][b]\, dV = EA \int \begin{bmatrix} -\dfrac{1}{L} \\[2mm] \dfrac{1}{L} \end{bmatrix} \left[-\frac{1}{L}\quad\frac{1}{L}\right] dx = \frac{EA}{L}\begin{bmatrix} 1 & -1 \\ -1 & 1 \end{bmatrix} \tag{4.6.12}$$

which is exactly what we got in Equation 3.3.31. Thus, the line element stiffness matrix we have been using all along is isoparametric.

While we are on the subject of line elements, we shall use the occasion to formulate an element with an extra node. This allows a higher-degree polynomial to be used. Such an element is shown in Figure 4.6.2 in both regular and natural coordinates.

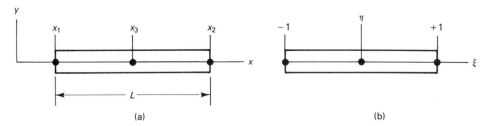

Figure 4.6.2 Three-Node Line Element in Regular and Natural Coordinates.

In this case,

$$\{u_e\} = [\varnothing]\{a\} = [1\ \xi\ \xi^2]\begin{bmatrix} a_1 \\ a_2 \\ a_3 \end{bmatrix} \tag{4.6.13}$$

then, by the same steps,

$$[n] = \left[\frac{-\xi + \xi^2}{2}\quad\frac{\xi + \xi^2}{2}\quad 1 - \xi^2\right] \tag{4.6.14}$$

and the Jacobian is

$$J = \left[\frac{-1 + 2\xi}{2}\quad\frac{1 + 2\xi}{2}\quad -2\xi\right]\{g\} \tag{4.6.15}$$

A check of

$$\{c\} = [n]\{g\} = x = \left[\frac{-\xi + \xi^2}{2}\quad\frac{\xi + \xi^2}{2}\quad 1 - \xi^2\right]\begin{bmatrix} x_1 \\ x_2 \\ x_3 \end{bmatrix} \tag{4.6.16}$$

confirms that this will be an isoparametric element. Continuing with the derivation of the element stiffness matrix, we note that

$$[b] = [D][n] = \frac{1}{J} \frac{d}{d\xi} [n] = \frac{1}{J} \left[\frac{-1 + 2\xi}{2} \quad \frac{1 + 2\xi}{2} \quad -2\xi \right] \qquad (4.6.17)$$

When x_3 is at the midpoint of the element, $J = L/2$ and

$$[k] = \int [b]^T [G][b] \, dV = \frac{EA}{3L} \begin{bmatrix} 7 & -8 & 1 \\ -8 & 16 & -8 \\ 1 & -8 & 7 \end{bmatrix} \qquad (4.6.18)$$

When x_3 is not at the midpoint, the matrix $[k]$ must be found by numerical integration, because the integral does not then have a known closed-form solution.

The most accepted form of numerical integration is called *Gauss quadrature*. A summary of Gauss quadrature is given in Reference 15. We shall not pursue this point here because the three-node line element with the mid-edge node off center is not that important to us. In the next section, numerical integration becomes much more important, and we shall develop the subject further there.

4.6.2 Plane Stress Elements

Generally, nonrectangular quadrilateral elements for plane stress have been generated as isoparametric elements. Such a quadrilateral is shown in Figure 4.6.3.

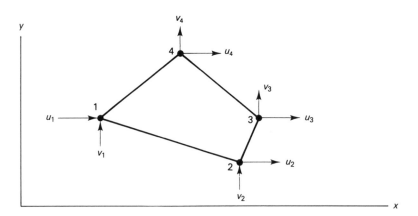

Figure 4.6.3 General Plane Quadrilateral Element.

We wish to find $[n]$ such that

$$\{u_e\} = [n]\{r_e\}, \qquad \{c\} = [n]\{g\} \qquad (4.6.19)$$

where

$$\{u_e\} = \begin{bmatrix} u \\ v \end{bmatrix}, \qquad \{c\} = \begin{bmatrix} x \\ y \end{bmatrix}, \qquad \{r_e\} = \begin{bmatrix} u_1 \\ v_1 \\ u_2 \\ v_2 \\ u_3 \\ v_3 \\ u_4 \\ v_4 \end{bmatrix}, \qquad \{g\} = \begin{bmatrix} x_1 \\ y_1 \\ x_2 \\ y_2 \\ x_3 \\ y_3 \\ x_4 \\ y_4 \end{bmatrix} \qquad (4.6.20)$$

The quadrilateral shown in Figure 4.6.3 is presented in regular and natural coordinates in Figure 4.6.4. We see that in the ξ-η plane the element is a square, as shown in Figure 4.6.4(b).

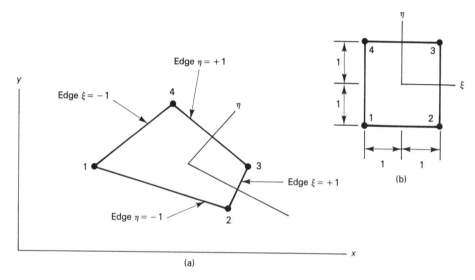

Figure 4.6.4 Quadrilateral Element in Natural Coordinates.

Our next task is to find a suitable polynomial representation of the displacements with respect to the ξ-η axes. The values chosen for the rectangular element in Section 4.5 would appear to be suitable. In fact, if we use the polynomial form of Equation 4.5.11, we get exactly the shape functions given in Equation 4.5.14. From this point on, the derivations for the stiffness matrices in the two cases differ. In the case of the quadrilateral, there is a Jacobian to deal with. The Jacobian relating x-y space to ξ-η space in this case is

$$[J] = \begin{bmatrix} \dfrac{\partial x}{\partial \xi} & \dfrac{\partial y}{\partial \xi} \\ \dfrac{\partial x}{\partial \eta} & \dfrac{\partial y}{\partial \eta} \end{bmatrix} \qquad (4.6.21)$$

where for a plane element

$$[k] = \iint [b]^T [G][b] h \; dx \; dy = \int [b]^T [G][b] h \; |J| \; d\xi \; d\eta \qquad (4.6.22)$$

where h is the plate thickness and $|J|$ is the determinant of the Jacobian matrix. To go into this in detail is well beyond the scope of this text. The point is that $|J|$ puts polynomial terms in the denominator of the integrand, $[b]$ puts polynomial terms in the numerator, and the two together are impossible to integrate in closed form. Numerical integration via Gauss quadrature saves the day. See References 13 through 17 for the details.

In any case, the result is a stiffness matrix of higher order, and it can be shown that it is a conformable element. The quadrilateral, consequently, is the most used plane stress element. In the next example, we shall illustrate a solution of a typical plane stress problem using a quadrilateral element that is very difficult to solve by analytical means.

Example 4.6.1

Find the stresses in a plate with a hole that is loaded by uniform edge stresses, as shown in figure (a). The hole has a diameter of 300 mm, the plate is 10 mm thick, and it is made of aluminum. Both the deflections and the stresses are sought. The stress is of special interest because of the phenomenon of stress concentration near the hole; that is, there is a sharp increase in the value of the normal stress along the y-axis near the hole compared to the same place in a plate with the same loading but without the hole.

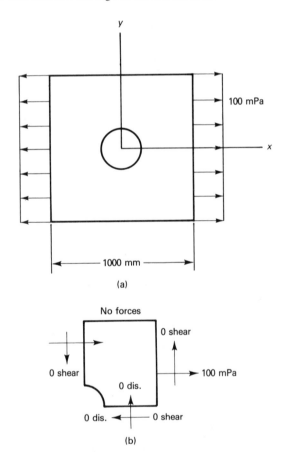

(a)

No forces

(b)

Solution: The problem can be reduced in size by one-fourth by using double symmetry. The boundary conditions using just the one quadrant are identified in figure (b).

Once again we shall set up the problem using the seven items of input data.

1. Divide the structure into suitable elements and number the nodes and elements: Let us use four-node quadrilateral elements arranged and with nodes numbered as shown in figure (c).

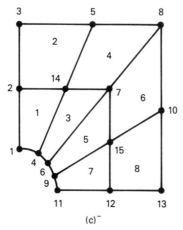

(c)⁻

2. Record the coordinates of the nodes:

Node Number	x	y
1	0.0	150.0
2	0.0	325.0
3	0.0	500.0
4	57.4	138.6
5	250.0	500.0
6	106.1	106.1
7	303.0	303.0
8	500.0	500.0
9	138.6	57.4
10	500.0	250.0
11	150.0	0.0
12	325.0	0.0
13	500.0	0.0
14	191.2	325.0
15	325.0	191.2

3. Identify the nodes that go with each element:

Element	Nodes (counterclockwise)
1	1, 4, 14, 2
2	2, 14, 5, 3
3	4, 6, 7, 14
4	14, 7, 8, 5
5	6, 9, 15, 7
6	7, 15, 10, 8
7	9, 11, 12, 15
8	15, 12, 13, 10

4. Specify the material properties of each element:

$$E = 68{,}950.0 \text{ MPa}$$

$$G = 26{,}520.0 \text{ MPa}$$

$$\upsilon = 0.3$$

5. Specify the geometric properties of each element: All the geometrical data are given by the nodal coordinates except the thickness, which is 10 mm.

6. Specify geometric constraints: In plane stress, there are no z-displacements and no rotations about any axis. Additional constraints in the x- and y-directions are specified according to the symmetry conditions. From symmetry, we conclude that nodes 1, 2, and 3 do not displace in the x-direction, and nodes 11, 12, and 13 do not displace in the y-direction. Thus,

$$u_1 = u_2 = u_3 = 0, \qquad v_{11} = v_{12} = v_{13} = 0$$

7. Specify equivalent nodal loads: The following equivalent nodal loads are suggested for this case by manual conversion of the distributed load into lumped loads, rather than by using the formal integral process:

Node	R_x	R_y
8	125,000.0	0.0
10	250,000.0	0.0
13	125,000.0	0.0

This problem has been executed on MSC/NASTRAN; the input file is shown in Appendix J, and output data on stresses and deflections are summarized next. The displacements in millimeters are

Node	u	v	Node	u	v
1	0.0	−0.3344	9	0.6604	−0.0862
2	0.0	−0.4094	10	0.9049	−0.0395
3	0.0	−0.4610	11	0.7419	0.0
4	0.2667	−0.2756	12	0.0897	0.0
5	0.2812	−0.3050	13	1.0965	0.0
6	0.4909	−0.1804	14	0.3316	−0.2699
7	0.5375	−0.1514	15	0.7195	−0.0711
8	0.6491	−0.0659			

The stresses in megapascals are

Element	σ_x	σ_y	τ_{xy}
1	157.71	29.14	40.08
2	95.72	5.67	132.73
3	95.87	35.24	73.93
4	86.74	12.77	40.09

continued

Element	σ_x	σ_y	τ_{xy}
5	49.13	35.99	69.03
6	21.71	87.05	43.33
7	−9.61	43.30	37.89
8	24.82	90.36	17.91

Several comments about the stresses are necessary. First, note that only element stresses are given. It is characteristic of the particular element used in this particular code to only report a single value for the whole element. This is not a constant strain element, however. Clearly, this does not provide good accuracy for the number of elements used. The values shown may be regarded as a kind of average across the element.

The answer can be improved by using more elements. The problem was redone using the nodes and elements shown in figure (d). We shall not give all the results, but only note that, in the element containing node 1,

$$\sigma_x = 205.99 \text{ MPa}$$

A plot of the displacements is given in figure (e). This plot was generated by the MSC/NASTRAN plot routine. It shows both the deflected and undeflected structure and is normalized to show the maximum deflection as a fraction of a characteristic length of the structure. Note the elongation of the hole. Note that the NASTRAN plotter shows straight lines between nodes in both the undeflected and deflected positions.

This solution is substantially more accurate, but to get very good accuracy a much denser mesh of elements would be necessary; nevertheless, the phenomenon of stress concentration around a hole is illustrated. This phenomenon is well known, and is studied in some detail in, for example, References 6 through 8.

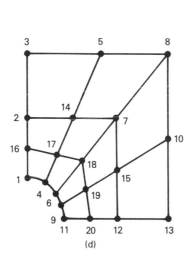

(d)

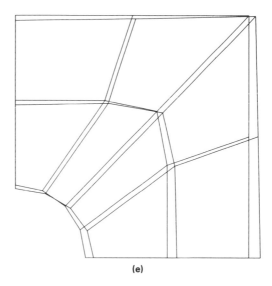

(e)

An element can be further improved by using more nodes. Since polygons more complex than quadrilaterals are not particularly good for modeling, the way to add more nodes is to place them on edges between the corner nodes. These are usually placed at midpoints between the corner nodes, but they may be biased toward one corner or the other. These are called *mid-edge nodes*. Among the interesting results is that with mid-edge nodes the polynomial for a quadrilateral permits curved edges; that is, the three nodes along an edge need not lie in a straight line but can follow a curve. Another feature of the isoparametric formulation ensures that elements are compatible or conformable. These are the elements that find their way into major commercial FEM codes. In Figure 4.6.5, we show the four most popular elements used in plane stress analysis. The four- and eight-node quadrilaterals are most used, while the three- and six-node triangles are used, often mixed with quadrilaterals, when geometric requirements make it convenient to use triangles.

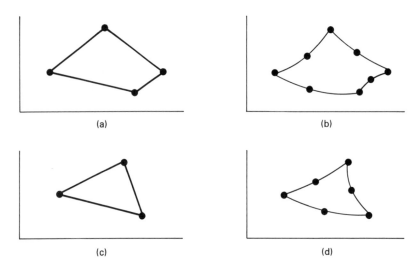

Figure 4.6.5 Popular Plane Stress Elements.

A further step in improving the elements might be to add even more nodes and, thereby, more degrees of freedom to each element. The additional degrees of freedom permit higher-order polynomial representation of the shape functions. Higher-order polynomials may mean that each element has more capacity to take on a shape closely approximating the true shape. Thus, while we have more degrees of freedom per element, we may need fewer elements per structure, so the amount of computation as represented by the number of equations to be solved may actually go down; nevertheless, more nodes than shown in Figure 4.6.5 are not widely used.

4.6.3 Solid Elements

A fairly simple extension of the plane stress case to three dimensions will give us the equations for a solid element. The solid element will provide the means for finding answers to full three-dimensional cases of elasticity, the subject studied in Chapter 1 and for which we met with little success in getting answers. In fact, if we could have only one element, the solid element could be used to solve all the problems in this text; however, it is computationally expensive for many problems compared to using other appropriate elements so that we use it, generally, only when necessary. It is normal practice to divide the structure into tetrahedral or hexahedral elements or a combination of both. We shall briefly illustrate both, but it is much beyond the scope of this text to go into the necessary detail to fully derive these elements. The reader is referred to References 14 through 17.

An element in the shape of a tetrahedron is shown in Figure 4.6.6. It was the first solid element developed and it proved very useful. There are three degrees of freedom at each node and, therefore, 12 terms in the nodal displacement matrix. This leads to

$$u_e(x, y, z) = a_1 + a_2x + a_3y + a_4z$$

$$v_e(x, y, z) = a_5 + a_6x + a_7y + a_8z \tag{4.6.23}$$

$$w_e(x, y, z) = a_9 + a_{10}x + a_{11}y + a_{12}z$$

which in matrix form is

$$\{u_e\} = [\emptyset]\{a\} = \begin{bmatrix} 1 & x & y & z & 0 & 0 & 0 & 0 & 0 & 0 & 0 & 0 \\ 0 & 0 & 0 & 0 & 1 & x & y & z & 0 & 0 & 0 & 0 \\ 0 & 0 & 0 & 0 & 0 & 0 & 0 & 0 & 1 & x & y & z \end{bmatrix} \begin{bmatrix} a_1 \\ a_2 \\ a_3 \\ a_4 \\ a_5 \\ a_6 \\ a_7 \\ a_8 \\ a_9 \\ a_{10} \\ a_{11} \\ a_{12} \end{bmatrix} \tag{4.6.24}$$

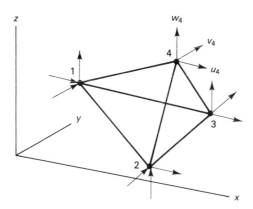

Figure 4.6.6 Tetrahedral Element.

From this we can form [h] and invert it. Once [h] is inverted, we can use Equations 4.5.5 to obtain the stiffness matrix. The edges remain straight between any two nodes, thus ensuring compatibility.

A general hexahedral element is shown in Figure 4.6.7. This element is even more widely used. An isoparametric formulation is necessary. Both the tetrahedral and hexahedral elements have been extended to include mid-edge nodes as shown in Figure 4.6.8. The mid-edge nodes allow curved edges, as illustrated.

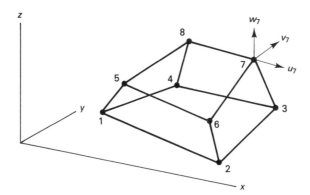

Figure 4.6.7 Hexahedral Element.

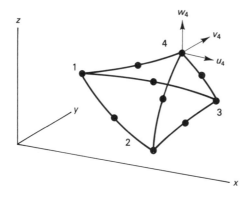

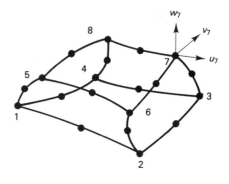

Figure 4.6.8 Higher-Order Elements with Mid-edge Nodes.

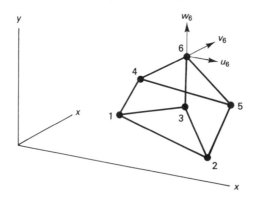

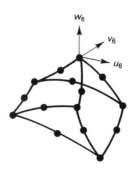

Figure 4.6.9 Wedge Elements.

Applications of the Finite Element Method   Chap. 4

Still a third element has been created for modeling purposes. It is illustrated in Figure 4.6.9 with and without mid-edge nodes and, appropriately, is called a *wedge element*.

It is more than we have space for to derive the stiffness matrices for these elements; however, we shall give a simple example solved on MSC/NASTRAN using hexahedral and wedge elements. Again we refer the reader to References 14 through 17.

Example 4.6.2

A bracket, shown in figure (a), is welded to a rigid wall, as shown in figure (b), resulting in a fixed edge constraint. It has a uniform pressure applied on the one surface as shown in figure (a).

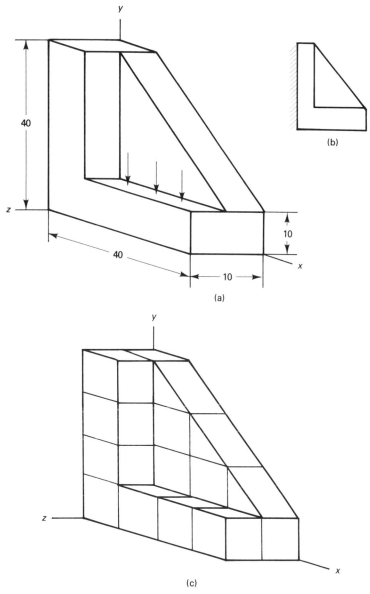

(a)

(b)

(c)

Solution: In this case we shall not set up the seven steps in solving an FEM problem. Suffice it to say that this has been done, and the NASTRAN input file is given in Appendix I.

We have modeled this structure with hexadedral and wedge elements as shown in figure (c). When hexahedral and wedge elements with only corner nodes are used, there are 54 nodes and 20 elements. Greater accuracy would be obtained if we used more elements or if we used elements with mid-edge nodes, or both.

The problem has been executed on MSC/NASTRAN and the output data for displacements is shown in figure (d) as generated by the NASTRAN plot routines. Our purpose in this example has not been to present a complete solution; rather, it has been to illustrate briefly the kind of problem that can be solved with solid elements. The MSC/NASTRAN input file is listed in Appendix J. Those interested can retrieve the input data from the input file.

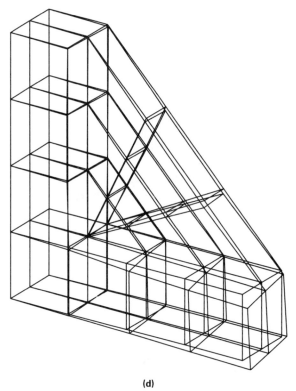

(d)

The development of solid elements opened up a large new world of stress analysis. Before that time, structures that could not be reduced to simple forms such as beams or plates were rarely solved by either analytical or numerical methods. More likely, experimental methods had to be used. Building the part and testing it, even in scale-model form, was very expensive and time consuming; then, rather suddenly, there was a numerical method that could be pushed to limits of high accuracy. Much of the empirical approach to stress and deflection analysis of linear elastic structures with small deflections has now been replaced with FEM solutions.

4.7 PLATE APPLICATIONS

We need to recall the basic assumptions for thin plates in bending. If we neglect in-plane deformations, we have, from Section 2.4, the classical assumption for displacements:

$$u = -z \frac{\partial w}{\partial x}, \qquad v = -z \frac{\partial w}{\partial y} \tag{4.7.1}$$

The strains are seen to be

$$\epsilon_x = -z \frac{\partial^2 w}{\partial x^2}, \qquad \epsilon_y = -z \frac{\partial^2 w}{\partial y^2}, \qquad \gamma_{xy} = -2z \frac{\partial^2 w}{\partial x \, \partial y} \tag{4.7.2}$$

or in matrix form

$$\{\epsilon\} = [D_c]\{u_c\} = \begin{bmatrix} \epsilon_x \\ \epsilon_y \\ \gamma_{xy} \end{bmatrix} = \begin{bmatrix} -z \dfrac{\partial^2}{\partial x^2} \\ -z \dfrac{\partial^2}{\partial y^2} \\ -2z \dfrac{\partial^2}{\partial x \, \partial y} \end{bmatrix} [w] \tag{4.7.3}$$

The stress–strain matrix for a linearly elastic, homogeneous, isotropic material is

$$[G] = \frac{E}{1 - v^2} \begin{bmatrix} 1 & v & 0 \\ v & 1 & 0 \\ 0 & 0 & \lambda \end{bmatrix} \tag{4.7.4}$$

where

$$\lambda = \frac{1 - v}{2} \tag{4.7.5}$$

We are now ready to develop a plate element stiffness matrix.

4.7.1 Simple Plate Elements

Simple triangular and rectangular plate elements have been derived using the now familiar polynomial representation. Consider the elements in Figure 4.7.1. All nodes lie in the x-y plane. At each node, there are three degrees of freedom. These are the lateral deflection w, the rotation θ about the x-axis, and the rotation φ about the y-axis.

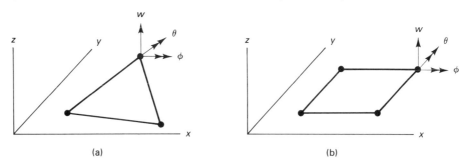

(a)   (b)

Figure 4.7.1 Two Examples of Plate Elements.

We note that

$$\varphi = \frac{\partial w}{\partial y}, \qquad \theta = -\frac{\partial w}{\partial x} \tag{4.7.6}$$

Thus, at a given node,

$$\{r_e\} = \begin{bmatrix} w \\ \theta \\ \varphi \end{bmatrix} = \begin{bmatrix} w \\ -\dfrac{\partial w}{\partial x} \\ \dfrac{\partial w}{\partial y} \end{bmatrix} \tag{4.7.7}$$

Polynomial terms are often chosen from an array known as Pascal's triangle, shown here in Figure 4.7.2 with terms up to the fourth power. For the triangular element with nine total degrees of freedom, we need a nine-term polynomial. All polynomial terms through the third power add up to ten; thus, the question of which third power term to neglect arises. As it turns out, neglecting any one of the terms leads to an unsatisfactory stiffness matrix. See References 14 or 15 for a discussion of the problems.

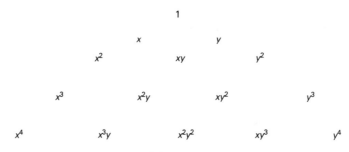

Figure 4.7.2 Pascal's Triangle.

Perhaps we would have better luck with the rectangular element. One early success was with a four-node rectangular element with 12 degrees of freedom. The chosen polynomial (with commas added to show separation of the components of the matrix [∅]) was

$$\{u_e\} = [\emptyset]\{a\} = [1, x, y, x^2, xy, y^2, x^3, x^2y, xy^2, y^3, x^3y, xy^3] \{a\} \tag{4.7.9}$$

where all the terms to the fourth power in Pascal's triangle, as shown in Figure 4.7.2, are included except x^4, x^2y^2, and y^4. Actually, the element turns out to be nonconforming because slope continuity normal to the edges of adjacent elements is not maintained. The solution based on an assembly of such elements, however, does converge to the correct answer as the elements become smaller. Thus, the element has enjoyed some success.

In time, better elements were derived. Those based on isoparametric formulations are especially useful. The most popular elements used in commercial codes are the triangular and quadrilateral elements, either with just corner nodes or with corner and mid-edge nodes. These have the same shape as the elements shown in Figure 4.5.5 for plane stress; however, they have different degrees of freedom at the nodes.

The full development of the plate element is a lengthy process and is well beyond the space available in this text. Suffice it to say that the principles for element stiffness matrix derivation, assembly of stiffnesses, specification of equivalent nodal loads, specification of constraints, and so on, are precisely as outlined for other simpler elements we have studied along the way. So is the process of preparing a plate problem for solution. This may be ascertained by studying References 14 through 17. In the next example, we shall take the very same structure used in Example 4.5.1 and constrain and load it for plate behavior and prepare the data for use in a commercial FEM code.

Example 4.7.1

Find the stresses in the plate shown in figure (a), which is fixed on two opposite edges and free on the other two opposite edges and which is loaded by a uniform lateral (z-direction) ring load around the edge of the hole. The total load is 100 newtons. The deflection, stress, and support reaction results for larger or smaller loads can be found by multiplying the results obtained here by the ratio of the load divided by 100. The hole has a diameter of 300 mm; the plate is 10 mm thick and is made of aluminum. Both the deflections and the stresses are sought.

Solution: From symmetry, the problem can be reduced in size by one-fourth. The boundary conditions for one quadrant are shown in figure (b). Once again we shall set up the problem using the seven items of input data.

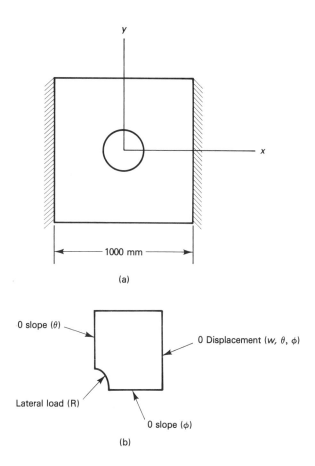

(a)

(b)

1. Divide the structure into suitable elements and number the nodes and elements: Let us use four-node quadrilateral elements arranged and with nodes numbered as shown in figure (c).

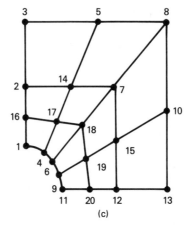

(c)

2. Record the coordinates of the nodes:

Node Number	x	y
1	0.0	150.0
2	0.0	325.0
3	0.0	500.0
4	57.4	138.6
5	250.0	500.0
6	106.1	106.1
7	303.0	303.0
8	500.0	500.0
9	138.6	57.4
10	500.0	250.0
11	150.0	0.0
12	325.0	0.0
13	500.0	0.0
14	191.2	325.0
15	325.0	191.2
16	0.0	237.5
17	123.3	231.8
18	204.55	204.55
19	231.8	123.3
20	237.5	0.0

3. Identify the nodes that go with each element:

Element	Nodes (counterclockwise)
1	1, 4, 17, 16
2	2, 14, 5, 3
3	4, 6, 18, 17

continued

Element	Nodes (counterclockwise)
4	14, 7, 8, 5
5	6, 9, 19, 18
6	7, 15, 10, 8
7	9, 11, 20, 19
8	15, 12, 13, 10
9	16, 17, 14, 2
10	17, 18, 7, 14
11	18, 19, 15, 7
12	19, 20, 12, 15

4. Specify the material properties of each element:

$$E = 68,950.0 \text{ MPa}$$

$$G = 26,520.0 \text{ MPa}$$

$$\upsilon = 0.3$$

5. Specify the geometric properties of each element: All the geometrical data are given by the nodal coordinates except for the thickness, which is 10 mm.

6. Specify the geometric constraints. The nodes on the right vertical edge are completely constrained in deflection and rotation, or

$$w_8 = \theta_8 = \varphi_8 = 0, \qquad w_{10} = \theta_{10} = \varphi_{10} = 0, \qquad w_{13} = \theta_{13} = \varphi_{13} = 0$$

The slopes on the bottom horizontal edge are

$$\varphi_{11} = \varphi_{12} = \varphi_{13} = \varphi_{20} = 0$$

and on the left vertical edge are

$$\theta_1 = \theta_2 = \theta_3 = \theta_{16} = 0$$

7. Specify the equivalent nodal loads. The following equivalent nodal loads are suggested for this case by manual conversion of the distributed load into lumped loads, rather than by using the formal integral process. There would be 16 nodes around the full circumference of the circle, and at each node we would place 1/16 of the total load. Using just one quadrant, we would place 1/16 of the load on each of nodes 4, 6, and 9, and 1/32 of the load on each of nodes 1 and 11; thus,

Node	R_x	R_y	R_z
1	0	0	−3.125
4	0	0	−6.25
6	0	0	−6.25
9	0	0	−6.25
11	0	0	−3.125

This problem has been executed on MSC/NASTRAN, and the input file is listed in Appendix J. The output data have been obtained. We shall report here only the deflections in millimeters at nodes around the hole.

Node Number	z-Deflection
1	−0.1305
4	−0.1287
6	−0.1246
9	−0.1206
11	−0.1189

From the NASTRAN plot routine we have the plot shown in figure (d).

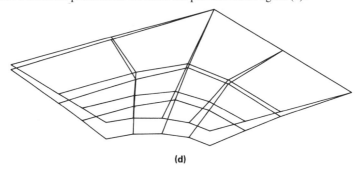

(d)

4.7.2 Combined Plate and Membrane Elements

The plane stress, or membrane, forces and displacements in the elements discussed in Section 4.5 can be combined with plate force and displacements in the elements in Section 4.7.1 in much the same way that the axial line element is combined with the beam bending element. When combined, we have an element with five degrees of freedom at each node. The most popular such elements are triangular and quadrilateral, both with and without mid-edge nodes, as shown in Figure 4.7.3.

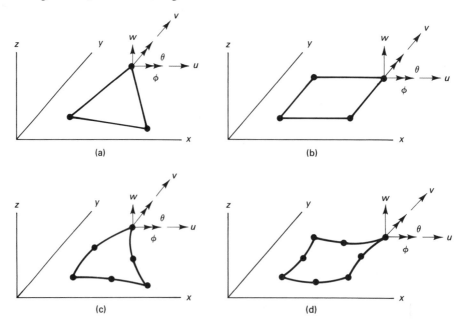

Figure 4.7.3 Four Popular Combined Plate/Membrane Elements.

4.8 BUILT-UP STRUCTURES

Complex built-up structures, such as airplanes and spacecraft, automobiles, ships, and bridges, are made of a variety of structural types (beams, plates, solids) all in one structure. The finite element model of all or part of such structures combines several elements all in the same analysis. The combining of several element types in the same structure often requires careful consideration of the compatibility of different elements. For example, beam and solid elements cannot be joined at a single node because the beam element supports rotation, while the solid element does not. Other elements are easily joined at nodes. We cannot go into many cases in the limited space available here. We shall show an example of a stiffened plate.

Example 4.8.1

It is proposed to strengthen the plate in Example 4.7.1 by reinforcing the hole with a ring as shown in figures (a) and (b). The ring has a cross section of 20-mm depth and 10-mm width and is made of the same material as the plate. The loading, constraints, and so on, are unchanged.

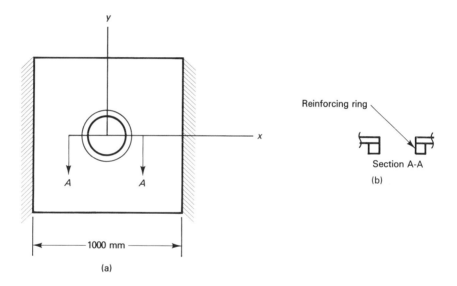

(a)

Section A-A
(b)

Solution: The seven steps are given next.

1.-2. Divide the structure into suitable elements, number the nodes and elements, and record the coordinates of the nodes: We divide the plate portion of the structure into nodes and elements and number them exactly as before. We add four beam elements between the pairs of nodes 1-4, 4-6, 6-9, and 9-11. Although the nodes of beam elements are properly at the centroid of the cross section, we can use an offset feature of the BAR element in MSC/NASTRAN to use the same nodes for the beam as for the plate. Thus we need add no additional nodal coordinates. Observe that the centroid of the cross section is 15 mm below the mid-surface of the plate. It is also 5 mm out along the radius of the hole.

3. Identify the nodes that go with each element: The connectivity of the new elements is

Element	Nodes
13	1, 4
14	4, 6
15	6, 9
16	9, 11

4. Specify the material properties: Same as Example 4.7.1.

5. Specify the geometric properties of each element: For the beam elements,

$$A = 200 \text{ mm}^2, \qquad I_1 = 6666.667 \text{ mm}^4, \qquad I_2 = 1666.667 \text{ mm}^4$$

6. Specify the constraints: No change.

7. Specify the equivalent nodal loads: No change.

This problem has been run on MSC/NASTRAN, and the input file is listed in Appendix J. The output data have been obtained. We shall report here only the deflections in millimeters at nodes around the hole.

Node Number	z-Deflection
1	−0.0775
4	−0.0762
6	−0.0737
9	−0.0714
11	−0.0705

When compared to the plate without the reinforcing ring, this represents a reduction in deflection of over 40% around the hole. The normalized plot does not look substantially different from the plot in Example 4.7.1, so it will not be shown.

4.9 CONCLUSIONS AND SUMMARY

We have just extended the formulation of the finite element method to problems of space trusses, beams, frames, bodies in plane stress, three-dimensional solid bodies, flat plates and, very briefly, to built-up structures. In each case, an example was given that was solved by a commercial computer code. This list constitutes a large number of the total categories of practical problems of static structural analysis, including some examples, for example, of three-dimensional solid bodies, which were seldom solved with much accuracy before the advent of FEM and large-capacity, low-cost digital computers. This truly represents a revolution in structural analysis.

Referring back to the symbolic representation of the FEM equations in Figures 4.2.1 and 4.2.2, we have developed the two-part process of selecting shape functions for several cases, as depicted by the equations

$$\{u\} = [\emptyset]\{a\}, \qquad [r] = [h]\{a\} \longrightarrow \{a\} = [h]^{-1}\{r\} \qquad (4.2.1)$$

$$[n] = [\emptyset] [h]^{-1} \qquad (4.2.2)$$

And from the element equations (in Figure 4.2.2)

$$[k]\{r_e\} = \{V_e\}$$

we have assembled the global equations (in Figure 4.2.1)

$$[K]\{r\} = \{R\}$$

for several example problems.

In our first example, we show the use of coordinate transformation matrices for use in assembling the element equations for truss, or axial line elements derived in Chapter 3, into global equations for space, or three-dimensional truss structures. With this example, we introduce the seven steps for preparing the information for an input file for an FEM computer code.

Next we develop the beam equations. The shape functions and the uncoupled torsional stiffness matrix are found to be

$$[n]_i = \left[1 - \frac{s}{L_i} \quad \frac{s}{L_i} \right] \tag{4.4.8}$$

$$[k]_i = \frac{G_i J_i}{L_i} \begin{bmatrix} 1 & -1 \\ -1 & 1 \end{bmatrix} \tag{4.4.10}$$

For simple beams in bending and shear, the shape functions are

$$[n]_i = [\emptyset][h]^{-1} = \frac{1}{L_i^3}[L_i^3 - 3L_i s^2 + 2s^3, \; L_i^3 s - 2L_i^2 s^2 + L_i s^3, \tag{4.4.23}$$

$$3L_i s^2 - 2s^3, \; -L_i^2 s^2 + L_i s^3]$$

where 1 has been replaced with i to generalize for different lengths of elements.

In the x-y plane, where

$$\theta_1 = \left(\frac{dv}{ds} \right)_1, \qquad \theta_2 = \left(\frac{dv}{ds} \right)_2 \tag{4.4.17}$$

the stiffness matrix is

$$[k]_i = \frac{(EI_{zz})_i}{L_i^3} \begin{bmatrix} 12 & 6L_i & -12 & 6L_i \\ 6L_i & 4L_i^2 & -6L_i & 2L_i^2 \\ -12 & -6L_i & 12 & -6L_i \\ 6L_i & 2L_i^2 & -6L_i & 4L_i^2 \end{bmatrix} \tag{4.4.24}$$

In the x-z plane, where

$$\varphi_1 = -\left(\frac{dw}{dx} \right)_1, \qquad \varphi_2 = -\left(\frac{dw}{dx} \right)_2 \tag{4.4.27}$$

the stiffness matrix is given by

$$[k]_i = \int EI_{yy}[n'']_i^T[n'']_i \; ds = \frac{(EI_{yy})_i}{L_i^3} \begin{bmatrix} 12 & -6L_i & -12 & -6L_i \\ -6L_i & 4L_i^2 & 6L_i & 2L_i^2 \\ -12 & 6L_i & 12 & 6L_i \\ -6L_i & 2L_i^2 & 6L_i & 4L_i^2 \end{bmatrix} \tag{4.4.28}$$

The stiffnesses for axial, torsional, and bending in two planes are combined into a single matrix as given in Equation 4.4.31. The use of these in both beams and frames is illustrated by examples.

Beginning with the plane stress elements, the complete development of the shape functions and the stiffness matrices becomes complex, and we resort to outlining the methods for finding each, but we do not always write them down. In real life, these may be found in research publications and more advanced books, and, of most importance to us, they are embedded in computer codes. All developments start with a polynomial representation of the distributed displacements within an element. The constant strain triangle has the polynomial form

$$\{u_e\} = [\emptyset]\{a\} = \begin{bmatrix} 1 & x & y & 0 & 0 & 0 \\ 0 & 0 & 0 & 1 & x & y \end{bmatrix} \begin{bmatrix} a_1 \\ a_2 \\ a_3 \\ a_4 \\ a_5 \\ a_6 \end{bmatrix} \tag{4.5.2}$$

and the rectangular element in nondimensional coordinates has the form

$$\{u_e\} = [\emptyset]\{a\} = \begin{bmatrix} 1 & \xi & \eta & \xi\eta & 0 & 0 & 0 & 0 \\ 0 & 0 & 0 & 0 & 1 & \xi & \eta & \xi\eta \end{bmatrix} \begin{bmatrix} a_1 \\ a_2 \\ a_3 \\ a_4 \\ a_5 \\ a_6 \\ a_7 \\ a_8 \end{bmatrix} \tag{4.5.11}$$

In this case, we can show the shape functions, which are

$$[n] = [\emptyset][h]^{-1} = \frac{1}{4} \begin{bmatrix} N_1 & 0 & N_2 & 0 & N_3 & 0 & N_4 & 0 \\ 0 & N_1 & 0 & N_2 & 0 & N_3 & 0 & N_4 \end{bmatrix} \tag{4.5.14}$$

where

$$N_1 = (1 - \xi)(1 - \eta), \qquad N_2 = (1 + \xi)(1 - \eta),$$
$$N_3 = (1 + \xi)(1 + \eta) \qquad N_4 = (1 - \xi)(1 + \eta) \tag{4.5.15}$$

This leads us to the isoparametric formulations in which the shape functions serve to connect both the nodal displacements with the distributed displacements and the nodal coordinates with the coordinates, or

$$\{u_e\} = [n]\{r_e\}, \qquad \{c\} = [n]\{g\} \tag{4.6.1}$$

After noting that the line element already derived is isoparametric, we develop the general quadrilateral element for plane stress. This is done by mapping the quadrilateral into natural coordinates, where the element is a square and its polynomial representation and shape functions are given by Equations 4.5.11 and 4.5.14.

At this point, the solid tetrahedral and hexahedral elements are discussed. This is followed by the discussion of the plate elements and the combined membrane bending elements. Because of the complexity of these elements and the limited space available in this text, the equations are not carried very far, but examples using them in computer codes are given.

PROBLEMS

1. What is the element stiffness matrix for element 1 in Example 4.2.1 in terms of the global coordinates of the structure?

2. It is desired to stiffen the structure in Example 4.2.1 to obtain less deflection under the torsional load without increasing the total weight. How would you do this? What effect would this have on the deflection under the shear load?

3. The axially loaded bar in Problem 6 of Chapter 3 has dimensions and constraints as shown. The cross section of the left half is 2A and the right half is A. A uniform distributed load acts over the second quarter of the bar, and there is a concentrated load at the right end. In addition, there is a uniform distributed torque of magnitude t_0 applied over the left half of the beam. It has a solid circular cross section in each half.

   (a) Set up the FEM equations using three elements.

   (b) Show how the constraints are applied.

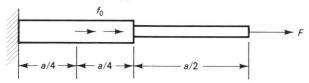

4. For the beam shown, set up the FEM equations using four elements of equal length. Apply constraints and partition the equation into the two sets for finding the deflection and the support reactions. What is the force in the spring? Would you expect the answer to be better or worse than the one obtained by the Rayleigh–Ritz method in Problem 3 of Chapter 3?

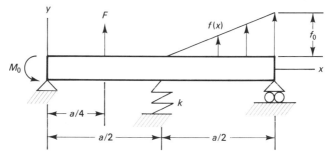

5. A uniform beam that was initially straight is given a deflection first by the prescribed loads as shown. Then a prescribed displacement at the middle support is added by raising the support an amount Δ by means of a screw jack under the support. Note that in all previous examples all prescribed support deflections were zero but that a nonzero prescribed deflection is permissible.

   (a) Write down the virtual work for this structure.

   (b) Set up the appropriate FEM equations. Identify each term in the virtual work expression with its counterpart in the FEM equations.

   (c) Partition the equations.

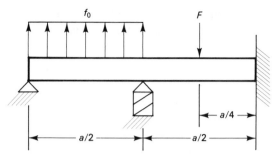

6. Given the following structure, as shown.
   (a) Write down the virtual work for this structure.
   (b) Set up the appropriate FEM equations. Identify each term in the virtual work expression with its counterpart in the FEM equations.
   (c) Partition the equations and solve for the deflections.

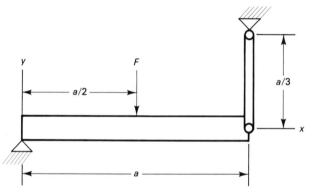

7. Consider a constant strain triangular element with global coordinates and nodal displacements as shown.

$$\{c\} = \begin{bmatrix} x_1 \\ y_1 \\ x_2 \\ y_2 \\ x_3 \\ y_3 \end{bmatrix} = \begin{bmatrix} 2 \\ 4 \\ 6 \\ 1 \\ 7 \\ 5 \end{bmatrix}, \qquad \{r_e\} = \begin{bmatrix} 0 \\ 0 \\ 0.001 \\ 0.002 \\ 0 \\ 0.001 \end{bmatrix}$$

   (a) What is the displacement of the centroid of the element?
   (b) If a concentrated force is applied at the centroid of the element as shown, what are the element equivalent nodal loads?

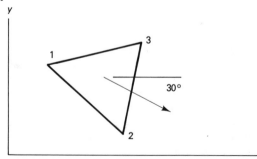

8. A rectangular element has the nodal coordinates and displacements shown.

$$\{c\} = \begin{bmatrix} x_1 \\ y_1 \\ x_2 \\ y_2 \\ x_3 \\ y_3 \\ x_4 \\ y_4 \end{bmatrix} = \begin{bmatrix} 0 \\ 0 \\ 5 \\ 0 \\ 5 \\ 3 \\ 0 \\ 3 \end{bmatrix}, \qquad \{r_e\} = \begin{bmatrix} 0.010 \\ 0.013 \\ 0.015 \\ 0.012 \\ 0.020 \\ 0.011 \\ 0.009 \\ 0.014 \end{bmatrix}$$

(a) What are the stresses at the nodes of the element?

(b) What are the stresses at the centroid of the element?

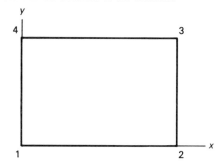

9. An axial line element has three nodes, two end nodes and one off the midpoint as shown in figure (a). It is shown in natural coordinates in figure (b).

(a) Evaluate the Jacobian in terms of L. Show that it is not constant in this case.

(b) Find the element stiffness matrix.

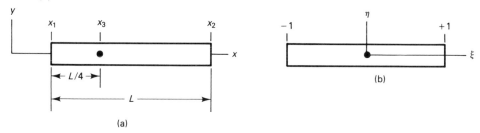

10. The structure shown is modeled in two different ways: (1) with ten beam elements each with two nodes, and (2) with ten plane stress elements each with four nodes. The beam has length a, width b, and depth c. The loading at the right end is $p_x(y, z) = p_0(1 + y/c)^2$ with units of pressure (force per unit area).

(a) For each case, what are the equivalent nodal loads?

(b) From your knowledge of shape functions, for each case what is the form (shape) of the displacement and stress distribution in the element at the right end? Point out similarities and differences.

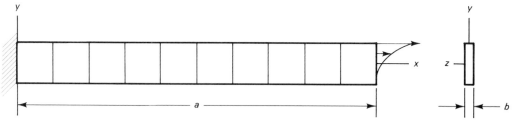

11. For the tetrahedral element in Figure 4.6.6, find the matrix $[h]$. Do not invert $[h]$, but show symbolically how you would define the stiffness matrix.

12. Beam theory applies to a structure with one dimension large compared to the other two; plate theory applies when one dimension is small compared to the other two. Consider a set of beams with rectangular cross sections and loaded with a uniform pressure, p_0, as shown. For this set of beams, a and c are the same, but b increases from one beam to the next.

(a) Construct a series of solutions using plate elements that would show when b gets too large for beam theory to be used. Show the number and kinds of elements and the constraints. Sketch the element arrangement for each case.

(b) Would plate elements of width b, that is, one row of plate elements, enable you to answer part (a)? Would your answer depend on the type of plate element selected?

(c) Suppose you wished to use a rectangular element with mid-edge nodes to solve this problem. Assuming simple plate degrees of freedom, that is, three at each node, what is the polynomial you would use for $[\emptyset]$?

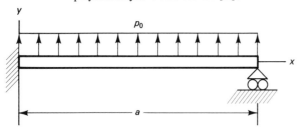
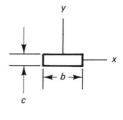

STRUCTURAL INSTABILITY

5

5.1 INTRODUCTION

Statically stable structures with linear elastic material properties have unique solutions for deflection, strain, and stress. Furthermore, the principle of superposition holds; that is, the sum of the solutions for two different loading states is equal to the solution for the sum of the two loading states. It is assumed that the two structures are identical in every way, including geometric boundary constraints; only the loading may differ. It happens that for certain conditions of geometry, constraint, and loading, a statically loaded structure may no longer have a unique solution, or an increase in loading may cause a disproportionately large deflection. Under these circumstances, the structure may be *unstable*. The large deflection is often catastrophic, resulting in the destruction of the structure and is referred to as *buckling*. In this chapter, we consider the conditions for instability and the methods of solution for unstable structures.

Instability is of primary concern for structures that are slender or thin-walled, such as beam, plates, and shells, and which are loaded with compressive loads. We first look at slender bars loaded with axial compressive loads. When loaded with both axial compressive and lateral loads, these bars are called *beam-columns*; when loaded with only axial compressive loads they are called *columns*. We next look at assemblies of slender bars, or *trusses* and *frames,* and examine the conditions under which they will buckle. Then we consider thin flat plates with compressive membrane forces and introduce the buckling of plates. Finally, we mention briefly other structures for which buckling is of concern.

The problems are formulated and solutions found for each structural type first by the differential equation method and then by virtual work and stationary potential energy. After exact and approximate analytical methods are briefly outlined, the finite element method is emphasized, since it is fast becoming the dominant method of analysis for unstable as well as stable structures.

The material presented here related to classical buckling analysis is expanded on in several references, including References 19, 20, and 21, among others. Finite element approaches to buckling are discussed in References 13, 14, 15, 18, 20, and 22, among others.

5.2 BUCKLING OF SLENDER BARS

Slender bars, or beams, are called beam-columns or columns when compressive axial loads are present of sufficient magnitude to affect the lateral deflection. These structures are chosen typically to introduce the phenomena of buckling, as we shall do here. Both differential equation and work and energy (integral equation) methods will be illustrated in the following paragraphs.

5.2.1 Differential Equations for Buckling

The various assumptions and definitions introduced in Chapter 2 to define slender bars and to develop a theory of slender bars are all included in this analysis; thus, the symbolic representation of the equations shown in Figure 2.2.1 may be used without modification. While the symbolic form of the various equations given in Equation 2.2.1 may also be adopted without change, in the usual derivation of the differential equations for buckling of slender bars there is a change in the way equilibrium is imposed. In Chapter 2, as noted for example in Figure 2.2.9, equilibrium is imposed in the undeflected position. In the derivation for buckling given next, equilibrium is imposed in the deflected position.

Consider the slender bar shown in Figure 5.2.1(a). It has all the same geometric and material properties of the beams studied in Chapter 2. There is both a lateral loading $f(x)$ and an axial load F applied at the free end. The bar is shown as fixed at one end and free at the other, but any geometric constraints used for beam analysis are acceptable as long as an applied axial load F is permitted. We shall consider only the two-dimensional case with deflections occurring in the x-z plane. For convenience, we have dropped the subscripts identifying the direction of the forces and moments and moments of inertia, since in a single plane this causes no confusion.

A differential element of the deflected beam is shown in Figure 5.2.1(b). We show the normal and shearing forces on each face of the element, as well as the moments on each face about the y-axis. We have

$$N(x + dx) = N(x) + \frac{\partial N}{\partial x}dx, \qquad S(x + dx) = S(x) + \frac{\partial S}{\partial x}dx$$

$$M(x + dx) = M(x) + \frac{\partial M}{\partial x}dx \tag{5.2.1}$$

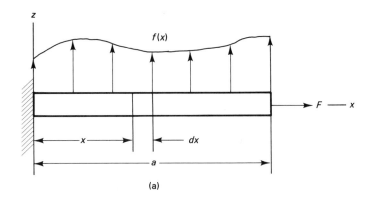

(a)

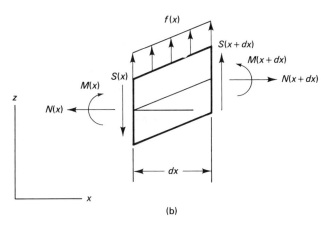

(b)

Figure 5.2.1 The Beam Element in the Deflected Position.

From summation of forces in the x-direction, we get

$$N(x) + \frac{\partial N}{\partial x}dx - N(x) = 0 \tag{5.2.2}$$

which reduces to

$$\frac{dN}{dx} = 0 \tag{5.2.3}$$

or N is constant. From equilibrium in the z-direction,

$$S(x) + \frac{\partial S}{\partial x}dx - S(x) + f(x) = 0 \tag{5.2.4}$$

which reduces to

$$\frac{dS}{dx} = -f(x) \tag{5.2.5}$$

and from summation of moments

$$M + \frac{\partial M}{\partial x}dx - M + [f(x)\,dx]\frac{dx}{2} - S\,dx - N\frac{dw}{dx}dx = 0 \qquad (5.2.6)$$

which reduces to

$$\frac{dM}{dx} - N\frac{dw}{dx} = -S \qquad (5.2.7)$$

Note the new term, $N\,dw/dx$, which does not appear in the general beam theory of Chapter 2.

If all the usual assumptions of beam theory are retained and we repeat the steps in Section 2.2.1 that lead to the bending moment curvature relation, we find that it is unaltered, or

$$EIw'' = M \qquad (5.2.8)$$

It follows that

$$(EIw'')' - Nw' = -S \qquad (5.2.9)$$

and it also follows that

$$(EIw'')'' - Nw'' = f(x) \qquad (5.2.10)$$

where once again primes denote differentiation with respect to x. The boundary conditions are stated in terms of deflection, slope, moment, or shear, just as in Chapter 2.

What we have is a set of two differential equations, Equations 5.2.3 and 5.2.10, that govern the behavior of the structure under combined axial and lateral load. The two dependent variables, N and w, are functions of the one independent variable x. This is a nonlinear set of equations because of the product of the two dependent variables in the second term in Equation 5.2.10. Since Equation 5.2.3 contains only one unknown, it can be solved independently and the result substituted into Equation 5.2.10; thus, the nonlinearity presents no mathematical difficulties in this case.

When we solve Equation 5.2.3 for N, we get

$$\frac{dN}{dx} = 0 \longrightarrow N(x) = a_1 \qquad \text{(a constant)} \qquad (5.2.11)$$

and since

$$N(a) = a_1 = F \longrightarrow N(x) = F \qquad (5.2.12)$$

Note that N is the internal stress resultant, which in this special case happens to be equal to the applied load F. When the external applied load, and hence the internal stress resultant, is a compressive force P, we let

$$N = -P \qquad (5.2.13)$$

and the equation becomes

$$(EIw'')'' + Pw'' = f(x) \qquad (5.2.14)$$

We are interested primarily in the case where the internal axial stress resultant is compressive; this is the form of the equation we will use for most of the subsequent work.

When P in Equation 5.2.14 is compressive, this is called the *beam-column equation*. It is a fourth-order ordinary differential equation. If the beam is uniform, it becomes an ordinary differential equation with constant coefficients, which can be written in the convenient form

$$w'''' + \lambda^2 w'' = \frac{f(x)}{EI} \qquad (5.2.15)$$

where

$$\lambda^2 = \frac{P}{EI}$$

for which general solutions are available. The solutions of this equation are in the form of

$$w(x) = w_h(x) + w_p(x) \qquad (5.2.16)$$

where $w_h(x)$ is the solution of the homogeneous equation

$$w'''' + \lambda^2 w'' = 0 \qquad (5.2.17)$$

and $w_p(x)$ is a particular solution of the full equation. From elementary differential equation theory, the homogeneous solution is

$$w_h(x) = c_1 \sin\lambda x + c_2 \cos\lambda x + c_3 x + c_4 \qquad (5.2.18)$$

and the particular solution awaits the specification of a particular lateral loading. Before we do that, let us see what we can glean from the homogeneous equation.

To complete the statement of the problem, we must know the boundary conditions. With one exception, these are the same as discussed in Section 2.2.4 and applied to various problems in Chapter 2. The exception is the case for a shear force applied at the boundary. Equation 2.2.50 is replaced by Equation 5.2.9, or

$$(EIw'')' + Pw' = -S \qquad (5.2.19)$$

This change carries over to the case where springs are employed; thus, Equation 2.2.55 is replaced by

$$EIw'''(a) + Pw'(a) - kw(a) = 0, \qquad EIw''(a) = 0 \qquad (5.2.20)$$

and Equation 2.2.59 by

$$EIw'''(a) + Pw'(a) - kw(a) = 0, \qquad EIw''(a) + \mu w'(a) = 0 \qquad (5.2.21)$$

Solutions for problems with various boundary conditions will be illustrated in the examples that follow in the next section.

5.2.2 Euler Column

When a beam has an axial compressive load and no lateral load, it is called a *column*. The equation for columns is simply the homogeneous part of the beam-column equation, or

$$w'''' + \lambda^2 w'' = 0 \qquad (5.2.22)$$

and the complete solution is

$$w(x) = c_1 \sin\lambda x + c_2 \cos\lambda x + c_3 x + c_4 \qquad (5.2.23)$$

When this solution is substituted into the four boundary conditions, four simultaneous linear homogeneous equations are obtained for determining the values of c_1, c_2, c_3, and c_4. In matrix form,

$$[\Lambda]\{c\} = \{0\} \qquad (5.2.24)$$

One obvious solution is $\{c\} = \{0\}$, which concludes that there is no deflection at all and is called the *trivial solution*. It has been found, however, that under certain conditions other solutions are possible. It is a characteristic of homogeneous linear algebraic equations that solutions other than the trivial solution may be possible if the determinant of the coefficients is zero, or if

$$|\Lambda| = 0 \qquad (5.2.25)$$

This leads to very interesting results in this case. To understand this, an example is needed.

Example 5.2.1

Find the deflection of a simply supported column as shown in figure (a).

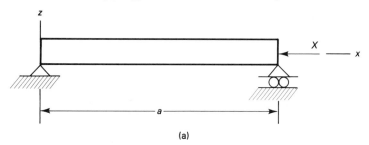

(a)

Solution: As noted previously, an axial applied load X as shown would produce an internal stress resultant P equal to X. The general solution is given by Equation 5.2.23, and the boundary conditions on deflection and moment at the ends are

$$w(0) = w(a) = EIw''(0) = EIw''(a) = 0 \qquad (a)$$

When the general solution is substituted into the boundary conditions, we obtain the matrix equation

$$[\Lambda]\{c\} = \{0\} \qquad (b)$$

or

$$\begin{bmatrix} 0 & 1 & 0 & 1 \\ 0 & \lambda^2 & 0 & 0 \\ \sin\lambda a & \cos\lambda a & a & 1 \\ \lambda^2 \sin\lambda a & \lambda^2 \cos\lambda a & 0 & 0 \end{bmatrix} \begin{bmatrix} c_1 \\ c_2 \\ c_3 \\ c_4 \end{bmatrix} = \begin{bmatrix} 0 \\ 0 \\ 0 \\ 0 \end{bmatrix} \qquad (c)$$

As noted, letting

$$\{c\} = \{0\} \qquad (d)$$

is an obvious solution, called the *trivial solution*, which concludes that there is no deflection at all. However, $\{c\}$ need not be zero if

$$|\Lambda| = 0 \qquad (e)$$

It has been discovered that significant information about the physical behavior of the column is found from this determinant. We must ask if there are conditions for which equation (e) is true. The answer is that if we allow P, and hence λ, to vary, there are values of P for which the determinant is zero. At this point, you can expand the determinant and set the result to zero; however, it is easier in this case to reduce the number of equations by substitution first. We note that the second of equations (c) is

$$\lambda^2 c_2 = 0 \longrightarrow c_2 = 0 \tag{f}$$

and the first of equations (c) is

$$c_2 + c_4 = 0 \longrightarrow c_4 = 0 \tag{g}$$

Thus, the equations reduce to

$$\begin{bmatrix} \sin\lambda a & a \\ \lambda^2 \sin\lambda a & 0 \end{bmatrix} \begin{bmatrix} c_1 \\ c_3 \end{bmatrix} = \begin{bmatrix} 0 \\ 0 \end{bmatrix} \tag{h}$$

The expansion of the determinant of this reduced coefficent matrix results in

$$\sin\lambda a = 0 \tag{i}$$

or

$$\lambda a = n\pi, \qquad n = 0, 1, 2, 3, \ldots \tag{j}$$

Thus,

$$P = \left(\frac{n\pi}{a}\right)^2 EI \tag{k}$$

This suggests that there are an infinite number of discrete values of P, one for each integer value of n, for which a solution other than $\{c\} = \{0\}$ exists.

If a deflection does exist at one of these discrete values, we wish to find its shape. When P is equal to one of the values in equation (k), we can examine equations (c) once again. From the second equation, we have $c_2 = 0$. With this information and the first equation, we have also that $c_4 = 0$. Now let us write out the last two equations:

$$(\sin\lambda a)c_1 + ac_3 = 0$$
$$(\lambda^2 \sin\lambda a)c_1 = 0 \tag{l}$$

Since $\sin \lambda a = 0$, we can conclude from these equations that $c_3 = 0$ and that c_1 is indeterminate. With $c_2 = c_3 = c_4 = 0$, we can conclude that the general solution becomes

$$w(x) = c_1 \sin\lambda x = c_1 \sin\frac{n\pi x}{a} \tag{m}$$

We now know that the shape of the deflection is a sine curve consisting of n half-sine waves according to the particular value of P chosen. The lowest value of P,

$$P_{cr} = \frac{\pi^2 EI}{a^2} \tag{n}$$

is of the greatest physical significance and is called the *critical* or *Euler buckling load* and has been given the subscript *cr* to emphasize this. The shape of the deflection is

$$w(x) = c_1 \sin\frac{\pi x}{a} \tag{o}$$

and is called the *buckling mode*. Note that from this analysis we cannot determine the value of c_1; however, that does not prevent us from knowing the shape of the deflection.

The higher loads and modes are of limited practical interest, because as the load is increased from a small value the structure will fail when the lowest critical load is reached.

From this example, we have learned some things that are true for all columns and for elastic structures in general. Under certain conditions, the equations governing the behavior of structures have an infinite number of solutions for the deflected shapes, each corresponding to one of an infinite number of discrete loading values. These equations are found to be in the class called *eigenvalue equations*; the discrete values of the parameter for which there is a solution other than the trivial one are called *eigenvalues,* and the functions that represent a solution to the equation for each of these eigenvalues are called *eigenfunctions*.

The column equation, Equation 5.2.22, is an eigenvalue equation, the discrete values of the parameter P or λ^2 are eigenvalues, and the buckling mode is an eigenfunction. Note that the eigenvalues are found precisely, but the eigenfunctions are indeterminate in magnitude although known in shape.

This fact can be illustrated by the plot of the deflection of a column shown in Figure 5.2.2. Consider the deflection of some point on the column, which we shall call Δ. As long as the axial compressive load P is below P_{cr}, the deflection Δ will be zero, the trivial solution, as shown in Figure 5.2.2 by the vertical line at the origin of Δ. Once P is equal to P_{cr}, the deflection can have any value plus or minus, as shown by the horizontal line at P_{cr}. The equations are valid only for deflections that are geometrically small and within the linear elastic range of the material; thus, large values of Δ are not valid in practical column analysis. Nevertheless, in the vicinity of the origin, an infinite number of deflections are possible.

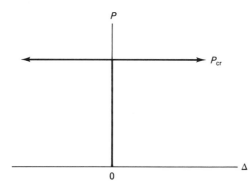

Figure 5.2.2 Deflection of a Column.

The first examination of this mathematical model of the column is attributed to the great mathematician Euler; therefore, it is often called the *Euler column* and the critical load is called the *Euler buckling load*. Perhaps more examples will be helpful.

Example 5.2.2

Find the buckling load and mode for a uniform column with both ends fixed.

Solution: The internal stress resultant is $P = X$. The solution is given by Equation 5.2.18 as

$$w(x) = c_1 \sin\lambda x + c_2 \cos\lambda x + c_3 x + c_4 \qquad (a)$$

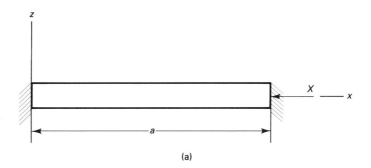

(a)

and the boundary conditions on deflection and slope at the ends are

$$w(0) = w(a) = w'(0) = w'(a) = 0 \tag{b}$$

When the solution is substituted into the boundary conditions, the following matrix equation is formed:

$$\begin{bmatrix} 0 & 1 & 0 & 1 \\ \lambda & 0 & 1 & 0 \\ \sin\lambda a & \cos\lambda a & a & 1 \\ \lambda\cos\lambda a & -\lambda\sin\lambda a & 1 & 0 \end{bmatrix} \begin{bmatrix} c_1 \\ c_2 \\ c_3 \\ c_4 \end{bmatrix} = \begin{bmatrix} 0 \\ 0 \\ 0 \\ 0 \end{bmatrix} \tag{c}$$

The first two equations can be used to reduce the number of unknowns to two, forming a 2×2 determinant, or exanding the 4×4 determinant of the coefficients directly gives us

$$\lambda a \sin\lambda a + 2(\cos\lambda a - 1) = 0 \tag{d}$$

What we seek now is the value of λa that is the lowest root of this transcendental equation. It may be done by plotting the function on the left-hand side versus λa and looking for where the plot crosses the axis. The first intersection occurs at $\lambda a = 0$, or the trivial case $P = 0$. The next intersection is at $\lambda a = 2\pi$, or

$$P_{cr} = \left(\frac{2\pi}{a}\right)^2 EI \tag{e}$$

A formal numerical method for finding the roots of equations readily adapted to computer solution is presented in Appendix K. This method will be used in Chapter 6 as well.

Substituting $\lambda a = 2\pi$ back into the equations, we obtain the results

$$c_1 = c_3 = 0, \qquad c_2 = -c_4 \tag{f}$$

and the mode shape is found to be

$$w(x) = c_2\left(\cos\frac{2\pi x}{a} - 1\right) \tag{g}$$

Example 5.2.3

Find the buckling load and mode for a column fixed on one end and free on the other end as shown in figure (a).

Solution: Once again the internal stress resultant is $P = X$. The boundary conditions on deflection and slope at the left end and on moment and shear at the right end are

$$w(0) = w'(0) = 0, \qquad EIw''(a) = 0, \qquad w'''(a) + \lambda^2 w'(a) = 0 \tag{a}$$

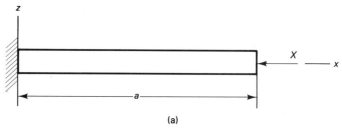

(a)

The matrix equation is

$$
\begin{bmatrix}
0 & 1 & 0 & 1 \\
\lambda & 0 & 1 & 0 \\
\sin\lambda a & \cos\lambda a & 0 & 0 \\
0 & 0 & 1 & 0
\end{bmatrix}
\begin{bmatrix}
c_1 \\
c_2 \\
c_3 \\
c_4
\end{bmatrix}
=
\begin{bmatrix}
0 \\
0 \\
0 \\
0
\end{bmatrix}
\tag{b}
$$

The determinant of the coefficients gives us

$$
\cos\lambda a = 0 \tag{c}
$$

or

$$
\lambda a = \frac{(2n + 1)\pi}{2}, \qquad n = 0, 1, 2, \ldots \tag{d}
$$

In this case, $n = 0$ is not trivial, and for $n = 0$ we get

$$
P_{cr} = \left(\frac{\pi}{2a}\right)^2 EI \tag{e}
$$

For $\lambda a = \pi/2$, we may conclude from the equations that

$$
c_1 = c_3 = 0, \qquad c_2 = -c_4 \tag{f}
$$

and, therefore,

$$
w(x) = c_2\left(\cos\frac{\pi x}{2a} - 1\right) \tag{g}
$$

From the foregoing analyses, we can observe that a general expression for the Euler buckling load for uniform columns with arbitrary boundary conditions may be written

$$
P_{cr} = c\left(\frac{\pi}{a}\right)^2 EI \tag{5.2.26}
$$

where the constant c is called the *end fixity coefficient*. The value of this constant depends only on the boundary conditions, and its value for specific cases can be obtained from the preceding examples.

If a new quantity

$$
a' = \frac{a}{\sqrt{c}} \tag{5.2.27}
$$

is defined, the buckling load is divided by the cross-sectional area, and the area moment of inertia is presented in the form

$$
I = A\rho^2 \tag{5.2.28}
$$

Structural Instability   Chap. 5

where ρ is the radius of gyration; then Equation 5.2.26 can be rewritten

$$\frac{\sigma_{cr}}{E} = \frac{\pi^2}{(a'/\rho)^2} \tag{5.2.29}$$

The length a' can be interpreted as the length of a simply supported column having the same cross section and buckling load as the column under consideration; consequently, it is termed the *equivalent* or *effective* length. A plot of Equation 5.2.29 is given in Figure 5.2.3.

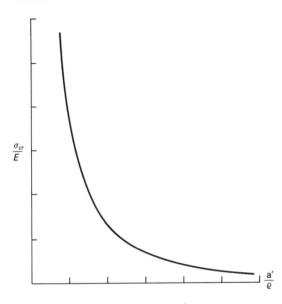

Figure 5.2.3 Buckling Stress versus Effective Length.

Solutions for other boundary conditions and other lateral loadings are readily obtained if the boundary conditions and loading are kept fairly simple. As the problems become more complicated, exact solutions are more difficult to obtain. Later in this chapter we shall deal with that problem.

5.2.3 Solutions for Beam-Columns

Now we shall deal with the case where both axial and lateral loads are present and particular solutions, as well as homogenous solutions to the differential equation, must be found. We refer back to Equations 5.2.14 through 5.2.18. Finding a particular solution is an exercise that depends on the particular form of the lateral load, so let us consider a specific example.

Example 5.2.4

Find the deflection of a uniform, simply supported beam-column with a uniform lateral load f_0 and an axial load X.

Solution: The particular solution must satisfy the equation

$$w'''' + \lambda^2 w'' = \frac{f_0}{EI} \tag{a}$$

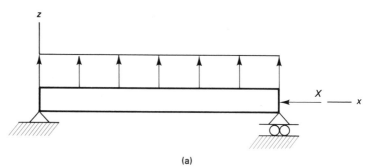

(a)

By the method of undetermined coefficients, explained at length in books on ordinary differential equations with constant coefficients, we know that when the right-hand side is a polynomial the particular solution is a polynomial of higher order, and it can be found by inspection. In this case, the right-hand side is a polynomial of order zero, that is, a constant, and the particular solution is a polynomial involving x^2, or

$$w_p(x) = \frac{f_0}{2P} x^2 \tag{b}$$

Thus, the general solution is

$$w(x) = c_1 \sin \lambda x + c_2 \cos \lambda x + c_3 x + c_4 + \frac{f_0}{2P} x^2 \tag{c}$$

The boundary conditions are

$$w(0) = w(a) = 0, \qquad w''(0) = w''(a) = 0 \tag{d}$$

When the general solution is substituted into these boundary conditions, we obtain

$$c_2 + c_4 = 0, \qquad \lambda^2 c_2 = \frac{f_0}{P}$$

$$c_1 \sin \lambda a + c_2 \cos \lambda a + c_3 a + c_4 = -\frac{f_0 a^2}{2P} \tag{e}$$

$$c_1 \lambda^2 \sin \lambda a + c_2 \lambda^2 \cos \lambda a = \frac{f_0}{P}$$

This can be put in the matrix form

$$[\Lambda]\{c\} = \{f\} \tag{f}$$

or

$$\begin{bmatrix} 0 & 1 & 0 & 1 \\ 0 & \lambda^2 & 0 & 0 \\ \sin \lambda a & \cos \lambda a & a & 1 \\ \lambda^2 \sin \lambda a & \lambda^2 \cos \lambda a & 0 & 0 \end{bmatrix} \begin{bmatrix} c_1 \\ c_2 \\ c_3 \\ c_4 \end{bmatrix} = \begin{bmatrix} 0 \\ \dfrac{f_0}{P} \\ -\dfrac{f_0 a^2}{2P} \\ \dfrac{f_0}{P} \end{bmatrix} \tag{g}$$

Equations (g) are four linear algebraic equations in terms of the four unknowns c_1, c_2, c_3, and c_4. When these are solved and the values put into equation (c), we obtain

$$w(x) = \frac{f_0}{P} \left(\frac{(1 - \cos \lambda a) \sin \lambda x}{\lambda^2 \sin \lambda a} + \frac{\cos \lambda x}{\lambda^2} - \frac{ax}{2} - \frac{1}{\lambda^2} + \frac{x^2}{2} \right) \tag{h}$$

It is instructive to study the nature of this solution as P, and consequently λ, takes on different values as it increases from 0. At $\lambda = 0$, when the member is simply a uniformly loaded beam, the solution is seen to be indeterminate; however, by a limiting process it does, in fact, reduce to the solution of the equation

$$w'''' = \frac{f_0}{EI} \tag{i}$$

which is

$$w(x) = \frac{f_0}{24EI} x(a^3 - 2ax^2 + x^3) \tag{j}$$

For the purposes described later, equation (h) can be given in the following alternative form by using some trigonometric identities and a nondimensional form, $\xi = x/a$, for the independent variable:

$$w(\xi) = \frac{f_0 a^3}{EI} \left(\frac{[\sin \lambda a \xi + \sin \lambda a (1 - \xi) - \sin \lambda a]}{\lambda^4 a^4 \sin \lambda a} - \frac{\xi(1 - \xi)}{2\lambda^2 a^2} \right) \tag{k}$$

A plot of Δ, the deflection of the midpoint of the beam, versus λa for various values of $f_0 a^3/(EI)$ is given in figure (b). As λ increases, the deflection increases and the deflection approaches ∞ as λa approaches the value π, that is, as P approaches the value $\pi^2 EI/a^2$. Thus, at this load, on the basis of the present analysis, the member will collapse even for vanishingly small values of f_0.

One may ask what happens when the axial load is tensile. Then, if we replace P by minus P, or N, the governing equation is

$$w'''' - \lambda^2 w'' = \frac{f_0}{EI} \tag{l}$$

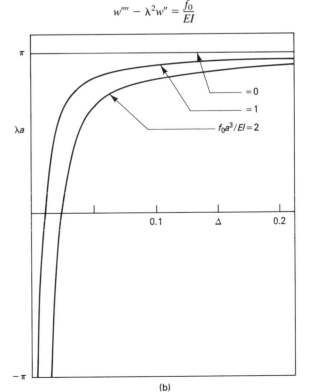

(b)

and the general solution is

$$w(x) = c_1 \sinh \lambda x + c_2 \cosh \lambda x + c_3 x + c_4 - \frac{f_0}{2P} x^2 \qquad \text{(m)}$$

When substituted into the boundary conditions and c_1, c_2, c_3, and c_4 are found, the solution becomes

$$w(\xi) = \frac{f_0 a^3}{EI} \left(\frac{[\sinh \lambda a \xi + \sinh \lambda a (1 - \xi) - \sinh \lambda a]}{\lambda^4 a^4 \sinh \lambda a} + \frac{\xi(1 - \xi)}{2\lambda^2 a^2} \right) \qquad \text{(n)}$$

The plots in figure (b) are extended for increasing tensile load as obtained from equation (n). Note that the tensile load has a slight stiffening effect; that is, the lateral deflection decreases as the axial tensile load increases.

We see that the presence of an axial compressive load on a beam can have a profound effect on the lateral displacement. The deflection is not proportional to the loading P; in fact, this relationship is highly nonlinear. Note that we are still using a linear relationship between stress and strain, or Hooke's law. Note also that we are solving a linear differential equation. Nevertheless, geometric nonlinearities give rise to the nonlinear relation between load and deflection in this case.

It must be remembered that, because of the assumptions made, the analysis is strictly correct only for small deflections; therefore, as the deflections increase, there comes a value at which the equations are no longer valid. Nevertheless, this analysis is very helpful in showing the importance of the axial load in producing lateral deflection and is accurate in the small-deflection region.

Had we used the equations of Chapter 2, we would have arrived at the conclusion that there was no effect at all. It should be noted that the more slender the beam, that is, the smaller the value of I/a^2, the lower the axial load at which the effect is significant. If the beam is not so slender, P may give rise to a high stress value, even up to the level of the yield stress, at a fraction of the value at which the effect on lateral deflection is significant. Thus, what we have learned in Chapter 2 is still relevant; it just becomes less accurate as the beam becomes more slender and as the axial compressive load becomes larger. A point is finally reached at which this nonlinear effect is dominant, and eventually buckling occurs.

The extension of the differential equation method to more complex cases becomes difficult. Consider, for example, a tapered beam-column, such as is shown in Figure 5.2.4(a). The governing equation is

$$[EI(x)w'']'' + Pw'' = f(x) \qquad (5.2.30)$$

an ordinary differential equation no longer with constant coefficients. There is no general solution for such an equation and few special solutions for particular values of the coefficients.

Or consider the beam column with additional constraints and concentrated lateral loads at intermediate points as shown in Figure 5.2.4(b). In this case, the beam must be broken into segments, with a different equation solved for each segment. This results in multiple sets of constants of integration, four constants for each segment, and a corresponding increase in the number of boundary conditions.

Still another case is encountered when the axial load itself varies over different segments of the beam-column, such as shown in Figure 5.2.4(c). Again the structure must

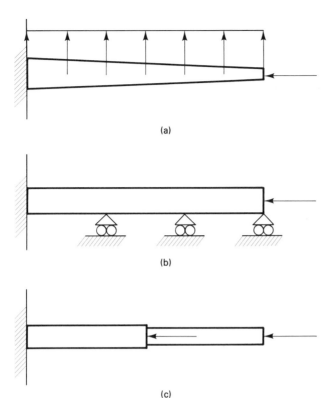

(a)

(b)

(c)

Figure 5.2.4 More Difficult Beam-Column Problems.

be divided into segments, with a different equation prevailing over each segment. We soon learn the various approximate methods are much more productive in obtaining solutions to many practical problems; therefore, we turn to the virtual work and potential energy methods in the next section. Nevertheless, the exact solutions found previously are of immense value, both in understanding the mathematical modeling of the buckling phenomenon and in providing test cases against which to measure the effectiveness of the approximate methods that follow.

5.2.4 Virtual Work for Beam-Columns

Appropriate equations for the beam-column can be obtained from the principle of virtual work. The virtual work expression for a statically stable beam is given in Section 3.2.2 and is repeated:

$$\delta W = \delta W_e + \delta W_i = \delta W_e - \delta U$$
$$= \int \delta w f(x)\, dx + \Sigma\, F_{zi}\, \delta w_i + \Sigma\, M_{yi}(\delta w_j)' - \int EI\, \delta w''w''dx = 0 \qquad (5.2.31)$$

We must modify this expression to add the mechanism for instability. Two alternative derivations are found in the literature. The traditional one takes account of the effect of the axial load on the external work term. As in the derivation of the differential equation, the beam-column is examined in the deflected position. In moving from the undeflected to the deflected position, we note that the axial load moves through an axial distance due to the lateral deflection.

Designating the axial deflection of the axial load F by $u(a)$, as shown in Figure 5.2.5, we now have for the external virtual work an additional contribution of

$$\Delta \delta W_e = F \, \delta u(a) \tag{5.2.32}$$

To determine the value of $u(a)$, we note that the developed length of the deflected beam is the same as the length of the undeflected beam. This ignores the effect of axial compression due purely to uniform axial strain. As we learned in Chapter 3, this effect can usually be neglected in beam theory when studying the lateral deflections. In Figure 5.2.5(c), we show an enlarged section ds of the deflected beam. The contribution of this element to the change in length is

$$du = dx - ds \tag{5.2.33}$$

but

$$ds = \sqrt{dx^2 + dw^2} = dx \sqrt{1 + (w')^2} \tag{5.2.34}$$

It is reasonable to assume that w' is small enough that only the first term of the binomial expansion need be retained. Thus,

$$ds = dx \left[1 + \frac{(w')^2}{2} \right] \tag{5.2.35}$$

so that

$$du = - \frac{(w')^2}{2} dx \tag{5.2.36}$$

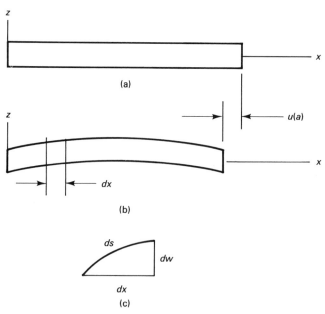

(a)

(b)

(c)

Figure 5.2.5 Beam-Column in Deflected Position.

and

$$u(a) = -\frac{1}{2} \int (w')^2 \, dx \qquad (5.2.37)$$

From a virtual displacement δw, we have

$$\delta u(a) = -\int (\delta w')(w') \, dx \qquad (5.2.38)$$

and, finally, since $F = N$

$$\Delta \delta W_e = F \, \delta u(a) = -N \int \delta w' w' \, dx \qquad (5.2.39)$$

The total virtual work is now

$$
\begin{aligned}
\delta W &= \int \delta w f(x) \, dx + \Sigma \, F_{bzi} \, \delta w_i + \Sigma \, M_{yj}(\delta w_j)' \\
&\quad - N \int \delta w' w' \, dx - \int EI \, \delta w'' w'' \, dx = 0 \\
&= \int \delta w f(x) \, dx + \Sigma \, F_{bzi} \, \delta w_i + \Sigma \, M_{yj}(\delta w_j)' \\
&\quad + P \int \delta w' w' \, dx - \int EI \, \delta w'' w'' \, dx = 0
\end{aligned}
\qquad (5.2.40)
$$

where it is recognized that the external load F produces an internal axial stress resultant $-P$.

In recent years, an alternative derivation has been favored, which approaches the problem from a view of the effect on the internal virtual work that results from the externally applied axial load. To invoke this derivation, we include a term in the strain–displacement equations that has been neglected up to now as higher ordered; that is, the expression for strain in terms of displacement includes a term from Equation 1.2.15 that involves the square of the lateral displacement, as follows:

$$\epsilon_x = \frac{\partial u}{\partial x} + \frac{1}{2}\left(\frac{\partial w}{\partial x}\right)^2 = u_0' - zw'' + \frac{1}{2}(w')^2 \qquad (5.2.41)$$

It has been found that this term contains the mechanism for structural instability. For convenience in notation, let

$$\epsilon_x = \epsilon_1 + \epsilon_2 \qquad (5.2.42)$$

where

$$\epsilon_1 = u_0' - zw'', \qquad \epsilon_2 = \frac{1}{2}(w')^2 \qquad (5.2.43)$$

and

$$\sigma_x = \sigma_1 + \sigma_2 \qquad (5.2.44)$$

where the initial axial stress in the beam is

$$\sigma_1 = \frac{N}{A} \qquad (5.2.45)$$

and the bending stress is

$$\sigma_2 = -\frac{M}{I}z \qquad (5.2.46)$$

We note that

$$\delta \epsilon_1 = \delta u_0' - z \, \delta w'', \qquad \delta \epsilon_2 = w' \, \delta w' \qquad (5.2.47)$$

The internal virtual work for a beam-column is then

$$\delta U = \int \{\delta\epsilon\}^T \{\sigma\} \, dV$$
$$= \int (\delta\epsilon_1 + \delta\epsilon_2)(\sigma_1 + \sigma_2) \, dV \qquad (5.2.48)$$
$$= \int (\delta\epsilon_1\sigma_1 + \delta\epsilon_1\sigma_2 + \delta\epsilon_2\sigma_1 + \delta\epsilon_2\sigma_2) \, dV$$

The term $\delta\epsilon_1\sigma_2$ is the contributor already identified in Chapter 3. The terms $\delta\epsilon_2\sigma_2$ and $\delta\epsilon_1\sigma_1$ will not contribute. This may be seen by writing the integrals out as

$$\int \delta\epsilon_2\sigma_2 \, dV = -\int \delta\epsilon_2 \frac{Mz}{I} \, dA \, dx = 0$$
$$\int \delta\epsilon_1\sigma_1 \, dV = \int (u_0 - z\,\delta w'') \frac{N}{A} \, dA \, dx = 0 \qquad (5.2.49)$$

The first integral and the second part of the second integral contain a term $\int z \, dA = 0$ if centroidal axes are used. And the first part of the second integral in the axial strain energy term decouples from the bending energy and can be neglected. We are left with one new term of the form

$$\Delta\delta U = \int \delta\epsilon_2\sigma_1 \, dV = \iint w' \, \delta w' \frac{N}{A} \, dA \, dx = \int Nw' \, \delta w' \, dx \qquad (5.2.50)$$

The total virtual work is now

$$\delta W = \delta W_e - (\delta U + \Delta\delta U)$$
$$= \int \delta w f(x) \, dx + \Sigma F_{zi} \, \delta w_i + \Sigma M_{yi}(\delta w_j)' \qquad (5.2.51)$$
$$- \int N\delta w' w' \, dx - \int EI\delta w'' w'' \, dx = 0$$

In a simple beam, when the axial internal force is compressive and is a constant equal to the axial load, we let $N = -P$ and can take the P outside the integral. This is exactly the expression for virtual work we got before with the external work method. In more complicated structures, finding the value of N is, itself, a problem to be solved that is not always simple. It may be, for example, a function of x. In such cases, the internal virtual work method is to be preferred.

The difference in the two derivations and whether the new term is introduced via the internal or external work term apparently depend on a subtle difference in how you view the problems. In the first approach, you view the beam in the deflected position and examine the external work done by the axial load. In the second approach, you examine the internal work in the presence of an initial stress field that is caused by the external load set. Both arrive at the same answer and that this is the correct expression is verified by experiment.

5.2.5 Rayleigh–Ritz Method

In the past, much use of the Rayleigh–Ritz method has been made in converting the principle of virtual work into a workable method of analysis for buckling problems, as well as for statically stable problems. If we choose to use the Rayleigh–Ritz method, where we introduce a set of global admissible functions and generalized coordinates of the form

$$w = [\Omega]\{q\} \qquad (5.2.52)$$

the equations of Section 3.2.3 must also be modified accordingly. In this case, Equation 3.2.52 for the generalized force must have an additional term that arises from the new term in the virtual work given in Equation 5.2.50. Upon substitution of Equation 5.2.52 into this expression, we get

$$\delta \Delta U = \{\delta q\}^T (\int N[\Omega']^T[\Omega']\{q\} \, dx) = \{\delta q\}^T[K^d]\{q\} \tag{5.2.53}$$

The algebraic matrix equation, which is Equation 3.2.50 for the stable structure, now takes on the form

$$([K] + [K^d])\{q\} = \{Q\} \tag{5.2.54}$$

where, as before in Equation 3.2.50,

$$[K] = \int [B]^T[G][B] \, dV = \int EI[\Omega'']^T[\Omega''] \, dx$$
$$\{Q\} = \int [\Omega]^T f(x) \, dx + \sum [\Omega(x_i)]^T F_{zi} + \sum [\Omega'(x_j)]^T M_{yj} \tag{5.2.55}$$

The expression for $[K^d]$ supposes that N is known. In the case of simple beams with a single axial load at one end, as in Examples 5.2.1 through 5.2.4, we have $N = -P$ and

$$[K^d] = -P \int [\Omega']^T[\Omega'] \, dx \tag{5.2.56}$$

In cases of multiple or variable axial loading, the N may vary with x and must be found before $[K^d]$ may be formulated.

The matrix $[K^d]$ has been given a variety of names, chiefly, the *differential stiffness,* the *geometric stiffness*, or the *initial stress* matrix. We shall adopt the name differential stiffness.

Equation 5.2.54 is the matrix equation for a beam-column. Setting the lateral loads to zero gives us

$$([K] + [K^d])\{q\} = \{0\} \tag{5.2.57}$$

which gives us the matrix eigenvalue equation for a column. We find the critical load by setting

$$|[K] + [K^d]| = 0 \tag{5.2.58}$$

In the simple column case, we look for a value of the lowest value of the load P, which is contained in the matrix $[K^d]$, that will cause the determinant to be zero. This is the critical or buckling load. Often the quantity $-P$ is factored from $[K^d]$, and Equation 5.2.57 is rewritten in the form

$$([K] - P[K^D])\{q\} = \{0\} \tag{5.2.59}$$

where

$$[K^D] = \int [\Omega']^T[\Omega'] \, dx \tag{5.2.60}$$

Thus,

$$|[K] - P[K^D]| = 0 \tag{5.2.61}$$

This is the more familiar form with the quantity P, which becomes the eigenvalue, shown explicitly. It may be seen that expansion of this determinant will produce a polynomial in P, and the lowest root of this polynomial equation is the critical value of P.

The value of the Rayleigh–Ritz method is in working problems that are difficult by the differential equation method. In the next example, we find the deflection of a tapered beam-column and the critical buckling load and mode of a tapered column. Exact analytical solutions of tapered beam-columns, that is, of differential equations with nonconstant coefficients, are usually very difficult except in special cases.

Example 5.2.5

Find the deflection of a tapered column fixed on one end and simply supported on the other, which has a uniform lateral load. Also, find the buckling load and mode. The member has a uniform thickness h and a linear taper in width so that the width b of the cross section is given by

$$b = b_0(1 + cx), \qquad c = -\frac{b_0 - b_1}{b_0 a} \tag{a}$$

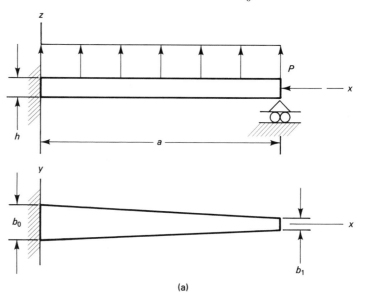

(a)

Thus, the stiffness is given by

$$EI = EI_0(1 + cx) \tag{b}$$

where

$$EI_0 = \frac{Eb_0 h^3}{12} \tag{c}$$

Solution: The geometric boundary conditions are

$$w(0) = w'(0) = 0, \qquad w(a) = 0 \tag{d}$$

An earlier user of the Rayleigh–Ritz method discovered that each of the following polynomial functions is admissible:

$$\Omega_n(x) = (n + 2)\left(\frac{x}{a}\right)^{n+1} - (2n + 3)\left(\frac{x}{a}\right)^{n+2} + (n + 1)\left(\frac{x}{a}\right)^{n+3} \tag{e}$$

We can construct a solution of the form

$$w(x) = \{\Omega\}^T\{q\} \tag{f}$$

Let us use just a single term:

$$\Omega_1(x) = 3\left(\frac{x}{a}\right)^2 - 5\left(\frac{x}{a}\right)^3 + 2\left(\frac{x}{a}\right)^4 \tag{g}$$

It follows that

$$K_{11} = \int EI(\Omega_1'')^2\, dx = \frac{EI_0}{a^3}(7.2 + 3ca)$$

$$K_{11}^D = \int (\Omega_1')^2\, dx = \frac{0.343}{a} \tag{h}$$

$$Q_1 = \int f_0 \Omega_1\, dx = 0.15 f_0 a$$

Thus, since

$$([K_{11}] - P[K_{11}^D])q_1 = Q_1 \tag{i}$$

$$q_1 = \frac{f_0 a^4}{8EI_0(6 + 2.5ca) - 2.287Pa^2} \tag{j}$$

and, therefore,

$$w(x) = q_1 \Omega_1$$

$$= \frac{f_0 a^4}{8EI_0(6 + 2.5ca) - 2.287Pa^2}\left[3\left(\frac{x}{a}\right)^2 - 5\left(\frac{x}{a}\right)^3 + 2\left(\frac{x}{a}\right)^4\right] \tag{k}$$

The buckling load may be found from

$$|K_{11} - PK_{11}^D| = 0 \tag{l}$$

which is

$$\frac{EI_0}{a^3}(7.2 + 3ca) - 0.343\frac{P}{a} = 0 \tag{m}$$

or

$$P_{cr} = \frac{EI_0(7.2 + 3ca)}{0.343a^2} \tag{n}$$

and the buckling mode is

$$w(x) = q_1\left[3\left(\frac{x}{a}\right)^2 - 5\left(\frac{x}{a}\right)^3 + 2\left(\frac{x}{a}\right)^4\right] \tag{o}$$

We can estimate the accuracy of this answer by comparing it with the exact answer for the uniform ($c = 0$) column. The exact answer from solving the differential equation is

$$P_{cr} = 20.19\frac{EI}{a^2} \tag{p}$$

From equation (n) for a uniform column,

$$P_{cr} = 20.99\frac{EI}{a^2} \tag{q}$$

The answer can be improved by choosing more terms in equation (g).

Sec. 5.2   Buckling of Slender Bars

5.2.6 Stationary Potential Energy

We can also write the potential energy expression for the beam-column as

$$\Pi = (U + \Delta U) - W_e$$

$$= \frac{1}{2} \int EI \frac{\partial^2 \delta w}{\partial x^2} \frac{\partial^2 w}{\partial x^2} \, dx - \frac{P}{2} \int (w')^2 \, dx \tag{5.2.62}$$

$$- \int wf(x) \, dx - \Sigma F_{zi} w_i - \Sigma M_{yj}(w_j)'$$

Minimization of the potential energy proceeds as follows:

$$\frac{\partial \Pi}{\partial q_i} = 0 \tag{5.2.63}$$

produces the matrix equation

$$([K] + [K^d])\{q\} = \{Q\} \tag{5.2.64}$$

which is identical to Equation 5.2.54 obtained from virtual work.

We shall now show that the potential energy is a minimum only for the stable configuration. Furthermore, for the unstable configuration, it is a maximum, and in the borderline case between stability and instability, it has a stationary value. We shall use an argument similar to that in Section 3.3, which established the minimum principle for stable structures.

Consider two kinematically admissible functions, one the true displacement $w(x)$ and another, $w(x) + \Delta w(x)$, which differs from the true by a small amount. Now denote the potential energy for the true displacement by Π and the other by Π_0. We shall drop the concentrated loads and moments for convenience, but the result would be the same if they were included. Then

$$\Pi_0 - \Pi = \frac{1}{2} \int EI(w'' + \Delta w'')^2 \, dx - \frac{P}{2} \int (w' + \Delta w')^2 \, dx$$

$$- \int f_z(w + \Delta w) \, dx - \frac{1}{2} \int EI(w'')^2 \, dx + \frac{P}{2} \int (w')^2 \, dx + \int f_z w \, dx$$

$$= \int EIw''\Delta w'' \, dx - P \int w'\Delta w'fx - \int f_z \, \Delta w \, dx \tag{5.2.65}$$

$$+ \frac{1}{2} \int EI(\Delta w'')^2 \, dx - \frac{P}{2} \int (\Delta w)^2 \, dx$$

Since Δw is a small kinematically admissible function, it satisfies all the conditions of a virtual displacement, and we may say

$$\delta W = \int EIw''\Delta w'' \, dx - P \int w'\Delta w' \, dx - \int f_z\Delta w \, dx = 0 \tag{5.2.66}$$

and, therefore,

$$\Pi_0 - \Pi = \frac{1}{2} \int EI(\Delta w'')^2 \, dx - \frac{P}{2} \int (\Delta w)^2 \, dx \tag{5.2.67}$$

If P is sufficiently small, $\Pi_0 - \Pi > 0$, which proves that the potential energy is a minimum. Compare this with the earlier proof in Section 3.3. If P is sufficiently large, Π_0

$- \Pi < 0$, indicating the potential energy is a maximum. There is an intermediate value of P for which $\Pi_0 - \Pi = 0$, signaling the transition from minimum to maximum where the potential energy is stationary. It can be shown that this value of P is P_{cr}. In summary,

$$\Pi_0 - \Pi = \begin{cases} > 0 \text{ if } P < P_{cr} \\ = 0 \text{ if } P = P_{cr} \\ < 0 \text{ if } P < P_{cr} \end{cases} \qquad (5.2.68)$$

The steps for finding a stationary value are the same as for finding a minimum, as we have seen in Example 5.2.4.

5.2.7 Finite Element Method

The finite element method, which in the form given here is a special case of the Rayleigh–Ritz method, can be adapted for problems in instability much as we have adapted the traditional Rayleigh–Ritz method in the previous section. We need only recognize that we can represent the distributed displacement of the beam $w(x)$ in terms of shape functions $[N]$ and the nodal displacements $\{r\}$, or

$$w(x) = [N]\{r\} \qquad (5.2.69)$$

and add a contribution to the equations of the form

$$[K^d]\{r\} = -P \int [N']^T[N']\{r\} \, dx \qquad (5.2.70)$$

These are precisely the same shape functions used for stable static stress and deflection analysis in Chapter 4. The shape and material of the structure and its constraints have not changed, only its loading. The algebraic matrix equation for the structure now takes on the form

$$([K] + [K^d])\{r\} = \{R\} \qquad (5.2.71)$$

where

$$[K] = \int [B]^T[G] \, [B] \, dV = \int EI \, [N'']^T[N''] \, dx \qquad (5.2.72)$$

and

$$\{R\} = \int [N]^T f(x) \, dx + \Sigma \, [N(x_i)]^T F_{zi} + \Sigma \, [N'(x_j)]^T M_{yj} \qquad (5.2.73)$$

The matrix $[K^d]$, as in the case of the Rayleigh–Ritz method, has been given the same variety of names, chiefly, the *differential stiffness,* the *geometric stiffness,* or the *initial stress* matrix. We shall adopt the name differential stiffness for both this form and the traditional Rayleigh–Ritz form.

Equation 5.2.71 is the matrix equation for a beam-column. When the set of axial internal forces is below that which causes buckling the structure is statically stable, and Equation 5.2.71 can be solved by the methods described in Chapters 3 and 4. To find the set of internal forces that causes buckling, first set the lateral loads to zero.

$$([K] + [K^d])\{r\} = \{0\} \qquad (5.2.74)$$

This defines the matrix eigenvalue problem for buckling. We find the critical loads by setting

$$| [K] + [K^d] | = 0 \tag{5.2.75}$$

In the simpler problems, such as columns with uniform axial internal forces, we look for the lowest value of the internal stress resultant P, which is contained in the matrix $[K^d]$, that will cause the determinant to be zero. This is the critical or buckling load. Often the factor $-P$ is factored from $[K^d]$, and Equation 5.2.74 is rewritten in the form

$$([K] - P[K^D])\{r\} = \{0\} \tag{5.2.76}$$

where

$$[K^D] = \int [N']^T [N'] \, dx \tag{5.2.77}$$

Thus,

$$| [K] - P[K^D] | = 0 \tag{5.2.78}$$

This is the more familiar form with the quantity P, which becomes the eigenvalue, shown explicitly. As in the traditional Rayleigh–Ritz method, it may be seen that expansion of this determinant will produce a polynomial in P, and the lowest root of this polynomial equation is the critical value of P. We shall see that in more complicated problems P cannot be factored from the $[K^D]$ matrix since the axial compressive internal stress resultant will differ from element to element. We shall show an example of this shortly and show how the eigenvalue is extracted.

As with the beam using FEM, we develop the shape functions, the stiffness matrix, the differential stiffness matrix, and the equivalent nodal loads by working with individual elements, and then we assemble the equations of the elements to provide the equations for the whole structure. The element shape functions $[n]$ given in Equation 4.2.23 are perfectly good for our purposes here and are repeated:

$$[n] = \frac{1}{L^3} [L^3 - 3L_s^2 + 2s^3, \; L^3 s - 2L^2 s^2 + Ls^3, \; 3Ls^2 - 2s^3, \\ -L^2 s^2 + Ls^3] \tag{5.2.79}$$

where s is measured from a node at one end of the element and is positive in the direction of the other node. Each of these shape functions is plotted in Figure 4.4.3. The element stiffness matrix is given in Equation 4.4.28 to be

$$[k]_i = \frac{EI}{L_i^3} \begin{bmatrix} 12 & -6L_i & -12 & -6L_i \\ -6L_i & 4L_i^2 & 6L_i & 2L_i^2 \\ -12 & 6L_i & 12 & 6L_i \\ -6L_i & 2L_i^2 & 6L_i & 4L_i^2 \end{bmatrix} \tag{5.2.80}$$

Inserting the shape functions into the integral for the element differential stiffness matrix gives us

$$[k^D]_i = \int_0^{L_i} [n']_i^T [n]_i \, ds = \frac{1}{30L_i} \begin{bmatrix} 36 & -3L_i & -36 & -3L_i \\ -3L_i & 4L_i^2 & 3L_i & -L_i^2 \\ -36 & 3L_i & 36 & 3L_i \\ -3L_i & -L_i^2 & 3L_i & 4L_i^2 \end{bmatrix} \tag{5.2.81}$$

For our first example of FEM applied to a buckling problem, let us consider a simple fixed ended column.

Structural Instability   Chap. 5

Example 5.2.6

Find the buckling load by the FEM for the column shown in figure (a).

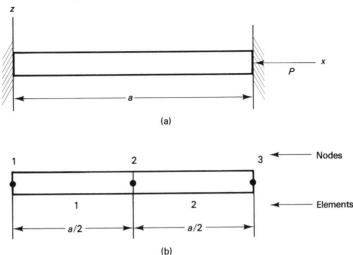

(a)

(b)

Solution: Let us divide the column into just two elements of equal length and number the nodes and elements as shown in figure (b). The assembled stiffness and differential stiffness matrices are then

$$
[K] - P[K^D] = \frac{EI}{\left(\frac{a}{2}\right)^3}
\begin{bmatrix}
12 & -6\left(\frac{a}{2}\right) & -12 & -6\left(\frac{a}{2}\right) & 0 & 0 \\
-6\left(\frac{a}{2}\right) & 4\left(\frac{a}{2}\right)^2 & 6\left(\frac{a}{2}\right) & 2\left(\frac{a}{2}\right)^2 & 0 & 0 \\
-12 & 6\left(\frac{a}{2}\right) & 24 & 0 & -12 & -6\left(\frac{a}{2}\right) \\
-6\left(\frac{a}{2}\right) & 2\left(\frac{a}{2}\right)^2 & 0 & 8\left(\frac{a}{2}\right)^2 & 6\left(\frac{a}{2}\right) & 2\left(\frac{a}{2}\right)^2 \\
0 & 0 & -12 & 6\left(\frac{a}{2}\right) & 12 & 6\left(\frac{a}{2}\right) \\
0 & 0 & -6\left(\frac{a}{2}\right) & 2\left(\frac{a}{2}\right)^2 & 6\left(\frac{a}{2}\right) & 4\left(\frac{a}{2}\right)^2
\end{bmatrix}
$$

$$
- \frac{P}{30\left(\frac{a}{2}\right)}
\begin{bmatrix}
36 & -3\left(\frac{a}{2}\right) & -36 & -3\left(\frac{a}{2}\right) & 0 & 0 \\
-3\left(\frac{a}{2}\right) & 4\left(\frac{a}{2}\right)^2 & 3\left(\frac{a}{2}\right) & -\left(\frac{a}{2}\right)^2 & 0 & 0 \\
-36 & 3\left(\frac{a}{2}\right) & 72 & 0 & -36 & -3\left(\frac{a}{2}\right) \\
-3\left(\frac{a}{2}\right) & -\left(\frac{a}{2}\right)^2 & 0 & 8\left(\frac{a}{2}\right)^2 & 3\left(\frac{a}{2}\right) & -\left(\frac{a}{2}\right)^2 \\
0 & 0 & -36 & 3\left(\frac{a}{2}\right) & 36 & 3\left(\frac{a}{2}\right) \\
0 & 0 & -3\left(\frac{a}{2}\right) & -\left(\frac{a}{2}\right)^2 & 3\left(\frac{a}{2}\right) & 4\left(\frac{a}{2}\right)^2
\end{bmatrix}
\qquad (a)
$$

The boundary conditions are

$$w_1 = \theta_1 = w_3 = \theta_3 = 0 \tag{b}$$

This eliminates the first two and last two equations and the first and last two columns of both stiffness matrices, leaving

$$([K] - P[K^D])\{r\} = \left(\begin{bmatrix} 24 & 0 \\ 0 & 8\left(\frac{a}{2}\right)^2 \end{bmatrix} - \beta \begin{bmatrix} 72 & 0 \\ 0 & 8\left(\frac{a}{2}\right)^2 \end{bmatrix}\right) \begin{bmatrix} w_2 \\ \theta_2 \end{bmatrix} = 0 \tag{c}$$

where

$$\beta = \frac{P}{30EI}\left(\frac{a}{2}\right)^2 \tag{d}$$

These are, actually, two independent equations

$$(24 - 72\beta)w_2 = 0, \qquad 8\left(\frac{a}{2}\right)^2(1 - \beta)\theta_2 = 0 \tag{e}$$

Thus, the two roots are

$$24 - 72\beta = 0 \longrightarrow \beta_1 = \frac{1}{3} = \frac{P}{30EI}\left(\frac{a}{2}\right)^2$$

$$1 - \beta = 0 \longrightarrow \beta_2 = 1 = \frac{P}{30EI}\left(\frac{a}{2}\right)^2 \tag{f}$$

or

$$P_1 = \frac{40EI}{a^2}, \qquad P_2 = \frac{120EI}{a^2} \tag{g}$$

Alternatively, we could have set the determinant of the coefficient matrix to zero or

$$|[K] - P[K^D]| = \left|\begin{bmatrix} 24 & 0 \\ 0 & 8\left(\frac{a}{2}\right)^2 \end{bmatrix} - \beta \begin{bmatrix} 72 & 0 \\ 0 & 8\left(\frac{a}{2}\right)^2 \end{bmatrix}\right| = 0 \tag{h}$$

When this determinant is expanded, we obtain

$$(24 - 72\beta)\left[8\left(\frac{a}{2}\right)^2(1 - \beta)\right] = 0 \tag{i}$$

This leads to the same two roots for P. The lowest root is designated P_{cr}. This may be compared with the exact value obtained in Example 5.2.2 of

$$P_{cr} = \frac{39.478EI}{a^2} \tag{j}$$

That is a quite good answer for such a crude FEM model.

To find the mode shape for the critical load, we have

$$(24 - 72\beta_1)w_2 = 0$$

$$(1 - \beta_1)\theta_2 = 0 \tag{k}$$

From this, $\theta_2 = 0$ and $w_2 \neq 0$. Thus,

$$w(s) = [n]_1 \begin{bmatrix} 0 \\ 0 \\ w_2 \\ 0 \end{bmatrix} = w_2\left(12\,\frac{s^2}{a^2} - 16\,\frac{s^3}{a_3}\right), \qquad \text{element 1} \tag{l}$$

$$w(s) = [n]_2 \begin{bmatrix} w_2 \\ 0 \\ 0 \\ 0 \end{bmatrix} = w_2\left(1 - 12\,\frac{s^2}{a^2} + 16\,\frac{s^3}{a_3}\right), \qquad \text{element 2}$$

Note that equations (a) can be used to solve many different boundary conditions. If both ends are simply supported,

$$w_1 = w_3 = 0 \tag{m}$$

and appropriate rows and columns of the equation are eliminated, and the remaining equations are solved for the buckling load. As an exercise, think of several more boundary conditions served by the same set of equations.

For problems with many degrees of freedom, the extraction of the eigenvalues is not done easily by expanding the determinant. Several methods are available and are discussed in References 23 and 24, among others. A program to extract both eigenvalues and eigenfunctions from equations of the form of Equations 5.2.77 is given in Appendix L. Note that this program is set up to serve us for eigenvalue problems in vibration as well as buckling, and so we will be using a subset of the program at this time. It is based on a method of matrix iteration often called the *power method* or the *inverse power method*, depending on the way the equations are organized. We shall briefly describe how this program solves the buckling problem in the following paragraphs.

The method we are about to describe always finds the highest eigenvalue first and then each succeeding one in descending order. If we divide by $-P$ and get

$$\left([K^D] - \frac{1}{P}\,[K]\right)\{r\} = ([K^D] - \psi[K])\,\{r\} = \{0\} \tag{5.2.82}$$

Thus, the highest eigenvalue for ψ will be the lowest for P.

The iteration method proceeds by writing the equation in the following form:

$$[K]^{-1}\,[K^D]\,\{r\} = \psi\,\{r\} \longrightarrow [Y]\,\{r\} = \psi\,\{r\} \tag{5.2.83}$$

We start with an assumed trial value $\{r\} = \{a\}^0$ and calculate a first approximation from Equations 5.2.83 as follows:

$$[Y]\,\{a\}^0 = \{C\}^1 = \psi^1\,\{a\}^1 \tag{5.2.84}$$

Since $[Y]$ is known and an assumed trial value $\{a\}^0$ is known, we can calculate $[C]^1$; then we normalize $\{C\}^1$, or extract a factor ψ^1 from $\{C\}^1$ so that the value of one of the elements of $\{a\}^1$ is equal to the value of the element of $\{a\}^0$ in a corresponding row. Note that we are using superscripts to denote different trial values; they are not exponents.

This is repeated several times until we converge on values of ψ and $\{a\}$. These are the first eigenvalue and first eigenfunction, respectively. The recursion formula for this operation is

$$[Y]\{a\}^i = \{C\}^{i+1} = \psi^{i+1}\{a\}^{i+1} \tag{5.2.85}$$

Perhaps an example will help understanding.

Example 5.2.7

Find the first eigenvalue and eigenfunction of the set of matrix equations given.

$$\begin{bmatrix} 2 & 4 \\ 4 & 6 \end{bmatrix} \begin{bmatrix} a_1 \\ a_2 \end{bmatrix} = \psi \begin{bmatrix} a_1 \\ a_2 \end{bmatrix} \tag{a}$$

Solution: The first trial value selected is shown, and the first approximation or second trial value is calculated.

$$\begin{bmatrix} 2 & 4 \\ 4 & 6 \end{bmatrix} \begin{bmatrix} 1.0 \\ 1.0 \end{bmatrix} = \begin{bmatrix} 6.0 \\ 10.0 \end{bmatrix} = 10.0 \begin{bmatrix} 0.6 \\ 1.0 \end{bmatrix} \tag{b}$$

Now repeat to find the next approximation.

$$\begin{bmatrix} 2 & 4 \\ 4 & 6 \end{bmatrix} \begin{bmatrix} 0.6 \\ 1.0 \end{bmatrix} = \begin{bmatrix} 5.2 \\ 8.4 \end{bmatrix} = 8.4 \begin{bmatrix} 0.619 \\ 1.0 \end{bmatrix} \tag{c}$$

And repeat again.

$$\begin{bmatrix} 2 & 4 \\ 4 & 6 \end{bmatrix} \begin{bmatrix} 0.619 \\ 1.0 \end{bmatrix} = \begin{bmatrix} 5.238 \\ 8.476 \end{bmatrix} = 8.476 \begin{bmatrix} 0.618 \\ 1.0 \end{bmatrix} \tag{d}$$

And again.

$$\begin{bmatrix} 2 & 4 \\ 4 & 6 \end{bmatrix} \begin{bmatrix} 0.618 \\ 1.0 \end{bmatrix} = \begin{bmatrix} 5.236 \\ 8.472 \end{bmatrix} = 8.472 \begin{bmatrix} 0.618 \\ 1.0 \end{bmatrix} \tag{e}$$

And we see that we have convergence. Note that this is the highest of the two eigenvalues for this problem.

The more important role of FEM in column and beam-column analysis is to solve problems that are difficult to solve by other methods. Columns and beam-columns with variable cross sections, multiple intermediate support, nonuniform loading including varying axial load, and so on, are the problems best served by FEM. In the next example, we shall consider such a case.

Example 5.2.8

Find the buckling load for the uniform beam shown in figure (a).

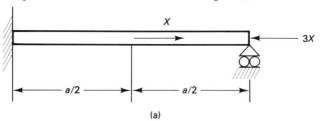

(a)

Solution: The same element matrices as used in Example 5.2.6 can be used here. In fact, the assembled stiffness matrix is identical; however, the assembled differential stiffness matrix differs. Remember that two sets of equations must be solved to find the buckling load. First, we must find the axial loads in the structure. In this case, we can easily see that in element 1 the internal axial stress resultant is a compressive force of $2X$ and in element 2 it is a compressive force of $3X$. Thus, in assembling the differential stiffness matrix, we must let

$$[k^d]_i = \frac{P_i}{30L_i} \begin{bmatrix} 36 & -3L_i & -36 & -3L_i \\ -3L_i & 4L_i^2 & 3L_i & -L_i^2 \\ -36 & 3L_i & 36 & 3L_i \\ -3L_i & -L_i^2 & 3L_i & 4L_i^2 \end{bmatrix} \tag{a}$$

where P_i is the axial compressive stress resultant in element i.

$$P_1 = 2X, \qquad P_2 = 3X, \qquad L_1 = L_2 = \frac{a}{2} \tag{b}$$

Thus,

$$[K] + [K^d] = \frac{EI}{\left(\frac{a}{2}\right)^3} \begin{bmatrix} 12 & -6\left(\frac{a}{2}\right) & -12 & -6\left(\frac{a}{2}\right) & 0 & 0 \\ -6\left(\frac{a}{2}\right) & 4\left(\frac{a}{2}\right)^2 & 6\left(\frac{a}{2}\right) & 2\left(\frac{a}{2}\right)^2 & 0 & 0 \\ -12 & 6\left(\frac{a}{2}\right) & 24 & 0 & -12 & -6\left(\frac{a}{2}\right) \\ -6\left(\frac{a}{2}\right) & 2\left(\frac{a}{2}\right)^2 & 0 & 8\left(\frac{a}{2}\right)^2 & 6\left(\frac{a}{2}\right) & 2\left(\frac{a}{2}\right)^2 \\ 0 & 0 & -12 & 6\left(\frac{a}{2}\right) & 12 & 6\left(\frac{a}{2}\right) \\ 0 & 0 & -6\left(\frac{a}{2}\right) & 2\left(\frac{a}{2}\right)^2 & 6\left(\frac{a}{2}\right) & 4\left(\frac{a}{2}\right)^2 \end{bmatrix}$$

$$- \frac{X}{30\left(\frac{a}{2}\right)^3} \begin{bmatrix} 72 & -6\left(\frac{a}{2}\right) & -72 & -6\left(\frac{a}{2}\right) & 0 & 0 \\ -6\left(\frac{a}{2}\right) & 8\left(\frac{a}{2}\right)^2 & 6\left(\frac{a}{2}\right) & -2\left(\frac{a}{2}\right)^2 & 0 & 0 \\ -72 & 6\left(\frac{a}{2}\right) & 180 & 3\left(\frac{a}{2}\right) & -108 & -9\left(\frac{a}{2}\right) \\ -6\left(\frac{a}{2}\right) & -2\left(\frac{a}{2}\right)^2 & 3\left(\frac{a}{2}\right) & 20\left(\frac{a}{2}\right)^2 & 9\left(\frac{a}{2}\right) & -3\left(\frac{a}{2}\right)^2 \\ 0 & 0 & -108 & 9\left(\frac{a}{2}\right) & 108 & 9\left(\frac{a}{2}\right) \\ 0 & 0 & -9\left(\frac{a}{2}\right) & -3\left(\frac{a}{2}\right)^2 & 9\left(\frac{a}{2}\right) & 12\left(\frac{a}{2}\right)^2 \end{bmatrix} \tag{c}$$

The boundary conditions are

$$w_1 = \theta_1 = w_3 = 0 \tag{d}$$

and the remaining equations are

$$\left\{ \frac{EI}{\left(\frac{a}{2}\right)^2} \begin{bmatrix} 24 & 0 & -6\left(\frac{a}{2}\right) \\ 0 & 8\left(\frac{a}{2}\right)^2 & 2\left(\frac{a}{2}\right)^2 \\ -6\left(\frac{a}{2}\right) & 2\left(\frac{a}{2}\right)^2 & 4\left(\frac{a}{2}\right)^2 \end{bmatrix} - \frac{X}{30\left(\frac{a}{2}\right)} \begin{bmatrix} 180 & 3\left(\frac{a}{2}\right) & -9\left(\frac{a}{2}\right) \\ 3\left(\frac{a}{2}\right) & 20\left(\frac{a}{2}\right)^2 & -3\left(\frac{a}{2}\right)^2 \\ -9\left(\frac{a}{2}\right) & -3\left(\frac{a}{2}\right)^2 & 12\left(\frac{a}{2}\right)^2 \end{bmatrix} \right\} \begin{bmatrix} w_2 \\ \theta_2 \\ \theta_3 \end{bmatrix} = \begin{bmatrix} 0 \\ 0 \\ 0 \end{bmatrix} \tag{e}$$

If we let

$$\beta = \frac{X(a/2)^2}{30EI} \tag{f}$$

the determinant of the coefficients becomes

$$\left| \begin{bmatrix} 24 & 0 & -6\left(\dfrac{a}{2}\right) \\ 0 & 8\left(\dfrac{a}{2}\right)^2 & 2\left(\dfrac{a}{2}\right)^2 \\ -6\left(\dfrac{a}{2}\right) & 2\left(\dfrac{a}{2}\right)^2 & 4\left(\dfrac{a}{2}\right)^2 \end{bmatrix} - \beta \begin{bmatrix} 180 & 3\left(\dfrac{a}{2}\right) & -9\left(\dfrac{a}{2}\right) \\ 3\left(\dfrac{a}{2}\right) & 20\left(\dfrac{a}{2}\right)^2 & -3\left(\dfrac{a}{2}\right)^2 \\ -9\left(\dfrac{a}{2}\right) & -3\left(\dfrac{a}{2}\right)^2 & 12\left(\dfrac{a}{2}\right)^2 \end{bmatrix} \right| = 0 \tag{g}$$

which may be solved for the lowest root of β and hence X. This becomes a bit of a tedious task; however, with computer codes now available, a numerical solution is readily obtained. For example, let us select an aluminum column with a length of 1000 mm. The matrix equation then becomes

$$\left(\begin{bmatrix} 24 & 0 & -3\text{ E3} \\ 0 & 2\text{ E6} & 5\text{ E5} \\ -3\text{ E3} & 5\text{ E5} & 1\text{ E6} \end{bmatrix} - \beta \begin{bmatrix} 180 & 1500 & -4500 \\ 1500 & 5\text{ E6} & -75\text{ E4} \\ -4500 & -75\text{ E4} & 3\text{ E6} \end{bmatrix} \right) \begin{bmatrix} w_2 \\ \theta_2 \\ \theta_3 \end{bmatrix} = \begin{bmatrix} 0 \\ 0 \\ 0 \end{bmatrix} \tag{h}$$

where

$$\beta = \frac{25000X}{3EI} \tag{i}$$

This problem may be solved by the method discussed in Appendix L. The critical load and mode found by that method are

$$X_{\text{cr}} = 7.956, \qquad \{r\} = \begin{bmatrix} w_2 \\ \theta_2 \\ \theta_3 \end{bmatrix} = w_2 \begin{bmatrix} 1.000000 \\ 0.001513 \\ -0.004410 \end{bmatrix} \tag{j}$$

This is the mode in terms of the nodal displacements. The distributed displacements are

$$w(x)_i = [n]_i \{r\}_i \tag{k}$$

Still better is to go directly to one of the FEM computer codes that does the assembly of the elements as well as the solution. This is done in the next two examples. Because, in general, the axial stress resultant in any element may vary from that in other elements, all the following problems require, first, a static stable stress analysis of the structure under a given small load system to determine the P_i's in each and every element. This load system is then ratioed up equally until the internal stress resultants produce buckling. Perhaps this is best understood by example. In this example, we have a tapered beam, multiple loads, and multiple constraints that produce a problem that would be very time consuming to solve by the other methods.

Example 5.2.9

Set up the problem for finding the buckling load for the tapered beam structure from Example 4.4.3 using a FEM computer code. The beam has exactly the same geometry, but different constraints. It is loaded with three axial loads: $-2X$ at $x = a/4$, X at $x = 3a/4$, and $-4X$ at $x = a$, as shown.

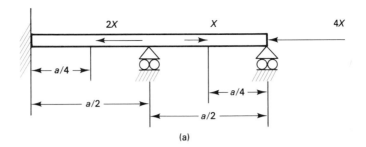

(a)

Solution: The dimensions from Example 4.4.3 are

$$a = 4000 \text{ mm}, \qquad b(0) = 150 \text{ mm}, \qquad c = 100 \text{ mm} \qquad \text{(a)}$$

Now let us set up the problem by identifying the several items of input data. We shall use the same division into elements, numbering of nodes and elements, nodal coordinates, material properties, and so on, as in Example 4.3.1. Items 1 through 5 in that example are taken unchanged. Item 6, the constraints, now include all six degrees of freedom at node 1 and the vertical displacement at nodes 3 and 5. We shall further restrict displacement to the x-z plane.

Item 7, the loading, will be different. We are looking for the critical loads; thus, the loads are not known. To proceed in the popular FEM code NASTRAN, for example, we would now specify a small reference load, well below the anticipated buckling load, in proper proportion at each node. For example,

7. Specify reference nodal loads.

Node	R_x	R_y	R_z
1	0.0	0.0	0.0
2	−2.0	0.0	0.0
3	0.0	0.0	0.0
4	1.0	0.0	0.0
5	−4.0	0.0	0.0

The first solution obtained is for the static stress analysis of the structure for this set of loads. From this analysis, the internal axial stress resultants are found for each element. Then the eigenvalue equation solver is invoked to find the value of internal stress resultants that cause the structure to buckle. This is reported in terms of multiples of the internal loads caused by the reference loads selected in item 7. This problem has been executed on MSC/NASTRAN, and the input file is listed in Appendix M. From the output, we note that the internal loads in each element from the reference loads are

Element	Internal Load
1	−5.0
2	−3.0
3	−3.0
4	−4.0

The eigenvalue solver has found that

$$X_{cr} = 517,907 \text{ N},$$

$$w_1 = 0.0$$
$$\theta_1 = 0.0$$
$$w_2 = -0.3107900$$
$$\theta_2 = 0.0002877$$
$$w_3 = 0.0$$
$$\theta_3 = -0.0009697$$
$$w_4 = 1.0$$
$$\theta_4 = -0.00037323$$
$$w_5 = 0.0$$
$$\theta_5 = 0.00185771$$

(b)

The internal loads in each element at the moment of buckling are

Element	Internal Load
1	$-2.590 \cdot 10^6$
2	$-1.554 \cdot 10^6$
3	$-1.554 \cdot 10^6$
4	$-2.072 \cdot 10^6$

Each of these loads contributes to the differential stiffness matrix.

The finite element method is particularly helpful for finding buckling loads for assemblies of beams called *frames*. The next example shows a simple frame with concentrated applied loads. In all the examples up to now, the axial internal stress resultants that cause the buckling are independent of the lateral loads. The homogeneous eigenvalue equation is obtained by setting the lateral loads on the beam-column to zero. The external axial loads that give rise to axial internal loads, which are a part of the homogeneous equations, are then increased, thus increasing the effect of the differential stiffness matrix until buckling occurs. In frame analysis, we shall see that lateral loads on one member can affect the axial internal stress resultants in another; thus, the axial internal stress resultants are no longer independent of the lateral loads.

In this case, we set up the full equations including all loads, find the internal loads in each element for a reference set of external loads, and then solve for the buckling load and mode in terms of multiples of the reference load. The next two examples should help make this clear.

Example 5.2.10

Set up the problem for a finite element solution using a FEM computer code for the buckling of the frame shown in figure (a) under the load case given.

Solution: First, the internal axial loads in each element of the frame are found. Because the applied loads cause compressive axial loads in at least one of the members of the frame, the

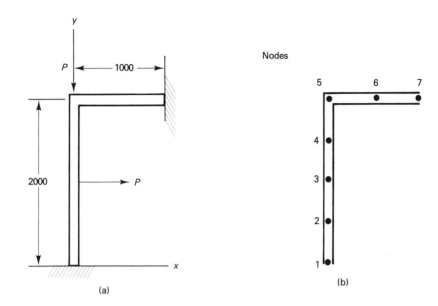

(a)

(b)

possibility of buckling occurs. Once these internal loads are found, the eigenvalue solver is invoked and the buckling load and mode are found.

The following seven steps are taken to prepare the FEM model for analysis:

1. Divide the structure into suitable elements and number the nodes and elements. Let us use six elements of equal length as shown in figure (b).

2. Record the coordinates of each node. They are

Node	x	y	z
1	0.0	0.0	0.0
2	0.0	500.0	0.0
3	0.0	1000.0	0.0
4	0.0	1500.0	0.0
5	0.0	2000.0	0.0
6	500.0	2000.0	0.0
7	1000.0	2000.0	0.0

3. Identify the nodes that go with each element.

Element	Nodes
1	1, 2
2	2, 3
3	3, 4
4	4, 5
5	5, 6
6	6, 7

4. Specify the material properties for all elements. Let us use steel, for which

$$E = 206,800.0 \text{ MPa}$$

$$G = 79,550.0 \text{ MPa}$$

$$v = 0.3$$

5. Specify the geometric properties of each cross section. Let us use a solid square cross section of 24 mm on a side. Then

$$A = 576 \text{ mm}^2$$

$$I_1 = I_2 = 27,648 \text{ mm}^4$$

If we choose the four corners of the cross section for stress output, the coordinates are ± 12 at each corner.

6. Specify geometric constraints. All degrees of freedom are constrained at nodes 1 and 7. The frame is also constrained to deflect only in the *x-y* plane.

7. Specify equivalent nodal loads. We shall select unit loads for easy interpretation of the eigenvalues.

Node	R_x	R_y	R_z
3	1.0	0.0	0.0
5	0.0	−1.0	0.0

This problem has been run on MSC/NASTRAN; the input file is listed in Appendix M, and output data on stress, deflections, and buckling loads have been obtained. They are summarized next.

The NASTRAN solution reports an eigenvalue of 12,078. Given the reference loads used, this provides for a buckling load of

$$P_{cr} = 12,077 \text{ N}$$

The buckling mode in terms of the nodal displacements is given in the following table:

Node	u_i	v_i	φ_i
1	0.0	0.0	0.0
2	0.04406	−0.00030	−0.0014439
3	1.0	−0.00026	−0.0003943
4	0.70235	−0.00039	0.0013998
5	−0.00002	−0.00052	0.0009485
6	−0.00001	0.12566	−0.0002448
7	0.0	0.0	0.0

A plot of the mode from the NASTRAN plotter is shown in figure (c). The plotter only connects straight lines between nodal displacements and therefore does not show correct rotations at the nodes; nevertheless, a measure of understanding of the mode shape is given. The actual mode shape is a set of cubic polynomials between nodes, as we have noted before.

(c)

It is instructive to study the internal axial forces in the elements when the external loads reach critical values, since they are the quantities that fixed the magnitude of the differential stiffness matrix. They are, in fact, the P_i terms in each $[k^d]_i$. For this particular external load set, the internal loads at the onset of buckling are as follows:

Element	Axial Force
1	−15,088
2	−15,088
3	−15,088
4	−15,088
5	−5,281
6	−5,281

It is these compressive axial forces that cause the frame to buckle.

For comparison, the problem was repeated without the horizontal load at node 3, but with the vertical load at node 5. In this case, the buckling load is

$$P_{cr} = 15,347$$

which is a 21% increase over the value obtained previously. The mode shape in terms of the nodal displacements is

Node	u_i	v_i	φ_i
1	0.0	0.0	0.0
2	0.44345	−0.00013	−0.0014503
3	1.0	−0.00026	−0.0003740
4	0.69111	−0.00038	0.0014111
5	−0.00002	−0.00051	0.0008898
6	−0.00001	0.11098	−0.0002217
7	0.0	0.0	0.0

Note how little the mode shape has changed; in fact, there is not enough change to need to repeat the plot. The difference in buckling load can be attributed to the different values of the differential stiffness resulting from the different axial forces at buckling in elements 5 and 6. These are listed next.

Element	Axial Force
1	−15,338
2	−15,338
3	−15,338
4	−15,338
5	−2
6	−2

The small values of the axial forces in elements 5 and 6 cause virtually no contribution to the differential stiffness from those elements. Note that the axial force in elements 1 through 4 is only slightly higher than in the other case.

One advantage of FEM is that the complexity of preparing a problem for solution by one of the commercial FEM computer codes does not go up proportionally to the complexity of the geometry, loading, and constraints, as is usually the case with exact and approximate analytical methods. It is clear why FEM is becoming the dominant analysis tool in instability analysis, as well as in the static stress and deflection analysis of stable structures.

5.3 BUCKLING OF PLATES

The next class of structures for which buckling is a major concern is thin, flat plates. We shall now examine them with all the same methods we used for columns and beam-columns.

5.3.1 Differential Equations for Buckling

The plate equations of Section 2.4 must be modified to account for the *membrane loads*, or loads with components in the *x-y* plane. We begin, once again, with the assumed deformations of Equations 2.4.1 and 2.4.2.

$$u = u_0 - z\frac{\partial w}{\partial x}, \qquad v = v_0 - z\frac{\partial w}{\partial y} \qquad (5.3.1)$$

but this time we retain the midplane displacements u_0 and v_0. The strains become

$$\epsilon_x = \frac{\partial u_0}{\partial x} - z\frac{\partial^2 w}{\partial x^2}, \qquad \epsilon_y = \frac{\partial v_0}{\partial y} - z\frac{\partial^2 w}{\partial y^2}$$

$$\gamma_{xy} = \frac{\partial u_0}{\partial y} + \frac{\partial v_0}{\partial x} - 2z\frac{\partial^2 w}{\partial x \, \partial y} \qquad (5.3.2)$$

The stresses are then

$$\sigma_x = \frac{E}{1 - v^2} (\epsilon_x + v\epsilon_y)$$

$$= \frac{E}{1 - v^2} \left(\frac{\partial u_0}{\partial x} + v \frac{\partial v_0}{\partial y} \right) - z \left(\frac{\partial^2 w}{\partial x^2} + v \frac{\partial^2 w}{\partial y^2} \right)$$

$$\sigma_y = \frac{E}{1 - v^2} (\epsilon_y + v\epsilon_x) \tag{5.3.3}$$

$$= \frac{E}{1 - v^2} \left(\frac{\partial v_0}{\partial y} + v \frac{\partial u_0}{\partial x} \right) - z \left(\frac{\partial^2 w}{\partial y^2} + v \frac{\partial^2 w}{\partial x^2} \right)$$

$$\tau_{xy} = G\gamma_{xy} = \frac{E}{2(1 + v)} \left(\frac{\partial u_0}{\partial y} + \frac{\partial v_0}{\partial x} - 2z \frac{\partial^2 w}{\partial x \, \partial y} \right)$$

It is convenient to work with stress resultants, such as the distributed moments defined in Equation 2.4.6, the distributed shear forces defined in Equation 2.4.8, and the distributed membrane forces defined next.

$$N_x = \int \sigma_x \, dz = \frac{Eh}{1 - v^2} \left(\frac{\partial u_0}{\partial x} + v \frac{\partial v_0}{\partial y} \right)$$

$$N_y = \int \sigma_y \, dz = \frac{Eh}{1 - v^2} \left(\frac{\partial v_0}{\partial y} + v \frac{\partial u_0}{\partial x} \right) \tag{5.3.4}$$

$$N_{xy} = \int \tau_{xy} \, dz = \frac{E}{2(1 + v)} \left(\frac{\partial u_0}{\partial y} + \frac{\partial v_0}{\partial x} \right)$$

Let us now consider the equilibrium of a small element acted on by the distributed forces and moments just defined. The membrane forces on an element are shown in Figure 5.3.1(a). Looking at the edge of the element along the y-axis in the x-z plane, as shown in Figure 5.3.1(b), we see that N changes in direction as well as magnitude if we look at the plate element in the deflected position. These are superimposed on the lateral forces and moments shown in Figure 2.4.3, which is repeated as Figure 5.3.2.

First, we note that the cosine of the angle of the forces N and $N + dN$ with respect to the x-y plane is small and therefore can be taken as 1. If we sum forces in the x- and y-directions, in the absence of in-plane body forces, we obtain

$$\frac{\partial N_x}{\partial x} + \frac{\partial N_{xy}}{\partial y} = 0, \qquad \frac{\partial N_{xy}}{\partial x} + \frac{\partial N_y}{\partial y} = 0 \tag{5.3.5}$$

These are identical to the equilibrium equations for plane stress, Equations 1.3.2, except they are expressed in terms of stress resultants, which have units of force per unit length. By eliminating u_0 and v_0 from Equations 5.3.4, we can obtain a third equation:

$$\frac{\partial^2}{\partial x^2} \left(\frac{N_y - vN_x}{h} \right) - 2(1 + v) \frac{\partial^2}{\partial x \, \partial y} \left(\frac{N_{xy}}{h} \right) + \frac{\partial^2}{\partial y^2} \left(\frac{N_x - vN_y}{h} \right) = 0 \tag{5.3.6}$$

This is identical to the compatibility equation for plane stress given in Equation 1.3.7, except it is expressed in terms of the stress resultants.

Sec. 5.3   Buckling of Plates

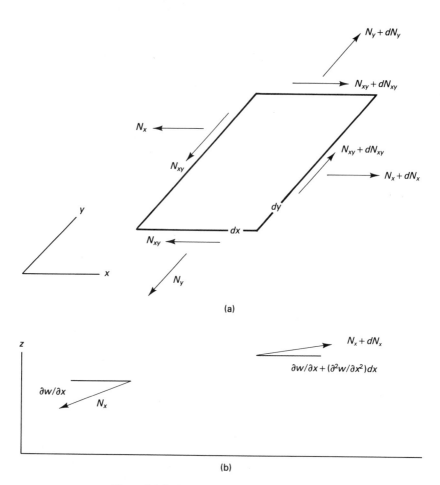

(a)

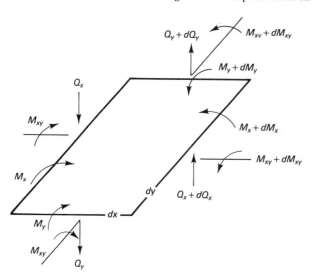

(b)

Figure 5.3.1 In-plane Forces on a Plate Element.

Figure 5.3.2 Plate Element in Bending and Shear.

Structural Instability   Chap. 5

Solving these three equations for the three unknown membrane stress resultants is the same as the plane stress problem described in Section 1.3.1, but is no small task if attempted analytically. Fortunately, one problem of major interest has been solved, that of uniform edge loads on a rectangular plate. See Example 1.3.1 for a solution with a uniform edge load in one coordinate direction. You can verify that any combination of constant values of normal and shear load on the edges will produce uniform stresses and stress resultants in the interior of the plate. We shall be using this fact in a moment.

Now we consider force equilibrium in the z-direction and moment equilibrium about the x- and y-axes. This time (see Equation 2.4.10) we look at the plate in the deflected position and include the effect of the in-plane loads. Summing forces in the z-direction, we note the contribution of the membrane forces. First, the z-components of N_x are

$$\left(N_x + \frac{\partial N_x}{\partial x}\, dx\right)\left(\frac{\partial w}{\partial x} + \frac{\partial^2 w}{\partial x^2}\, dx\right) dy - N_x \frac{\partial w}{\partial x}\, dy = N_x \frac{\partial w^2}{\partial x^2}\, dx\, dy$$
$$+ \frac{\partial N_x}{\partial x}\frac{\partial w}{\partial x}\, dx\, dy \tag{5.3.7}$$

Similarly, there are components of N_y forces

$$N_y \frac{\partial^2 w}{\partial x^2}\, dx\, dy + \frac{\partial N_y}{\partial y}\frac{\partial w}{\partial y}\, dx\, dy \tag{5.3.8}$$

and also components of N_{xy} forces

$$2N_{xy}\frac{\partial^2 w}{\partial x\, \partial y}\, dx\, dy + \frac{\partial N_{xy}}{\partial x}\frac{\partial w}{\partial y}\, dx\, dy + \frac{\partial N_{xy}}{\partial y}\frac{\partial w}{\partial x}\, dx\, dy \tag{5.3.9}$$

When these components are added to the other forces, already identified in Equations 2.4.10, we get

$$\frac{\partial Q_x}{\partial x} + \frac{\partial Q_y}{\partial y} + p + N_x \frac{\partial^2 w}{\partial x^2} + 2N_{xy}\frac{\partial^2 w}{\partial x\, \partial y} + N_y \frac{\partial^2 w}{\partial y^2}$$
$$\left(\frac{\partial N_x}{\partial x} + \frac{\partial N_{xy}}{\partial y}\right)\frac{\partial w}{\partial x} + \left(\frac{\partial N_{xy}}{\partial x} + \frac{\partial N_y}{\partial y}\right)\frac{\partial w}{\partial y} = 0 \tag{5.3.10}$$

With the help of equilibrium Equations 5.3.5, we get

$$\frac{\partial Q_x}{\partial x} + \frac{\partial Q_y}{\partial y} + p + N_x \frac{\partial^2 w}{\partial x^2} + 2N_{xy}\frac{\partial^2 w}{\partial x\, \partial y} + N_y \frac{\partial^2 w}{\partial y^2} = 0 \tag{5.3.11}$$

and then when we use Equations 2.4.11 we get, finally,

$$D\nabla^4 w = p + N_x \frac{\partial^2 w}{\partial x^2} + 2N_{xy}\frac{\partial^2 w}{\partial x\, \partial y} + N_y \frac{\partial^2 w}{\partial y^2} \tag{5.3.12}$$

where

$$D = \frac{Eh^3}{12(1 - v^2)} \tag{5.3.13}$$

All that remains to complete the mathematical model of the buckling of plates is to state the boundary conditions. These conditions for a simply supported, a fixed or clamped, and a free edge are given at the end of Section 2.4. As stated, the free edge must not be loaded. The other edges may be loaded or unloaded.

This elegant partial differential equation has few known exact analytical solutions. Before we turn to approximate methods, let us examine two of them: (1) a strip of an infinite plate, and (2) the rectangular, simply supported plate with uniform edge loads. We shall forego the problem of combined lateral and membrane loading for now and turn directly to the plate buckling problem.

Example 5.3.1

Consider a strip of a plate infinite in the $\pm y$-directions that lies between $y = 0$ and $y = 1$ and for which the edges $x = 0$ and $x = a$ are simply supported. There is no lateral load, and the two simply supported edges are loaded by a uniform compressive load, $N_x = -X$.

Solution: Equation 5.3.12 for such a plate reduces to

$$D \frac{d^4 w}{dx^4} + X \frac{d^2 w}{dx^2} = 0 \tag{a}$$

The boundary conditions are

$$w(0) = w(a) = 0, \qquad \frac{d^2 w(0)}{dx^2} = \frac{d^2 w(a)}{dx^2} = 0 \tag{b}$$

This has all the appearances of the simply supported column with D replacing EI and X replacing P; therefore, the eigenvalues are

$$X = \frac{m^2 \pi^2 D}{a^2} = \frac{m^2 \pi^2 E h^3}{12(1 - v^2)a^2} \tag{c}$$

The critical buckling load is

$$X_{cr} = \frac{\pi^2 E h^3}{12(1 - v^2)a^2} \tag{d}$$

Note that for a simply supported column with a rectangular cross section of unit width the critical buckling load is

$$P_{cr} = \frac{\pi^2 E h^3}{12a^2} \tag{e}$$

differing only by the absence of the $(1 - v^2)$ in the denominator. The corresponding eigenfunctions are

$$w_i(x, y) = A_i \sin \frac{i \pi x}{a} \tag{f}$$

the lowest of which is the critical buckling mode

$$w_1(x, y) = A_1 \sin \frac{\pi x}{a} \tag{g}$$

which is the same shape as for the simply supported column.

Example 5.3.2

Consider a rectangular plate simply supported on all four edges and loaded by uniform compressive edge loads on opposite pairs of edges as shown in figure (a). The two pairs of edge loads have values of X in the x-direction and Y in the y-direction.

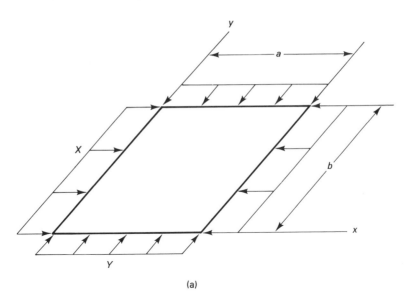

(a)

Solution: The governing equation is

$$D\nabla^4 w + X \frac{\partial^2 w}{\partial x^2} + Y \frac{\partial^2 w}{\partial y^2} = 0 \tag{a}$$

and the boundary conditions are

$$w(0, y) = w(a, y) = 0, \qquad \frac{\partial^2 w(0, y)}{\partial x^2} = \frac{\partial^2 w(a, y)}{\partial x^2} = 0$$

$$w(x, 0) = w(x, b) = 0 \qquad \frac{\partial^2 w(x, 0)}{\partial y^2} = \frac{\partial^2 w(x, b)}{\partial y^2} = 0 \tag{b}$$

Someone thought to try a solution of the following form:

$$w(x, y) = \Sigma\Sigma \, A_{ij} \sin\frac{i\pi x}{a} \sin\frac{j\pi y}{b} \tag{c}$$

Upon substitution of this into equation (a), we obtain

$$\Sigma\Sigma \left[D\pi^4 \left(\frac{i^2}{a^2} + \frac{j^2}{b^2} \right)^2 - X\pi^2 \frac{i^2}{a^2} - Y\pi^2 \frac{j^2}{b^2} \right] A_{ij} \sin\frac{i\pi x}{a} \sin\frac{j\pi y}{b} = 0 \tag{d}$$

One possible solution is $A_{ij} = 0$; however, this represents the trivial solution, $w(x, y) = 0$, and corresponds to equilibrium in the unbuckled state. Another possible solution is obtained by setting the quantity in square brackets equal to zero, or

$$X\pi^2 \frac{i^2}{a^2} + Y\pi^2 \frac{j^2}{b^2} = D\pi^4 \left(\frac{i^2}{a^2} + \frac{j^2}{b^2} \right)^2 \tag{e}$$

This result indicates that any one of the functions

$$w_{ij}(x, y) = A_{ij} \sin\frac{i\pi x}{a} \sin\frac{j\pi y}{b} \tag{f}$$

is a buckling mode, and the corresponding critical load is obtained from equation (e). If first we examine the case for $Y = 0$,

$$X\pi^2 \frac{i^2}{a^2} = D\pi^4 \left(\frac{i^2}{a^2} + \frac{j^2}{b^2}\right)^2 \tag{g}$$

by rearranging terms we get

$$X = \frac{D\pi^2}{b^2} \left(\frac{ib}{a} + \frac{j^2 a}{ib}\right)^2 \tag{h}$$

We can see that for all possible values for j, $j = 1$ provides the lowest value for X; thus,

$$X_{cr} = \frac{D\pi^2}{b^2} \left(\frac{ib}{a} + \frac{a}{ib}\right)^2 \tag{i}$$

or in terms of stress

$$\sigma_{xcr} = -K \frac{\pi^2 E}{12(1 - v^2)} \left(\frac{h}{b}\right)^2 \tag{j}$$

where

$$K = \left(\frac{ib}{a} + \frac{a}{ib}\right)^2 \tag{k}$$

For a given value of i, the parameter K depends only on the ratio a/b, called the *aspect ratio* of the plate. It is convenient to present this in graphical form, as shown in figure (b). Plots of K versus a/b are given for $i = 1, 2, 3,$ and 4.

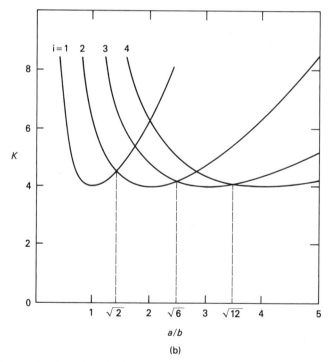

(b)

It is seen that the value of i yielding the minimum value of σ_{xcr} depends on a/b; thus, for plates that are approximately square, $i = 1$, and the buckled mode shape is

$$w_{11}(x, y) = A_{11} \sin \frac{\pi x}{a} \sin \frac{\pi y}{b} \tag{l}$$

In fact, it appears that for a/b close to any value of i, the buckled mode shape is

$$w_{i1} = A_{i1} \sin \frac{i\pi x}{a} \sin \frac{\pi y}{b} \tag{m}$$

We see that plate buckling has some interesting characteristics quite different from columns. For example, the lowest buckling stress, or load, does not always correspond to the same buckling mode, but changes with the aspect ratio of the plate.

For larger aspect ratios, say $a/b > 3$, $K \to 4$, and if the material is such that $v = 0.3$, equation (j) reduces to

$$\sigma_{xcr} = -3.62E \left(\frac{h}{b}\right)^2 \tag{n}$$

and as $(a/b)^2 \ll 1$, $K \to (b/a)^2$, and the buckling stress becomes

$$\sigma_{xcr} = -\frac{\pi^2 E}{12(1 - v^2)} \left(\frac{h}{a}\right)^2 = -0.904E \left(\frac{h}{a}\right)^2 \tag{o}$$

This latter is the same case as in Example 5.3.1.

Actually, finding an exact solution for a practical case that gives so much insight to the nature of plate buckling is fortuitous. We learn quickly that not much else can be done with the differential equations for plate buckling if exact analytical solutions are our goal. What can be done is outlined in the references cited at the beginning of the chapter. For our purposes, it is best to go on to approximate methods based on work and energy right away.

5.3.2 Virtual Work in Plate Buckling

The expression for the virtual work of a general elastic body given in Section 3.2.2 must be adapted to plates with both membrane and lateral loads and deflections. The expression for virtual work in Equation 3.2.21 still applies if we retain the higher-order terms in the strain–displacement relation. Thus we replace Equations 5.3.2 with

$$\epsilon_x = \frac{\partial u_0}{\partial x} - z \frac{\partial^2 w}{\partial x^2} + \frac{1}{2} \left(\frac{\partial w}{\partial x}\right)^2$$

$$\epsilon_y = \frac{\partial v_0}{\partial y} - z \frac{\partial^2 w}{\partial y^2} + \frac{1}{2} \left(\frac{\partial w}{\partial y}\right)^2 \tag{5.3.14}$$

$$\gamma_{xy} = \frac{\partial u_0}{\partial y} + \frac{\partial v_0}{\partial x} - 2z \frac{\partial^2 w}{\partial x \, \partial y} + \frac{1}{2} \left(\frac{\partial w}{\partial x}\frac{\partial w}{\partial y} + \frac{\partial w}{\partial y}\frac{\partial w}{\partial x}\right)$$

When these are substituted into the virtual work expression and reductions are done analogous to those for a beam in Equations 5.2.42 through 5.2.50, we obtain

$$\delta W = \delta U - \delta W_e = 0$$

$$= \iint ([D_c]\{\delta u_c\})^T [G][D_c]\{u_c\} h \, dx \, dy - \iint pw \, dx \, dy \tag{5.3.15}$$

$$- \frac{1}{2} \iint \begin{bmatrix} \dfrac{\partial \delta w}{\partial x} \\ \dfrac{\partial \delta w}{\partial y} \end{bmatrix} \begin{bmatrix} N_x & N_{xy} \\ N_{xy} & N_y \end{bmatrix} \begin{bmatrix} \dfrac{\partial w}{\partial x} \\ \dfrac{\partial w}{\partial y} \end{bmatrix} dx \, dy$$

where $[G]$ is given by Equation 4.7.4 and $[D_c]$ and $\{u_c\}$ are defined in Equation 4.7.3.

Much of the technical literature on plate buckling prior to about 1970 depended heavily on the Rayleigh–Ritz method, but now it is the finite element method that commands most of the attention. It is interesting, however, to reproduce a set of curves in Figure 5.3.3 that summarizes a large number of plate buckling solutions obtained largely by differential equation and Rayleigh–Ritz methods. Few other solutions were readily available, and many problems not covered in this figure, for example, nonrectangular plates, plates with holes, and plates with irregular boundary conditions, were solved by estimating the buckling load from values of these standard cases.

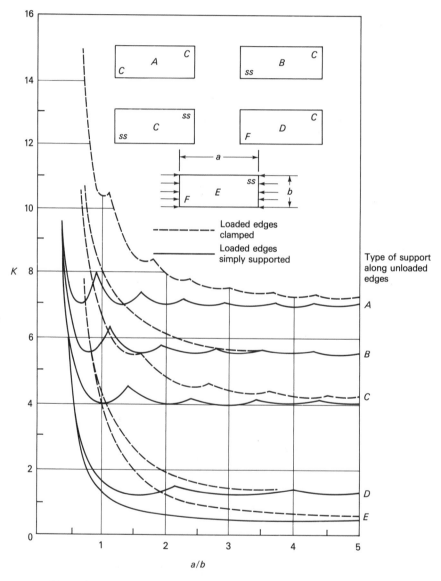

Figure 5.3.3 Buckling of Rectangular Plates for Various Boundary Conditions.

Fortunately, FEM provides us with a general approach to plate buckling. So let us pursue it here. First, we let

$$\{u_e\} = w = [n]\{r_e\} \tag{5.3.16}$$

and

$$[b] = [D][n] \tag{5.3.17}$$

using the same shape functions as in Section 4.7. The finite element equations for plate buckling become

$$([k] + [k^d])\{r_e\} = \{R_e\} \tag{5.3.18}$$

where

$$[k] = \iint [b]^T [G][b]h \, dx \, dy$$

$$[k^d] = -\iint [n']^T \begin{bmatrix} N_x & N_{xy} \\ N_{xy} & N_y \end{bmatrix} [n'] \, dx \, dy \tag{5.3.19}$$

$$\{R_e\} = \iint [n]p \, dx \, dy$$

The difficulty in selecting a set of shape functions for a plate element has been discussed in Sections 4.5.1 and 4.5.2. To pursue this further is beyond the scope of this text. Fortunately, suitable element stiffness and differential stiffness matrices have been developed and incorporated in FEM computer codes. The four most popular elements have the same characteristics as shown in Figure 4.5.2.

We shall now consider two example cases. The first is a finite element solution for the buckling load of one of the plates given in Figure 5.3.2. The next is of a plate with a hole in it. This latter is the kind of problem that FEM can do rather easily, but that other methods find difficult.

Example 5.3.3

Consider a rectangular plate that is simply supported on two loaded edges and clamped on the other two, as in case A in Figure 5.3.3. Let us consider a square plate, that is, one with aspect

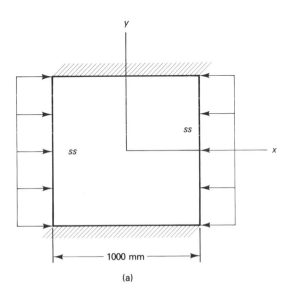

(a)

ratio of 1. We shall select a plate with the same dimensions and material properties as in Examples 4.4.1 and 4.5.1, but with no hole, as shown in figure (a).

Solution: Because we are not sure of the mode shape, let us not prejudge any planes of symmetry. Now the seven items of input data are determined.

1. Divide the structure into elements and number the nodes and elements: We shall use quadrilateral elements in the arrangement shown in figure (b), where the node numbers are in block numerals and the element numbers are in italic numerals.

(b)

2. Record the coordinates of the nodes:

Node Number	x	y
1	−500.0	−500.0
2	−250.0	−500.0
3	0.0	−500.0
4	250.0	−500.0
5	500.0	−500.0
6	−500.0	−250.0
7	−250.0	−250.0
8	0.0	−250.0
9	250.0	−250.0
10	500.0	−250.0
11	−500.0	0.0
12	−250.0	0.0
13	0.0	0.0
14	250.0	0.0
15	500.0	0.0
16	−500.0	250.0
17	−250.0	250.0
18	0.0	250.0
19	250.0	250.0

continued

(a)

Node Number	x	y
20	500.0	250.0
21	−500.0	500.0
22	−250.0	500.0
23	0.0	500.0
24	250.0	500.0
25	500.0	500.0

3. Identify the nodes that go with each element:

Element	Nodes (counterclockwise)			
1	1	2	7	6
2	2	3	8	7
3	3	4	9	8
4	4	5	10	9
5	6	7	12	11
6	7	8	13	12
7	8	9	14	13
8	9	10	15	14
9	11	12	17	16
10	12	13	18	17
11	13	14	19	18
12	14	15	20	19
13	16	17	22	21
14	17	18	23	22
15	18	19	24	23
16	19	20	25	24

(b)

4. Specify the material properties for aluminum:

$$E = 68,950.0 \text{ MPa}$$

$$G = 26,520.0 \text{ MPa}$$

$$\upsilon = 0.3$$

(c)

5. Specify other geometric properties: in this case, the thickness of 1 mm.
6. Specify the geometric constraints: These are chosen to correspond to the classical solution as reported in Figure 5.3.3. First, all edges are restrained in the z-direction.

$$w_1 = w_2 = w_3 = w_4 = w_5 = 0, \qquad w_{10} = w_{15} = w_{20} = w_{25} = 0$$
$$w_{24} = w_{23} = w_{22} = w_{21} = 0, \qquad w_{16} = w_{11} = w_6 = 0$$

(d)

The restraints that prevent rigid body motion in the x-y plane and about the z-axis are

$$u_{13} = v_{13} = 0, \qquad v_{15} = 0$$

(e)

These essentially attach the plate to the coordinate system and have no real effect on the displacements and buckling load. All other displacements in the x- and y-directions are permitted. There are, however, no rotations about the x-axis on the clamped top and bottom edges; thus,

$$\beta_1 = \beta_2 = \beta_3 = \beta_4 = \beta_5 = 0$$
$$\beta_{21} = \beta_{22} = \beta_{23} = \beta_{24} = \beta_{25} = 0$$

(f)

Furthermore, since the edges are straight,

$$\theta_1 = \theta_2 = \theta_3 = \theta_4 = \theta_5 = 0$$
$$\theta_{21} = \theta_{22} = \theta_{23} = \theta_{24} = \theta_{25} = 0$$
$$\beta_1 = \beta_6 = \beta_{11} = \beta_{16} = \beta_{21} = 0$$
$$\beta_5 = \beta_{10} = \beta_{15} = \beta_{20} = \beta_{25} = 0$$

(g)

It is a characteristic of the type of element used that all rotations about the z-axis are zero; that is,

$$\varphi_i = 0, \qquad i = 1, 2, \ldots, 25$$

(h)

7. Specify the equivalent nodal loads: The reference loads will be chosen to sum to 1.0 N on each loaded edge. This represents a uniform force per unit length on each loaded edge of 0.001 N.

Node	R_x
1	0.125
6	0.25
11	0.25
16	0.25
21	0.125
5	−0.125
10	−0.25
15	−0.25
20	−0.25
25	−0.125

(i)

Note that the reference loads on the corner nodes 1, 5, 21, and 25 are one-half the value of the loads on the interior nodes. This approximates the distributed uniform load on the edges as stated in the original problem. That this is the correct nodal representation of the distributed edge loading is verified by the fact that uniform membrane forces are generated by the loads in the FEM solution.

The FEM solution using MSC/NASTRAN proceeds in two parts. First, the membrane forces due to the listed reference loads are found. They are 0.001 N everywhere in the x-direction in the plate, zero in the y-direction, and there are no shearing membrane forces. This produces a uniform compressive deflection in the x-direction and, because of the Poisson's ratio effect, a uniform elongation of the plate in the y-direction, as shown in figure (c).

This deflection is depicted in a NASTRAN plot that shows the element boundaries in both the undeformed and deformed positions. The deflection is greatly exaggerated and is normalized to show the relative amounts of deflection in the two directions. It may be noted that the extension in the y-direction is 0.3 times the compression in the x-direction. This is exactly as is expected for a Poisson's ratio of 0.3.

Node	w_i	β_i	θ_i
14	1.0	0.0	0.0
15	0.0	0.0	0.0073972
16	0.0	0.0	0.0037778
17	−0.51741	0.0037563	0.0
18	0.54399	0.0	−0.0037778
19	0.51741	−0.0037563	0.0
20	0.0	0.0	0.0037778
21	0.0	0.0	0.0
22	0.0	0.0	0.0
23	0.0	0.0	0.0
24	0.0	0.0	0.0
25	0.0	0.0	0.0

We do not have the shape functions in the NASTRAN plate element used to find these solutions; therefore, we cannot produce the distributed displacement functions for each element. If we were to fit cubic polynomials to the displacement data along lines of constant x and constant y through the nodes, we would have a very close approximation to the shape along those lines. Methods are available for constructing functional forms of surfaces that fit the preceding data. The NASTRAN plotter does a crude approximation by drawing straight line segments between lateral nodal displacements.

The buckling mode for the lowest buckling load is plotted in figure (d) using the NASTRAN plotter. Element boundaries are plotted for both the undeformed and deformed positions. Note what appears to be a full wave in the x-direction, but only a half-wave in the y-direction. Normally, we think of the lowest buckling load for a square plate to be half-waves in both coordinate directions and that is what it is for all edges simply supported. The presence of clamped edges on the two opposite unloaded sides and simply supported edges on the two opposite loaded sides is enough in this case to cause the lowest buckling load to be associated with what we think of as a higher mode. Because of the more complicated mode shape, more elements would be desirable for greater accuracy.

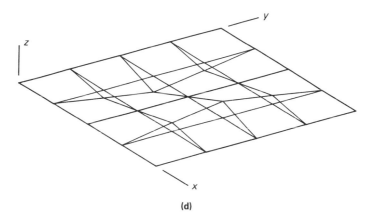

(d)

The preceding example is of a problem for which the answer is known. The advantage of the FEM, as we have stated so often, is in doing problems that were prohibitively difficult to do before. The fact is that these problems often are little more difficult to do than the simpler problems using FEM. The next example demonstrates this.

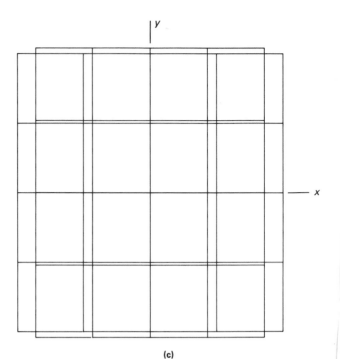

(c)

The NASTRAN solution finds a buckling stress of 0.5127 N/mm². From tion

$$\sigma_{x\ cr} = -K \frac{\pi^2 E}{12(1 - v^2)} \left(\frac{h}{b}\right)^2$$

we find a value of $K = 8.227$ which would plot on Figure 5.3.3 slightly above the there for $a/b = 1$. It is hard to read the figure accurately, but the value obtaine TRAN appears to be about 5% to 6% higher than the value shown there. Mor would lower the NASTRAN value slightly.

The buckling mode is given in terms of the nodal displacements, including the following table:

Node	w_i	β_i	θ_i
1	0.0	0.0	0.0
2	0.0	0.0	0.0
3	0.0	0.0	0.0
4	0.0	0.0	0.0
5	0.0	0.0	0.0
6	0.0	0.0	0.0037778
7	−0.51741	−0.0037563	0.0
8	0.54399	0.0	−0.0037778
9	0.51741	0.0037563	0.0
10	0.0	0.0	0.0037778
11	0.0	0.0	0.0073972
12	−1.0	0.0	0.0
13	0.0	0.0	−0.0073972

continued

Example 5.3.4

Consider the same rectangular plate with the same kind of loading and the same edge constraints, but now with a hole with a diameter of 300 mm cut in its center as shown in figure (a).

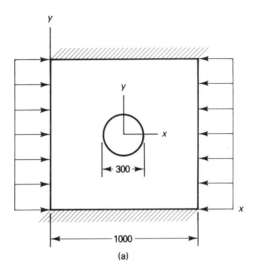

(a)

Solution: The seven items of input data are

1. Divide the plate into elements and number the nodes and elements: This is shown in figure (b). As much as possible is retained from Example 5.3.3 to simplify the preparation of the computer model. Note that the circular hole is modeled as a polygon because the element used in the analysis has straight edges between nodes. This introduces some error, which would lessen as the number of elements and nodes around the perimeter of the circle increases.

2. Record the coordinates of the nodes: Note that all the first 25 nodes have the same coordinates as in Example 5.3.3 except for number 13. That one plus the new ones are listed below:

Node	x	y
13	−150.0	0.0
26	−106.066	−106.066
27	0.0	−150.0
28	106.066	−106.066
29	150.0	0.0
30	106.066	106.066
31	0.0	150.0
32	−106.066	106.066

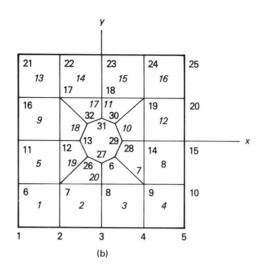

(b)

3. Identify the nodes that go with each element. Those that are different from Example 5.3.3 are given next:

Element	Nodes (counterclockwise)			
6	8	9	28	27
7	28	9	14	29
10	29	14	19	30
11	31	30	19	18
17	32	31	18	17
18	12	13	32	17
19	7	26	13	12
20	7	8	27	26

4. Specify the material properties for aluminum: Same as Example 5.3.3.

5. Specify other geometrical properties: Same as Example 5.3.3.

6. Specify the geometric constraints: Same as Example 5.3.3.

7. Specify the equivalent nodal loads: Same as Example 5.3.3.

The NASTRAN solution finds a buckling stress of 0.4824 N/mm². This produces a value about 5% lower than for the same plate without a hole; thus, the effect of the hole is to reduce the stiffness and, therefore, the buckling load by this amount.

The buckling mode is given in terms of the nodal displacements, including rotations, in the following table:

Node	w_i	β_i	θ_i
1	0.0	0.0	0.0
2	0.0	0.0	0.0
3	0.0	0.0	0.0

continued

Node	w_i	β_i	θ_i
4	0.0	0.0	0.0
5	0.0	0.0	0.0
6	0.0	0.0	−0.0008582
7	0.33214	0.0022512	−0.0018400
8	0.57455	0.0040333	0.0
9	0.33214	0.0022512	0.0018400
10	0.0	0.0	0.0008582
11	0.0	0.0	−0.0019057
12	0.61039	0.0	−0.0030019
13	0.96677	0.0	−0.0037769
14	0.61039	0.0	0.0030019
15	0.0	0.0	0.0019057
16	0.0	0.0	−0.0008582
17	0.33214	−0.0022512	−0.0018400
18	0.57455	−0.0040333	0.0
19	0.33214	−0.0022512	0.0018400
20	0.0	0.0	0.0008582
21	0.0	0.0	0.0
22	0.0	0.0	0.0
23	0.0	0.0	0.0
24	0.0	0.0	0.0
25	0.0	0.0	0.0
26	1.0	0.0022510	−0.0026470
27	0.99876	0.0042327	0.0
28	1.0	0.0022510	0.0026470
29	0.96677	0.0	0.0037769
30	1.0	−0.0022510	0.0026470
31	0.99876	−0.0042327	0.0
32	1.0	−0.0022510	−0.0026470

The first buckling mode is plotted in figure (c) using the NASTRAN plotter with its straight line segments between nodal displacements. Note the large change in buckling mode, even though there has been a small change in the buckling load. The release of stiffness by the hole has changed the mode to the form of a half-wave in each coordinate direction.

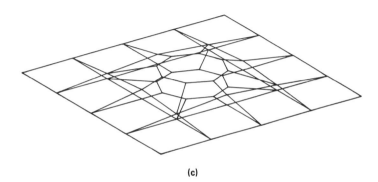

(c)

The two solutions obtained in Examples 5.3.3 and 5.3.4 use boundary conditions that correspond to those used in classical solutions to plate buckling. Although particular attention is not called to it, the fact is that the clamped edges are not truly clamped in the x-y plane; that is, the clamped edges are free to move in the x-y plane. A look at the boundary conditions shows that the constraints on the clamped edges are stated entirely in terms of deflection w_i in the z-direction and rotations β_i, θ_i, and φ_i about the x-, y-, and z-axes, respectively. Furthermore, the loaded, simply supported edges are also permitted to move in the x-y plane. Movement of the simply supported edges is necessary for the in-plane loads to develop in the plate. Free movement of the clamped edges is necessary to prevent in-plane loads from developing in the y-direction. In the classical analysis, we find the in-plane loads as part of an uncoupled plane stress problem. We then use that state of stress to find the buckling load.

It is just as easy in the FEM formulation of the plate buckling problem to let the clamped edges be restrained. In the next example, we repeat Examples 5.3.3 and 5.3.4 for a case of restrained edges.

Example 5.3.5

Repeat Examples 5.3.3 and 5.3.4, but with the clamped edges restrained from movement in the y-direction.

Solution: Most everything in both problems is identical except we add the following constraints:

$$v_1 = v_2 = v_3 = v_4 = v_5 = 0$$

$$v_{21} = v_{22} = v_{23} = v_{24} = v_{25} = 0 \tag{a}$$

With this addition the solutions are obtained by NASTRAN. The buckling load for the square plate without a hole is 0.4773 N/mm². The mode shape is very like that in Example 5.3.3. This load is nearly 7% lower than the same plate without in-plane restraint in the y-direction. The cause is the Poisson's ratio effect, which produces in-plane forces in the y-direction; in fact, the membrane stresses in the y-direction are exactly υ times those in the x-direction.

For the plate with a hole, the buckling load is 0.3470 N/mm² and the mode shape is very similar to the same plate without in-plane restraint in the y-direction. This is a 28% reduction in buckling load.

These examples illustrate the importance of boundary conditions on determining buckling loads in plates, something that is rarely treated in depth by classical methods.

Thin curved plates, or shells, are another important class of structures for which buckling is of great concern. Again, by the finite element method, many useful solutions are possible. This topic, however important, is beyond the scope of this text, but may be found in several of the references.

5.4 BUCKLING OF GENERAL SOLID BODIES

General solid bodies may be subject to buckling even when they do not fall into the classes of slender bars, thin plates, or shells. Little has been written on this subject of practical use based on classical methods. The finite element method, however, can handle the problem routinely. Differential stiffness matrices for solid elements have been derived by incorporating the strain–displacements of Equations 1.2.18 in the strain energy part of the expression for virtual work. As in the case of other elements, the same shape functions can be used to find the differential stiffness matrix as were used to find the stiffness matrix. MSC/NASTRAN and other computer codes have included this feature.

5.5 CONCLUSIONS AND SUMMARY

Either by looking at the equilibrium of a structure in the deflected position or by including the higher-order terms in the strain–displacement equations, we have discovered an important new phenomenon that plays an important role in the behavior of structures. This behavior change is most prominent in slender bars, thin flat plates, and thin shells under loadings that produce internal compressive stresses.

The governing equation for a slender bar with both a lateral loading a compressive internal axial force, or a beam-column, is

$$(EIw'')'' + Pw'' = f(x) \tag{5.2.14}$$

This specializes to the equation for a uniform column when there is no lateral load and EI is a constant, or

$$w'''' + \lambda^2 w'' = 0 \tag{5.2.17}$$

where

$$\lambda^2 = \frac{P}{EI}$$

This latter equation is of a class known as eigenvalue equations.

Solutions of these equations are generally limited to uniform bars with reasonable simple loadings and constraints. When solutions to more complex structures are sought, approximate methods derived from virtual work are more often used. The virtual work for a beam-column is

$$\begin{aligned}
\delta W &= \delta W_e - (\delta U + \Delta \delta U) \\
&= \int \delta w f(x) \, dx + \Sigma F_{zi} \delta w_i + \Sigma M_{yi} (\delta w_j)' \\
&\quad - \int N \delta w' w' \, dx - \int EI \delta w'' w'' \, dx = 0
\end{aligned} \tag{5.2.51}$$

This may be turned into a practical analysis tool by either the Rayleigh–Ritz or the finite element methods. With Rayleigh–Ritz, we use the same admissible functions that we did for stable structure analysis, that is,

$$w = [\Omega]\{q\} \tag{5.2.52}$$

and obtain the linear algebraic equations

$$([K] + [K^d])\{q\} = \{Q\} \tag{5.2.54}$$

where

$$\begin{aligned}
[K] &= \int [B]^T [G][B] \, dv = \int EI[\Omega'']^T [\Omega''] \, dx \\
\{Q\} &= \int [\Omega]^T f(x) \, dx + \Sigma [\Omega(x_i)]^T F_{zi} + \Sigma [\Omega'(x_j)]^T M_{yj}
\end{aligned} \tag{5.2.55}$$

and

$$[K^d] = -P \int [\Omega']^T[\Omega'] \, dx \tag{5.2.56}$$

The buckling loads and modes are found from the homogeneous equations

$$([K] + [K^d])\{q\} = \{0\} \tag{5.2.57}$$

by first setting

$$| \, [K] + [K^d] \, | = 0 \tag{5.2.58}$$

The FEM equations are a repeat of the Rayleigh–Ritz equations with a simple change in notation. We let

$$w(x) = [N]\{r\} \tag{5.2.69}$$

and obtain the linear algebraic equations

$$([K] + [K^d])\{r\} = \{R\} \tag{5.2.70}$$

where

$$[K^d] = -P \int [N']^T[N'] \, dx \tag{5.2.71}$$

$$[K] = \int [B]^T[G] \, [B] \, dv = \int EI[N'']^T[N''] \, dx \tag{5.2.72}$$

$$\{R\} = \int [N]^T f(x) \, dx + \Sigma \, [N(x_i)]^T F_{zi} + \Sigma \, [N'(x_i)]^T M_{yj} \tag{5.2.73}$$

The buckling loads and modes are found from the homogeneous equations

$$([K] + [K^d])\{r\} = \{0\} \tag{5.2.74}$$

The differential equation for the buckling of a thin flat plate is

$$D\nabla^4 w = p + N_x \frac{\partial^2 w}{\partial x^2} + 2N_{xy} \frac{\partial^2 w}{\partial x \, \partial y} + N_y \frac{\partial^2 w}{\partial y^2} \tag{5.3.12}$$

where

$$D = \frac{Eh^3}{12(1 - v^2)} \tag{5.3.13}$$

and the boundary conditions are as given in Section 2.4. The expression for virtual work for plate buckling is

$$\delta W = \delta U - \delta W_e = 0$$

$$= \iint ([D_c]\{\delta u_c\})^T[G] \, [D_c]\{u_c\} h \, dx \, dy - \iint pw \, dx \, dy \tag{5.3.15}$$

$$- \frac{1}{2} \iint \begin{bmatrix} \frac{\partial \delta w}{\partial x} \\ \frac{\partial \delta w}{\partial y} \end{bmatrix} \begin{bmatrix} N_x & N_{xy} \\ N_{xy} & N_y \end{bmatrix} \begin{bmatrix} \frac{\partial w}{\partial x} \\ \frac{\partial w}{\partial y} \end{bmatrix} dx \, dy$$

The plate element matrices are

$$\{u_e\} = w = [n]\{r_e\} \tag{5.3.16}$$

and

$$[b] = [D][n] \tag{5.3.17}$$

from which we obtain

$$([k] + [k^d])\{r_e\} = \{R_e\} \tag{5.3.18}$$

where

$$[k] = \iint [b]^T[G][b]h \, dx \, dy$$

$$[k^d] = -\iint [n']^T \begin{bmatrix} N_x & N_{xy} \\ N_{xy} & N_y \end{bmatrix} [n'] \, dx \, dy \tag{5.3.19}$$

$$\{R_e\} = \iint [n]p \, dx \, dy$$

PROBLEMS

1. Find the critical buckling load and mode for the column shown. What is the value of the end fixity coefficient?

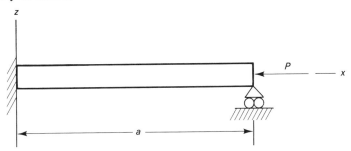

2. Find the critical buckling load and mode for the column shown for values of $ka^3/EI = \frac{1}{2}$. Then examine the solution for $k \longrightarrow \infty$ and $k \longrightarrow 0$.

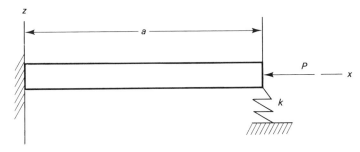

3. Find the deflection of the beam-column shown for $P < P_{cr}$.

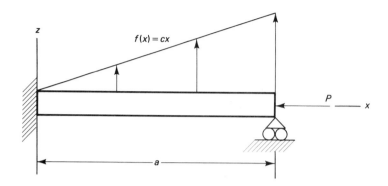

4. Use your knowledge of ordinary differential equations with constant coefficients and of beam-column boundary conditions to solve the following new problem for the deflection. The figure shows a uniform simply supported beam column on an elastic foundation. The governing differential equation is $EIw'''' + Pw'' + \alpha w = f_0$. Show how you would find the buckling load.

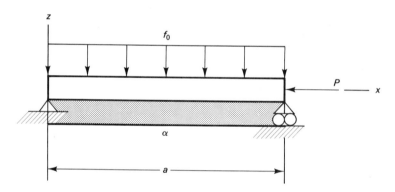

5. Write down the differential equation and boundary conditions for each of the columns or beam-columns shown. Do not solve. In figure (b), the value of the distributed load $f(x)$ at the origin is $f(0) = f_0$. Write down the expression for the virtual work for each of the two beams.

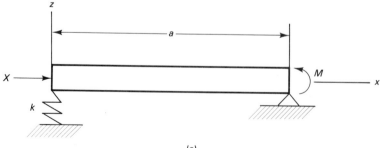

(a)

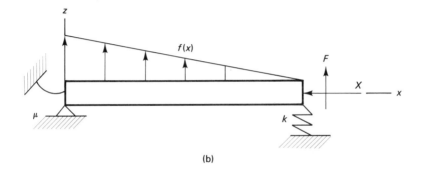

(b)

6. Use the R-R method to find an approximate value for the deflection of the beam-column shown. Let $EI_1 = 2EI_2$ and assume that $P < P_{cr}$. Find a one-term solution for each of the following functions that is admissible. Indicate which of these could not be used in any case.
   (a) $\Omega_i(x) = (x/a)[(x/a) - 1]$
   (b) $\Omega_i(x) = [(x/a) - 1]$
   (c) $\Omega_i(x) = \sin(\pi x/a)$
   (d) $\Omega_i(x) = (x/a)$
   Can you suggest a way to determine which of the solutions is most accurate without being able to compare it to a known exact solution?

   Then find and compare the critical buckling load and mode for the structure for the functions chosen.

   Finally, find the equations for a two-term Rayleigh–Ritz solution to this problem. Set up the equations but do not solve. Show how you would find the buckling load in this case.

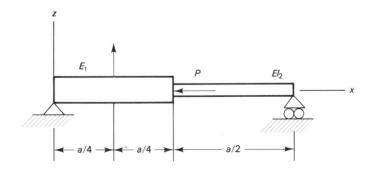

7. A beam element has a triangular load as shown. Find the equivalent nodal loads for this element.

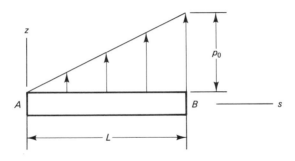

8. The finite element method is used on the problem shown. Divide into three elements, and number the nodes and elements. Assemble the equations and apply constraints. Explain how you would obtain the solution for the deflection when $P < P_{cr}$ and how you would find the buckling load and mode.

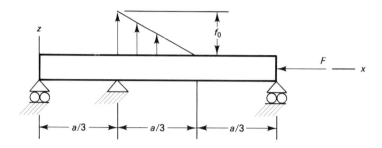

9. Suppose an axial compressive load F is added to the problem shown in Example 4.4.1 to make it a beam-column. Set up and solve the equations for finding the deflection when $P < P_{cr}$. Find the buckling load and compare with that found in Example 5.2.6. What is the buckling mode?

10. Find the buckling loads for the two plates shown and compare. All edges are simply supported. Sketch the modes.

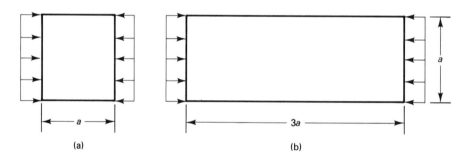

(a)   (b)

11. A rectangular plate of thickness h is clamped on all four edges and is loaded as shown. The total load is 500 lb. How thick must the plate be to avoid elastic buckling? $E = 10^7$ psi.

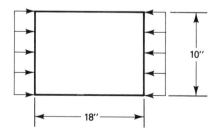

12. For the aluminum plate shown find the seven items of input data for an FEM solution for buckling. Select a convenient number of elements.

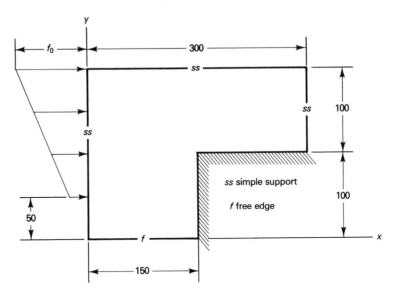

STRUCTURAL DYNAMICS

6

6.1 INTRODUCTION

In the analysis of structures in this text up to this point, we have assumed that all external forces are applied slowly, so slowly, in fact, that the loads and the resulting stresses and deformations are independent of time. We are all aware, however, that loads are often applied rapidly and motion of the structure occurs. The presence of such motion can have serious consequences, and the design of many structures must be based, in part, on dynamic considerations. The study of such motion is called *structural dynamics* or the *vibration of structures*.

The dynamics of continuous elastic systems can be modeled by partial differential equations based on Newton's laws or by integral equations based on considerations of virtual work. Both approaches will be used; in fact, we shall follow the same pattern as used in earlier chapters of looking first for exact analytical solutions by solving the differential equations, then looking for approximate solutions based on the Rayleigh-Ritz method, and, finally, looking for solutions by the finite element method.

The derivation of the appropriate time-dependent equations, in most cases, is a simple extension of the static case by adding effective forces to the system that result from accelerations of the mass of the structure. These are the *inertia forces*. For the most part, we take advantage of D'Alembert's principle to add the inertia forces as *reversed effective forces*. Other time dependent forces may also be considered. The most common of these are the *damping forces*, of which *viscous damping* is most commonly dealt with in the mathematical models of structures. In addition, we have time dependent applied loads.

The chapter begins with a very brief look at the equations of motion for a general linear elastic solid, but quickly turns to the analysis of beams and frames and then plates. For reasons that will become clear, various kinds of motion are singled out and named. There is *free motion,* which occurs in the absence of applied loads, but may be initiated by applying *initial conditions* to the structure. Then there is *forced motion,* which results from the application of loads. Forced motion comes in two kinds. There is *harmonic response,* which occurs when a periodic force is applied to the structure, and *transient response,* when the applied force is not periodic.

Many references are available, including References 22, which is representative of a book treating the vibration of both discrete and continuous systems. Finite element approaches to structural dynamics may be found in detail in References 23 and 24 and are introduced in less detail in References 13, 14, and 15, among others.

6.2 DYNAMICS OF A GENERAL ELASTIC SOLID

Consider the equations for the general elastic solid derived in Chapter 1. To adapt these equations to model undamped structural dynamics, we need only consider the dependent variables (displacements, strains, and stresses) as functions of time, let the applied loads be functions of time, and explicitly include inertia forces in the body forces according to D'Alembert's principle. Thus, in Equations 1.2.5, we let

$$f_x = f_{xo} - \rho \frac{\partial^2 u}{\partial t^2}, \qquad f_y = f_{yo} - \rho \frac{\partial^2 v}{\partial t^2} \qquad f_z = f_{zo} - \rho \frac{\partial^2 w}{\partial t^2} \tag{6.2.1}$$

where ρ is the mass density of the material and f_{xo}, f_{yo}, and f_{zo} are those body forces exclusive of the inertia forces.

Rarely is any attempt made to find exact analytical solutions to the full three-dimensional equations for the dynamic behavior of an elastic solid. It is even more pressing than in the static case to find simpler forms of the equations. Fortunately, the same forms (plane stress and strain, beams and plates, and so on) lend themselves well to dynamic analysis. Thus, we turn quickly to the dynamic analysis of beams in the next section.

6.3 BENDING VIBRATION OF SLENDER BARS BY DIFFERENTIAL EQUATION METHODS

The understanding of the behavior of beams in lateral vibration is not only useful knowledge in its own right, but also the methods learned carry over to most other structures of practical interest. In the next few paragraphs, we shall pursue this subject in great detail. First, we shall look at the differential equation approaches, which include such mathematical topics as separation of variables, eigenvalue analysis, and the use of infinite series in obtaining solutions. Then we shall look at integral equations approaches based on both virtual work and minimum potential energy, which use approximate analytical and numerical methods such as Rayleigh-Ritz, finite elements, and finite differences.

6.3.1 Derivation of the Differential Equations

The symbolic representation of the equations for beams shown in Figure 2.2.1 and the symbolic form of the equations given in Equation 2.2.1 may be used without modification. What changes is that now all the independent variables are functions of time as well as space, and the equations of equilibrium must also contain inertia force terms.

Consider the slender bar shown in Figure 6.3.1(a), which is acted on by a constant axial load F and a dynamic lateral load $f(x, t)$.

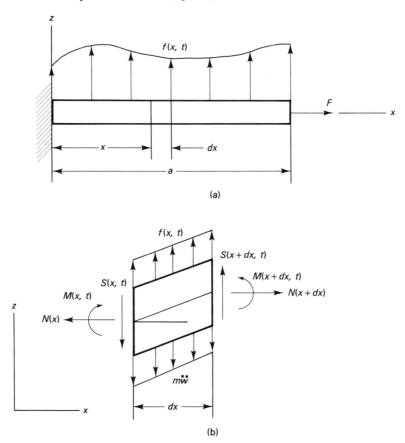

(a)

(b)

Figure 6.3.1 The Beam Element in Vibration.

Portions of the bar not constrained will be moving with a displacement $w(x, t)$, a velocity $\partial w(x, t)/\partial t$, and an acceleration $\partial^2 w(x, t)/\partial t^2$. An element of the bar is shown in Figure 6.3.1(b) with the time-dependent internal forces and moments acting on it. Only the axial load is considered independent of time at this stage. An inertia force $m(x)\ddot{w}(x, t)$ is included explicitly, where $m(x)$ is the mass per unit length of the bar, and derivatives of a function with respect to time are shown as dots over the function. By considering the inertia force just as any other distributed loading, by D'Alembert's principle we can say

$$(EIw'')'' - Nw'' = f(x, t) - m\ddot{w} \tag{6.3.1}$$

or

$$(EIw'')'' - Nw'' + m\ddot{w} = f(x, t) \tag{6.3.2}$$

where the prime now means the partial derivative with respect to x and the overdot means a partial derivative with respect to t. Note that when the axial load is compressive we can let $P = -N$.

This is a linear partial differential equation that is fourth order in derivatives with respect to x and second order in derivatives with respect to time. To complete the statement of the problem, we must specify four boundary conditions and two initial conditions. The boundary conditions are, generally, identical to those we used earlier in static stress and buckling analysis, except partial derivatives must replace the total derivatives used before. For example, the simply supported boundary conditions at $x = 0$ are

$$w(0, t) = 0, \qquad EIw''(0, t) = 0 \tag{6.3.3}$$

and for the free end at $x = a$ the boundary conditions are

$$EIw''(a, t) = 0, \qquad (EIw'')'(a, t) + Pw'(a, t) = 0 \tag{6.3.4}$$

and so on. Initial conditions are boundary conditions on time. The time may be specified at any time, but usually information is supplied on what is happening at time equal to zero (that is why they are called initial conditions) in the following form:

$$w(x, 0) = 0, \qquad \dot{w}(x, 0) = 0 \tag{6.3.5}$$

The solution of the preceding equation and its boundary conditions is most difficult except in the special case where EI and m are constants. In this case, the equation may be written

$$EIw'''' - Nw'' + m\ddot{w} = f(x, t) \tag{6.3.6}$$

This is the equation we are going to examine in detail in the next several paragraphs. It turns out that the solution case for $f(x, t) = 0$ is valuable not only in its own right, but also as a step in solving the problem for when $f(x, t) \neq 0$. This case is taken up in the next section.

6.3.2 Free Vibrations of Beams in Bending

In the static case, there is no deflection if there is no load. Not so in the dynamic case, as we shall see because of the initial conditions. So let us examine the equation when $f(x, t) = 0$ and $N = 0$, or

$$(EIw'')'' + m\ddot{w} = 0 \tag{6.3.7}$$

Note that this equation still requires four boundary conditions and two initial conditions.

It has been shown that a mathematical method known as *separation of variables* will work on this equation. Let

$$w(x, t) = W(x)T(t) \tag{6.3.8}$$

Substituting this into Equation 6.3.7, we have

$$(EIW'')'' T + mW\ddot{T} = 0 \qquad (6.3.9)$$

or, rearranging,

$$\frac{(EIW'')''}{mW} + \frac{\ddot{T}}{T} = 0 \qquad (6.3.10)$$

Now the first term in Equation 6.3.10 is a function of x only, and the second term is a function of t only. Since x and t are arbitrary, we can say that the only way a function of x plus a function of t can always be zero is if each function is a constant. This permits us to say

$$\frac{(EIW'')''}{mW} = -\frac{\ddot{T}}{T} = \omega^2 \qquad (6.3.11)$$

The choice of which function to assign the minus sign and the choice of calling the constant ω^2 are based strictly on the results obtained next; that is, it is convenient to do so.

These equations may be rewritten

$$\ddot{T} + \omega^2 T = 0, \qquad (EIW'')'' - m\omega^2 W = 0 \qquad (6.3.12)$$

The solution of these equations is found by assuming a solution of the form of an exponential. This is a well-known procedure that applies to all constant coefficient ordinary linear differential equations. It will be reviewed here briefly, but in subsequent applications it will be assumed the method is known. For example, let

$$T(t) = Ce^{rt} \qquad (6.3.13)$$

and substitute it into the first of Equations 6.3.12. We get

$$r^2 Ce^{rt} + \omega^2 Ce^{rt} = 0 \longrightarrow r^2 = -\omega^2 \qquad (6.3.14)$$

This latter equation is called the characteristic equation and we seek its roots. Thus, for $i = \sqrt{-1}$,

$$r = \pm i\omega \qquad (6.3.15)$$

and we may say

$$T(t) = a_1 e^{i\omega t} + a_2 e^{-i\omega t} \qquad (6.3.16)$$

which may be rewritten using trigonometric identities in the form

$$T(t) = c_1 \sin \omega t + c_2 \cos \omega t \qquad (6.3.17)$$

This particular equation defines the time behavior of all undamped, linear elastic structures with small deflections. It is so widely known that it has been named and is called the *harmonic equation* and its solution, given here, the *harmonic solution*. At this point, c_1, c_2, and ω are all unknown, but we can recognize that $T(t)$ is a periodic function with *frequency* ω. Now you know why we chose the constant in the separation of variables as $-\omega^2$. It is so convenient for use in this equation. But before we can find the three unknowns, we must take a look at the other equation.

The solution to the second equation is difficult to obtain in the general case, but when EI and m are constant, that is, the beam is uniform, it may be obtained readily. Then

$$EIW'''' - m\omega^2 W = 0 \qquad (6.3.18)$$

or

$$W'''' - \alpha^4 W = 0 \qquad (6.3.19)$$

where

$$\alpha^4 = \frac{m\omega^2}{EI} \qquad (6.3.20)$$

The general solution is found by assuming a solution of the exponential form. In this case, the characteristic equation is

$$r^4 = \alpha^4 \qquad (6.3.21)$$

and there are two pairs of roots. One pair, $\pm\alpha$ is real, and the other pair, $\pm i\alpha$, is imaginary. The solution may be written in exponential form,

$$W(x) = a_3 e^{\alpha x} + a_4 e^{-\alpha x} + a_5 e^{i\alpha x} + a_6 e^{-i\alpha x} \qquad (6.3.22)$$

or in the more convenient form of

$$W(x) = c_3 \sinh \alpha x + c_4 \cosh \alpha x + c_5 \sin \alpha x + c_6 \cos \alpha x \qquad (6.3.23)$$

Equation 6.3.23 contains the four unknown constants of integration and α, which is unknown because it contains the unknown ω.

Up to this point we have generated six constants of integration, c_1 through c_6, and the unknown quantity called the frequency, ω, for a total of seven unknowns. We have four boundary conditions and two initial conditions, a total of six conditions, still to satisfy. Interestingly enough, perhaps surprisingly so, this is enough to find the complete solution. First, we satisfy the boundary conditions. The variables in the boundary and initial conditions, too, must be separated. At this point, it is easier to follow what is going on in a specific example. We can continue the general discussion after the example.

Example 6.3.1

Apply the boundary conditions to Equation 6.3.23 for the uniform, simply supported beam shown in figure (a).

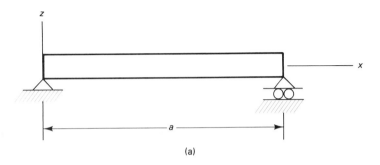

(a)

Solution: The boundary conditions are

$$w(0, t) = W(0)T(t) = 0 \qquad \longrightarrow \qquad W(0) = 0$$

$$EIw''(0, t) = EIW''(0)T(t) = 0 \qquad \longrightarrow \qquad W''(0) = 0$$

(a)

$$w(a, t) = W(a)T(t) = 0 \qquad \longrightarrow W(a) = 0$$

$$EIw''(a, t) = EIW''(a)T(t) = 0 \qquad \longrightarrow W''(a) = 0$$

Upon substitution, we have

$$W(0) = c_3 \sinh \alpha 0 + c_4 \cosh \alpha 0 + c_5 \sin \alpha 0 + c_6 \cos \alpha 0 = 0$$

$$= c_4 + c_6 = 0$$

$$W''(0) = c_3 \alpha^2 \sinh \alpha 0 + c_4 \alpha^2 \cosh \alpha 0 - c_5 \alpha^2 \sin \alpha 0 + c_6 \alpha^2 \cos \alpha 0 = 0$$

$$= c_4 - c_6 = 0 \tag{b}$$

$$W(a) = c_3 \sinh \alpha a + c_4 \cosh \alpha a + c_5 \sin \alpha a + c_6 \cos \alpha a = 0$$

$$W''(a) = c_3 \alpha^2 \sinh \alpha a + c_4 \alpha^2 \cosh \alpha a - c_5 \alpha^2 \sin \alpha a - c_6 \alpha^2 \cos \alpha a = 0$$

We may observe that what we have in equations (b) is a set of homogeneous linear algebraic equations of the form

$$[\Lambda]\{c\} = \{0\} \tag{c}$$

from which we may say that either

$$\{c\} = \{0\} \quad \text{or} \quad |\Lambda| = 0 \tag{d}$$

We have an eigenvalue problem, since the trivial solution $\{c\} = 0$ would proclaim that no motion ensues. But our experience says this is not so; thus, we look for the conditions where $|\Lambda| = 0$.

Rather than expand the determinant, it is easier to work directly with the equations in this case. We see from the first two boundary conditions that $c_4 = c_6 = 0$; thus, the second two equations become

$$c_3 \sinh \alpha a + c_5 \sin \alpha a = 0$$

$$c_3 \sinh \alpha a - c_5 \sin \alpha a = 0 \tag{e}$$

Adding the two, we get

$$c_3 \sinh \alpha a = 0 \longrightarrow c_3 = 0 \tag{f}$$

which leaves us with

$$c_5 \sin \alpha a = 0 \tag{g}$$

and either $c_5 = 0$, the trivial solution, or

$$\sin \alpha a = 0 \longrightarrow \alpha a = i\pi \longrightarrow \alpha = \frac{i\pi}{a} \tag{h}$$

where i may be any integer from 1 to ∞. From this, we can conclude that

$$\alpha^4 = \left(\frac{i\pi}{a}\right)^4 = \frac{m\omega^2}{EI} \longrightarrow \omega = \left(\frac{i\pi}{a}\right)^2 \sqrt{\frac{EI}{m}} \tag{i}$$

We have just found the frequency, which is called the *natural frequency* of the system, but, alas, we do not know the value of c_5. We do know that

$$W(x) = c_5 \sin \frac{i\pi x}{a} \tag{j}$$

so while we do not know all about $W(x)$, we do know its shape. It is a number of half-sine waves according to the value of i we select. This shape is called the natural *mode* of vibration for a uniform, simply supported beam. There is a distinct natural mode that goes with each natural frequency.

We can generalize what we have just learned here. For any free vibration of a beam, which is governed by a homogeneous differential equation and a set of homogeneous boundary conditions, the problem reduces to an eigenvalue problem. The solution to the homeogeneous equation for a uniform beam, Equation 6.3.23, when substituted into the boundary conditions provides us with an eigenvalue equation of the form

$$[\Lambda]\{c\} = \{0\} \tag{6.3.24}$$

from which we may find the eigenvalue, or frequency, by setting the determinant of the coefficient matrix equal to zero, or

$$|\Lambda| = 0 \tag{6.3.25}$$

The eigenfunctions may be found by substituting, in turn, each eigenvalue back into Equations 6.3.24 and then solving for all the constants but one in terms of that one. That is best understood by looking at an example.

Example 6.3.2

Find the natural frequencies and normal modes for a cantilever beam as shown in figure (a).

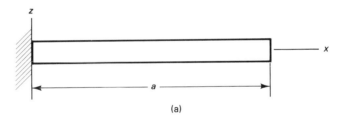

(a)

The boundary conditions are

$$W(0) = 0, \qquad W'(0) = 0, \qquad W''(a) = 0, \qquad W'''(a) = 0 \tag{a}$$

Solution: When Equation 6.3.23 is substituted into the boundary conditions, we get

$$c_4 + c_6 = 0$$
$$c_3 + c_5 = 0$$
$$c_3 \sinh \alpha a + c_4 \cosh \alpha a - c_5 \sin \alpha a - c_6 \cos \alpha a = 0$$
$$c_3 \cosh \alpha a + c_4 \sinh \alpha a - c_5 \cos \alpha a + c_6 \sin \alpha a = 0$$

$$\tag{b}$$

For these homogeneous equations to have a solution other than the trivial one, the determinant of the coefficients must be zero, or

$$\left| \begin{bmatrix} 0 & 1 & 0 & 1 \\ 1 & 0 & 1 & 0 \\ \sinh \alpha a & \cosh \alpha a & -\sin \alpha a & -\cos \alpha a \\ \cosh \alpha a & \sinh \alpha a & -\cos \alpha a & \sin \alpha a \end{bmatrix} \right| = 0 \tag{c}$$

When this is expanded, it yields

$$\cos \alpha a + \frac{1}{\cosh \alpha a} = 0 \tag{d}$$

Note that it may be more convenient to reduce the preceding set of equations to two by solving for two of the constants in terms of two others, as follows,

$$c_6 = -c_4, \qquad c_5 = -c_3 \tag{e}$$

and substitute this information into the other two equations to obtain

$$c_3 (\sinh \alpha a - \sin \alpha a) + c_4 (\cosh \alpha a + \cos \alpha a) = 0$$

$$c_3 (\cosh \alpha a + \cos \alpha a) + c_4 (\sinh \alpha a - \sin \alpha a) = 0 \tag{f}$$

Then the determinant becomes

$$\left| \begin{bmatrix} (\sinh \alpha a - \sin \alpha a) & (\cosh \alpha a + \cos \alpha a) \\ (\cosh \alpha a + \cos \alpha a) & (\sinh \alpha a - \sin \alpha a) \end{bmatrix} \right| = 0 \tag{g}$$

and when it is expanded we get equation (d).

It is not obvious what values of αa satisfy this equation. A numerical solution method and a computer program for finding the roots of this equation are described in Appendix K. Using that program, we have obtained the following for the ten lowest roots for αa.

$$
\begin{aligned}
(\alpha a)_1 &= 1.875 & &= 0.597\pi \\
(\alpha a)_2 &= 4.694 & &= 1.494\pi \\
(\alpha a)_3 &= 7.855 & &= 2.500\pi \\
(\alpha a)_4 &= 10.996 & &= 3.500\pi \\
(\alpha a)_5 &= 14.137 & &= 4.500\pi \\
(\alpha a)_6 &= 17.279 & &= 5.500\pi \\
(\alpha a)_7 &= 20.420 & &= 6.500\pi \\
(\alpha a)_8 &= 23.562 & &= 7.500\pi \\
(\alpha a)_9 &= 26.703 & &= 8.500\pi \\
(\alpha a)_{10} &= 29.845 & &= 9.500\pi
\end{aligned}
\tag{h}
$$

The pattern noted when the roots are presented as multiples of π may be explained by the fact that for higher values of αa

$$\frac{1}{\cosh \alpha a} \longrightarrow 0 \tag{i}$$

Thus, for these higher values,

$$\cos \alpha a = 0 \tag{j}$$

or

$$\alpha a = \frac{(2i - 1)\pi}{2} \tag{k}$$

It follows that

$$\omega_1 = \left(\frac{0.597\pi}{a} \right)^2 \sqrt{\frac{EI}{m}}, \qquad \omega_2 = \left(\frac{1.494\pi}{a} \right)^2 \sqrt{\frac{EI}{m}}$$

$$\omega_3 = \left(\frac{(2i - 1)\pi}{2a} \right)^2 \sqrt{\frac{EI}{m}}, \qquad \text{for } i < 2 \tag{l}$$

The mode shapes are obtained from equations (d) or from (e) and (f). We note from (f) and (e) that

$$c_4 = -c_3 \frac{\sinh \alpha_i a - \sin \alpha_i a}{\cosh \alpha_i a + \cos \alpha_i a}, \qquad c_5 = -c_3, \qquad c_6 = -c_4 \qquad \text{(m)}$$

Therefore,

$$W(x) = c_3 [(\sinh \alpha_i x - \sin \alpha_i x) - \left(\frac{\sinh \alpha_i a - \sin \alpha_i a}{\cosh \alpha_i a + \cos \alpha_i a} \right) (\cosh \alpha_i x - \cos \alpha_i x)] \qquad \text{(n)}$$

As discussed before, springs are sometimes included at the boundary to represent the elasticity of the supporting structure. For example, when cantilever beams are built into walls, the walls are not always perfectly rigid as the usual cantilever beam constraints suggest. Similarly, idealized simple supports permit no vertical deflection, whereas real supports may be elastic and be represented by springs. The next example shows how the elasticity of a support can change the natural frequencies of a beam.

Example 6.3.3

Find the equation for finding the natural frequencies of the elastically restrained cantilever beam of length a shown in figure (a).

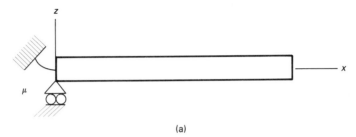

(a)

Solution: The boundary conditions are

$$W(0) = 0 \qquad EIW''(0) - \mu W'(0) = 0$$
$$EIW''(a) = 0 \qquad EIW'''(a) = 0 \qquad \text{(a)}$$

When Equation 6.3.23 is substituted into these boundary conditions, we get

$$\left| \begin{bmatrix} 0 & 1 & 0 & 1 \\ -1 & \dfrac{EI\alpha}{\mu} & -1 & \dfrac{EI\alpha}{\mu} \\ \sinh \alpha a & \cosh \alpha a & -\sin \alpha a & -\cos \alpha a \\ \cosh \alpha a & \sinh \alpha a & -\cos \alpha a & \sin \alpha a \end{bmatrix} \right| = 0 \qquad \text{(b)}$$

When expanded, this reduces to

$$\cos \alpha a + \frac{(1 + (EI/\mu a) \, \alpha a \, (-\cosh \alpha a \sin \alpha a + \sinh \alpha a \cos \alpha a)}{\cosh \alpha a} = 0 \qquad \text{(c)}$$

The program for finding roots may be used to find values of αa for given values of $EI/\mu a$.

This is all very fine, but we still have not completed the solution to the problem because we have not satisfied the initial conditions. What to do is suggested by the following example so, once again, before we generalize, let us consider a specific case.

Example 6.3.4

Find the motion for the simply supported beam in Example 6.3.3 when the initial conditions are

$$w(x, 0) = \frac{x}{a}\left(1 - \frac{x}{a}\right), \qquad \dot{w}(x, 0) = 0 \tag{a}$$

Solution: We know that one possible solution is

$$w_i(x, t) = W_i T_i = (c_{1i} \sin \omega_i t + c_{2i} \cos \omega_i t) c_{5i} \sin \frac{i\pi x}{a} \tag{b}$$

where the subscripts i have been added to emphasize that this solution holds for any given set of natural frequencies and modes. We may note that since c_1, c_2, and c_5 are all unknowns no generality is lost if we let

$$c_{1i} c_{5i} = A_i, \qquad c_{2i} c_{5i} = B_i \tag{c}$$

and let

$$w_i(x, t) = (A_i \sin \omega_i t + B_i \cos \omega_i t) \sin \frac{i\pi x}{a} \tag{d}$$

Unfortunately, if we set the time to zero and compare with the initial conditions in equations (a), we see that the second is satisfied if $A_i = 0$, but the first is never satisfied, as shown next:

$$w_i(x, 0) = B_i \sin \frac{i\pi x}{a} \neq \frac{x}{a}\left(1 - \frac{x}{a}\right)$$

$$\dot{w}_1(x, 0) = A_1 \omega_1 \sin \frac{i\pi x}{a} = 0 \longrightarrow A_1 = 0 \tag{e}$$

Fortunately, inspired by Fourier series analysis, someone thought to try the following. First, from the principle of superposition, if equation (d) is a solution to the equation and its boundary conditions for each frequency and mode pair, then a sum of all the w_i's will be a solution. So let

$$w(x, t) = \Sigma w_i(x, t) = \Sigma (A_i \sin \omega_i t + B_i \cos \omega_i t) \sin \frac{i\pi x}{a} \tag{f}$$

If the initial conditions are stated in terms of this series, we have

$$w(x, 0) = \Sigma B_i \sin \frac{i\pi x}{a} = \frac{x}{a}\left(1 - \frac{x}{a}\right)$$

$$\dot{w}(x, 0) = \Sigma A_i \omega_i \sin \frac{i\pi x}{a} = 0 \tag{g}$$

Next, multiply both sides of equations (g) by $\sin(j\pi x/a)$ and then integrate over the length of the beam; thus,

$$\int_0^a w(x, 0) \sin \frac{j\pi x}{a} dx = \int_0^a \Sigma B_i \sin \frac{i\pi x}{a} \sin \frac{j\pi x}{a} dx$$

$$= \int_0^a \frac{x}{a}\left(1 - \frac{x}{a}\right) \sin \frac{j\pi x}{a} dx \tag{h}$$

$$\int_0^a \dot{w}(x, 0) \sin \frac{j\pi x}{a} \, dx = \int_0^a \Sigma A_i \omega_i \sin \frac{i\pi x}{a} \sin \frac{j\pi x}{a} \, dx$$

$$= \int_0^a 0 \sin \frac{j\pi x}{a} \, dx$$

Then, if we interchange the integral and the summation signs and integrate term by term, we may note that

$$\int_0^a \sin \frac{i\pi x}{a} \sin \frac{j\pi x}{a} \, dx = 0, \qquad j \neq i$$

$$= \frac{a}{2}, \qquad j = i$$

(i)

Thus, for any given i, the summation reduces to only a single term. It follows that

$$A_i = \frac{2}{\omega_i a} \int_0^a 0 \sin \frac{i\pi x}{a} \, dx = 0$$

$$B_i = \frac{2}{a} \int_0^a \frac{x}{a} \left(1 - \frac{x}{a} \right) \sin \frac{i\pi x}{a} \, dx$$

(j)

With A_i and B_i known, equations (f) give us a complete solution to the problem:

$$w(x, t) = \Sigma B_i \cos \omega_i t \sin \frac{i\pi x}{a}$$

(k)

The equation, the boundary conditions, and the initial conditions are all satisfied.

What has just happened in a special case can be extended to the general case. For any beam, we can solve the set of homogeneous equation and homogeneous boundary conditions for the natural modes and frequencies. The natural mode is known in shape but is multiplied by an unknown constant. It is customary to assign an arbitrary value to this constant, say 1, and call the remaining shape the *normal mode*. Let us designate the normal mode for any beam by $\varphi_i(x)$. Then we conclude that

$$w(x, t) = \Sigma (A_i \sin \omega_i t + B_i \cos \omega_i t) \varphi_i(x)$$

(6.3.26)

We then satisfy the initial conditions by setting

$$w(x, 0) = \Sigma B_i \varphi_i(x), \qquad \dot{w}(x, 0) = \Sigma A_i \omega_i \varphi_i(x)$$

(6.3.27)

To obtain the constants A_i and B_i, we multiply both sides by $m\varphi_j(x)$ and integrate over the length of the beam, as follows:

$$\int \Sigma B_i \varphi_i(x) \varphi_j(x) m \, dx = \int w(x, 0) \varphi_j(x) m \, dx$$

$$\int \Sigma A_i \omega_i \varphi_i(x) \varphi_j(x) m \, dx = \int \dot{w}(x, 0) \varphi_j(x) m \, dx$$

(6.3.28)

The success of this method depends on a property of normal modes called *orthogonality*. It so happens that

$$\int \varphi_i(x) \varphi_j(x) m \, dx = 0, \qquad j \neq i$$

$$= M_i, \qquad j = i$$

(6.3.29)

Thus,

$$A_i = \frac{1}{M_i \omega_i} \int \dot{w}(x, 0)\varphi_i(x)m\,dx$$

$$B_i = \frac{1}{M_i} \int w(x, 0)\varphi_i(x)m\,dx \tag{6.3.30}$$

where

$$M_i = \int [\varphi_i(x)]^2 m\,dx \tag{6.3.31}$$

For proof of orthogonality, let us consider two dissimilar normal modes φ_i and φ_j. It follows that

$$(EI\varphi_i'')'' - m\omega_i^2 \varphi_i = 0, \qquad (EI\varphi_j'')'' - m\omega_j^2 \varphi_j = 0 \tag{6.3.32}$$

If we multiply the first equation by φ_j and the second by φ_i and integrate over the length of the beam, we get

$$\int (EI\varphi_i'')''\varphi_j\,dx = \omega_i^2 \int m\varphi_i \varphi_j\,dx$$

$$\int (EI\varphi_j'')''\varphi_i\,dx = \omega_j^2 \int m\varphi_j \varphi_i\,dx \tag{6.3.33}$$

Subtracting the first from the second of Equations 6.3.29 gives us

$$(\omega_j^2 - \omega_i^2) \int m\varphi_i \varphi_j\,dx = \int [(EI\varphi_j'')''\varphi_i\,dx$$

$$- \int (EI\varphi_i'')''\varphi_j\,dx] \tag{6.3.34}$$

Integrating the right-hand side by parts twice results in

$$(\omega_j^2 - \omega_i^2) \int m\varphi_i \varphi_j\,dx = [(EI\varphi_j'')'\varphi_i - (EI\varphi_i'')'\varphi_j$$

$$- (EI\varphi_j'')\varphi_i' + (EI\varphi_i'')'\varphi_j'] \Big|_0^a \tag{6.3.35}$$

The right-hand side of Equation 6.3.35 can be seen to vanish for typical boundary conditions, in fact, for all possible boundary homogeneous boundary conditions; thus, if $\omega_j \neq \omega_i$

$$\int m\varphi_i \varphi_j\,dx = 0 \tag{6.3.36}$$

which is a proof of orthogonality.

Let us now do another example of free motion started by initial conditions.

Example 6.3.5

Show how to find the motion of the cantilever beam in Example 6.3.2 imposed by applying a lateral load F at the free end and at time $t = 0$ releasing that load. The initial velocity is zero.

Solution: From the methods of Chapter 2, we can obtain the initial shape of the beam at time $t = 0$ and hence the initial conditions are

$$w(x, 0) = \frac{Fa^3}{6EI}\left(\frac{x}{a}\right)^2\left(\frac{x}{a} - 3\right), \qquad \dot{w}(x, 0) = 0 \tag{a}$$

From Equations 6.3.30 and equation (n) from Example 6.3.2, we obtain $A_i = 0$ and

$$B_i = \frac{1}{M_i} \int_0^a \frac{Fa^3}{6EI}\left(\frac{x}{a}\right)^2\left(\frac{x}{a} - 3\right)[(\sinh \alpha_i x - \sin \alpha_i x)$$

$$- \left(\frac{\sinh \alpha_i a - \sin \alpha_i a}{\cosh \alpha_i a + \cos \alpha_i a}\right)(\cosh \alpha_i x - \cos \alpha_i x)]m\,dx \tag{b}$$

where

$$M_i = \int_0^a [(\sinh \alpha_i x - \sin \alpha_i x)$$
$$- \left(\frac{\sinh \alpha_i a - \sin \alpha_i a}{\cosh \alpha_i a + \cos \alpha_i a}\right)(\cosh \alpha_i x - \cos \alpha_i x)]^2 m \, dx \qquad \text{(c)}$$

The complete solution is then obtained from Equation 6.3.26:

$$w(x, t) = \Sigma B_i \cos \omega_i t [(\sinh \alpha_i x - \sin \alpha_i x)$$
$$- \left(\frac{\sinh \alpha_i a - \sin \alpha_i a}{\cosh \alpha_i a + \cos \alpha_i a}\right)(\cosh \alpha_i x - \cos \alpha_i x)] \qquad \text{(d)}$$

where the α_i are found in equation (h) in Example 6.3.2.

The extension of the differential equation method for free vibration to more complex cases becomes difficult for the same reasons discussed at the end of Section 5.2.3. Before we turn to approximate methods based on work and energy, we shall continue the differential equation method to consider some additional features of structural dynamics. The value of these methods is in providing exact answers to use in comparison with the approximate methods.

6.3.3 Forced Vibration of Beams in Bending

When a distributed dynamic load is applied to the beam, the motion is described by Equation 6.3.2, which is repeated here for convenience:

$$(EIw'')'' - Nw'' + m\ddot{w} = f(x, t) \qquad (6.3.37)$$

or when there is no axial force

$$(EIw'')'' + m\ddot{w} = f(x, t) \qquad (6.3.38)$$

The applied load $f(x, t)$ is called the *forcing function*. We shall now show how natural frequencies and modes are used to find the forced motion. Let the forced motion be given in the form

$$w(x, t) = \Sigma \xi_i(t)\varphi_i(x) \qquad (6.3.39)$$

where $\xi_i(t)$ are as yet unknown functions of time, and $\varphi_i(x)$ are the known normal modes of a particular beam. We shall call the $\xi_i(t)$ *normal coordinates*.

Next we substitute Equation 6.3.39 into 6.3.38. We shall drop the axial force for the present to save writing it out each time, but it can easily be added back later. We get

$$m \Sigma \ddot{\xi}_i\varphi_i + \Sigma \xi_i(EI\varphi_i'')'' = f(x, t) \qquad (6.3.40)$$

If we now multiply both sides of this equation by $\varphi_j(t)$ and integrate over the length of the beam, we get

$$\Sigma \ddot{\xi}_i \int \varphi_i\varphi_j m \, dx + \Sigma \xi_i \int (EI\varphi_i'')'' \varphi_j \, dx = \int f(x, t) \varphi_j \, dx \qquad (6.3.41)$$

Since the normal modes individually satisfy

$$(EI\varphi_i'')'' = \omega_i^2 m\varphi_i \qquad (6.3.42)$$

we can substitute this into Equation 6.3.41 and get

$$\Sigma \, \ddot{\xi_i} \int \varphi_i \varphi_j m \, dx + \Sigma \, \xi_i \omega_i^2 \int \varphi_i \varphi_j m \, dx = \int f(x, t) \varphi_j \, dx \qquad (6.3.43)$$

Since $\int \varphi_i \varphi_j m \, dx = 0$ for $i \neq j$ from orthogonality, we can note that all the terms in the summation but one are zero, or

$$\ddot{\xi_j} \int \varphi_j^2 m \, dx + \xi_j \omega_j^2 \int \varphi_j^2 m \, dx = \int f(x, t) \varphi_j \, dx \qquad (6.3.44)$$

or

$$M_j \ddot{\xi_j} + M_j \omega_j^2 \xi_j = \Xi_j(t) \qquad (6.3.45)$$

where

$$M_j = \int \varphi_j^2 m \, dx, \qquad \Xi_j(t) = \int f(x, t) \varphi_j \, dx \qquad (6.3.46)$$

It is common to call M_j the *generalized mass* and $\Xi_j(t)$ the *generalized force*. Note that we can pick either i or j or, for that matter, choose another symbol to be the subscript.

This process has the interesting result of replacing the partial differential equation governing the behavior of a beam in forced motion by a set of second-order ordinary differential equations. Furthermore, they are uncoupled or independent differential equations that can be solved one by one.

The solution of each uncoupled equation is well known by analogy with the familiar mass-spring or *harmonic oscillator* system that is studied in great detail in books such as Reference 22. In Figure 6.3.2, we see such a system where m is the mass, k the spring

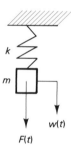

$F(t)$ **Figure 6.3.2**   Harmonic Oscillator.

constant, $F(t)$ the applied load, and $w(t)$ the deflection. The equation of motion of the harmonic oscillator is

$$m\ddot{w} + kw = F(t) \longrightarrow \ddot{w} + \omega^2 w = \frac{F(t)}{m} \qquad (6.3.47)$$

where

$$\omega^2 = \frac{k}{m} \qquad (6.3.48)$$

which may be compared with Equation 6.3.45 rewritten as

$$\ddot{\xi_j} + \omega_j^2 \xi_j = \frac{\Xi_j(t)}{M_j} \qquad (6.3.49)$$

Thus, what we have learned about the mass-spring system carries over directly to the uncoupled equations. We shall discuss this solution briefly next, but for a thorough understanding look to Reference 22 and similar books on the subject.

There are two main classifications of forced motion that derive their names from the response to the applied load. When the applied loading is periodic in time, particularly when it is sinusoidal, we refer to the deflection of the beam as *harmonic response*. When it is not periodic, we refer to it as *transient response*. Consider first the behavior of a beam when the load

$$\Xi_j(t) = Z_j \sin \Omega t \tag{6.3.50}$$

is applied and where Ω is a known frequency. Then Equation 6.3.6 can be written

$$\ddot{\xi}_j + \omega_j^2 \xi_j = \frac{Z_j}{M_j} \sin \Omega t \tag{6.3.51}$$

The particular solution will be of the form of $\sin \Omega t$ also. The complete solution is

$$\xi_j(t) = A_j \sin \omega_j t + B_j \cos \omega_j t + \frac{Z_j}{(\omega_j^2 - \Omega^2) M_j} \sin \Omega t \tag{6.3.52}$$

The initial conditions are given by

$$w(x, 0) = \Sigma \, \xi_i(0)\varphi_i(x), \qquad \dot{w}(x, 0) = \Sigma \, \dot{\xi}_i(0)\varphi_i(x) \tag{6.3.53}$$

Once again we multiply by $\varphi_j(x)m$ and integrate over the length of the beam. From this,

$$\Sigma \, \xi_i(0) \int \varphi_i(x)\varphi_j(x)m \, dx = \int w(x, 0)\varphi_j(x)m \, dx$$
$$\Sigma \, \dot{\xi}_i(0) \int \varphi_i(x)\varphi_j(x)m \, dx = \int \dot{w}(x, 0)\varphi_j(x)m \, dx \tag{6.3.54}$$

and we get

$$\xi_j(0) = \frac{1}{M_j} \int w(x, 0)\varphi_j(x)m \, dx, \qquad \dot{\xi}_j(0) = \frac{1}{M_j} \int \dot{w}(x, 0)\varphi_j(x)m \, dx \tag{6.3.55}$$

For the special case of a beam initially at rest, which is a frequently encountered case,

$$\xi_i(0) = \dot{\xi}_j(0) = 0 \tag{6.3.56}$$

For the beam initially at rest,

$$\xi_j(0) = B_j = 0$$
$$\dot{\xi}_j(0) = A_j\omega_j + \Omega \frac{Z_j}{\omega_j^2 M_j} = 0 \longrightarrow A_j = -\frac{\Omega}{\omega_j} \frac{Z_j}{\omega_j^2 M_j} \tag{6.3.57}$$

Thus,

$$\xi_j(t) = \frac{Z_j}{(\omega_j^2 - \Omega^2) M_j} \left(\sin \Omega t - \frac{\Omega}{\omega_j} \sin \omega_j t \right) \tag{6.3.58}$$

When this is substituted into Equation 6.3.39, we have a complete solution to the motion:

$$w(x, t) = \sum_{j=1}^{\infty} \frac{Z_j}{(\omega_j^2 - \Omega^2) M_j} \left(\sin \Omega t - \frac{\Omega}{\omega_j} \sin \omega_j t \right) \varphi_j(x) \tag{6.3.59}$$

This solution is seen to be indeterminate when $\Omega = \omega_j$. Applying a limiting procedure, for each normal coordinate when $\Omega = \omega_j$, we get

$$\xi_j(t) = \frac{Z_j}{2\omega_j^2 M_j}(\sin \Omega t - \omega_j t \sin \omega_j t) \tag{6.3.60}$$

This is an oscillation that increases indefinitely in amplitude with the passage of time. Such a response is known as *resonance* and occurs for any linear elastic structure of the type being studied when a periodic force is applied at a frequency that is equal to a natural frequency of the structure. Theoretically, the amplitude of resonance response will grow without bound.

It is instructive to plot the amplitude of motion depicted by Equation 6.3.57. Let

$$\frac{Z_j}{(\omega_j^2 - \Omega^2)M_j} = G_j = \frac{\Gamma_j}{[1 - (\Omega^2/\omega_j^2)]} \tag{6.3.61}$$

and let us plot the absolute value of G_j/Γ_j versus Ω/ω_j, as shown in Figure 6.3.3. This shows that as the applied frequency approaches a natural frequency, or $\Omega/\omega_j \longrightarrow 1$, the amplitude of the motion goes very large. This is called the *resonance peak*, and such a peak occurs for each natural frequency.

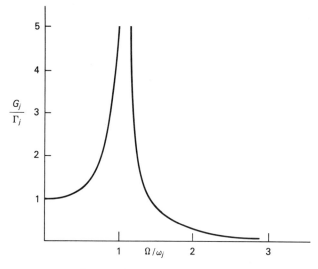

Figure 6.3.3 Resonance.

In a real structure, there is a dissipation of energy to prevent the motion from growing indefinitely. Instead, when the forcing frequency equals the natural frequency, the amplitude will build up to some large, but finite, value and remain constant at that value. The phenomenon that causes this is called *damping* in structures and will be considered in a subsequent section of this chapter.

Now for an example of resonance.

Example 6.3.6

Consider a simply supported beam initially at rest that has a uniform periodic force applied suddenly at time $t = 0$ as shown in figure (a).

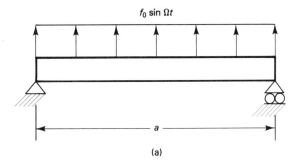

$f_0 \sin \Omega t$

(a)

Solution: The governing equation is

$$w'''' + \frac{m}{EI} \ddot{w} = \frac{f_0}{EI} \sin \Omega t \qquad (a)$$

and the boundary conditions are

$$w(0, t) = 0, \qquad EIw(0, t) = 0, \qquad w(a, t) = 0, \qquad EIw(a, t) = 0 \qquad (b)$$

The initial conditions for a beam at rest are

$$w(x, 0) = 0, \qquad \dot{w}(x, 0) = 0 \qquad (c)$$

The solution will be found in the form given in Equation 6.3.39, where $\xi_i(t)$ is the solution to Equation 6.3.45. From Example 6.3.1, we have the natural frequencies and normal modes for a simply supported uniform beam:

$$\omega_j = \left(\frac{j\pi}{a}\right)^2 \sqrt{\frac{EI}{m}}, \qquad \varphi_j = \sin\frac{j\pi x}{a} \qquad (d)$$

Then

$$M_j = \int_0^a \left(\sin\frac{j\pi x}{a}\right)^2 m \, dx = \frac{ma}{2}$$

$$\Xi_j = \int_0^a \frac{f_0}{EI} \sin \Omega t \sin\frac{j\pi x}{a} \, dx = Z_j \sin \Omega t \qquad (e)$$

where

$$Z_j = \frac{f_0 a}{j\pi EI} [1 - (-1)^j] \qquad (f)$$

The problem is now in the form of the harmonic response problem studied in Equations 6.3.49 through 6.3.58. The motion of the beam is

$$w(x, t) = \Sigma \, \xi_j(t) \sin\frac{j\pi x}{a} \qquad (g)$$

where

$$\xi_j(t) = \frac{Z_j}{M_j(\omega_j^2 - \Omega^2)} \left(\sin \Omega t - \frac{\Omega}{\omega_j} \sin \omega_j t\right) \qquad (h)$$

with Z_j, ω_j, and M_j given previously.

If the applied forcing function includes concentrated applied forces, $F(t)$, the generalized force is simply the concentrated force times the normal mode evaluated at the point where the load is applied. This is often stated in the mathematical form of an intergral using the Dirac delta function. Let $F_1(t)$ be a z-component of a lateral load on a beam at $x = a_1$ and $\delta(x - a_1)$ be the delta function; then

$$\Xi_j(t) = \int F_1(t)\delta(x - a_1)\varphi_j(x)\ dx = F_1(t)\varphi_j(a_1) \tag{6.3.62}$$

In a similar way, a concentrated moment, say, $M_2(t)$ at $x = a_2$ would contribute

$$\Xi_j(t) = \int M_2(t)\delta(x - a_2)\frac{\partial\varphi_j(x)}{\partial x}\ dx = M_2(t)\frac{\partial\varphi_j(a_2)}{\partial x} \tag{6.3.63}$$

The next example has a concentrated dynamic force.

Example 6.3.7

Consider a simply supported beam initially at rest that has a concentrated periodic force applied suddenly at time $t = 0$, as shown in figure (a).

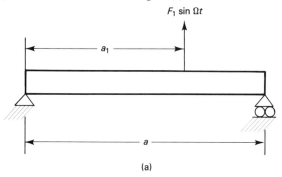

(a)

Solution: The generalized force for this case is

$$\Xi_j(t) = F_1 \sin \Omega t \sin \frac{j\pi a_1}{a} \tag{a}$$

The equations of forced motion are

$$\ddot{\xi}_j + \omega_j^2\xi_j = \frac{\Xi_j}{M_j} = \frac{2F_1}{ma} \sin \frac{j\pi a_1}{a} \sin \Omega t \tag{b}$$

which has for a solution

$$\xi_j(t) = \frac{2F_1}{ma(\omega_j^2 - \Omega^2)} \sin \frac{j\pi a_1}{a} \left(\sin \Omega t - \frac{\Omega}{\omega_j} \sin \omega_j t \right) \tag{c}$$

The complete motion of the beam is

$$w(x, t) = \sum_1^\infty \frac{2F_1}{ma(\omega_j^2 - \Omega^2)} \sin \frac{j\pi a_1}{a} \left(\sin \Omega t - \frac{\Omega}{\omega_j} \sin \omega_j t \right) \sin \frac{j\pi x}{a} \tag{d}$$

Once again, we observe resonance.

When the applied force is not periodic, we have a transient response. The method of solution is the same, as is shown in the next example, but the phenomenon of resonance does not appear. Each different forcing function requires a different particular solution.

Example 6.3.8

Find the motion of a simply supported beam that is at rest at time $t = 0$ and suddenly has a constant uniform load applied, as shown in the plot of load versus time in figure (a).

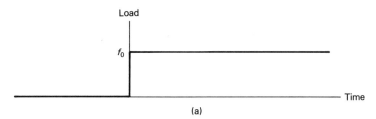

(a)

Solution: The generalized force is

$$\Xi_j(t) = f_0 \int_0^a \sin \frac{j\pi x}{a} \, dx = \frac{f_0 a}{j\pi} (\cos j\pi - 1) \tag{a}$$

The equations for forced motion are

$$\ddot{\xi}_j + \omega_j^2 \xi_j = \frac{\Xi_j}{M_j} = \frac{2f_0}{mj\pi} (\cos j\pi - 1) = C_j \tag{b}$$

where C_j is a constant for which

$$C_j = 0, \, j = \text{even}, \qquad C_j \neq 0, \, j = \text{odd} \tag{c}$$

A particular solution is

$$(\xi_j)_p = \frac{C_j}{\omega_j^2} \tag{d}$$

so that a general solution is

$$\xi_j(t) = A_j \sin \omega_j t + B_j \cos \omega_j t + \frac{C_j}{\omega_j^2} \tag{e}$$

The initial conditions are

$$\xi_j(0) = 0, \qquad \dot{\xi}_j(0) = 0 \tag{f}$$

Upon substitution of the general solution into the boundary conditions, we get

$$\xi_j(0) = B_j + \frac{C_j}{\omega_j^2} = 0, \qquad \longrightarrow B_j = -\frac{C_j}{\omega_j^2}$$

$$\dot{\xi}_j(0) = A_j \omega_j = 0, \qquad \longrightarrow A_j = 0 \tag{g}$$

so that

$$\xi_j(t) = \frac{C_j}{\omega_j^2} (1 - \cos \omega_j t), \qquad j = \text{odd} \tag{h}$$

Forced motion for the simply supported beam, both harmonic and transient, is relatively straightforward because the normal modes are simple sine functions. The normal modes for other beams are much more complicated and, therefore, are difficult to work with. This method is usually not used in the practical analysis of more complex structures, but it does serve as a norm for comparing approximate methods. The simply supported beam does have much more to offer us in understanding vibration via the differential equation method, so we shall continue with it for a bit longer.

6.3.4 Beams with Damping

As noted earlier, the motion of structures continues indefinitely once started by imposing initial conditions and/or applied loads. In practice, this motion often is modified or even dies out as a result of forces that resist the motion. These are the *damping forces*. The actual physical phenomena that cause damping are complex and generally hard to model mathematically. One idealization of the true damping forces that has a certain amount of mathematical simplicity is called *viscous damping*. This type generates a force proportional to the velocity and in opposition to the direction of the velocity. In its simplest form for a beam, it adds a force per unit length, $-c\dot{w}(x, t)$, to the beam element in Figure 6.3.1. The equation of motion given in Equation 6.3.2 now becomes

$$(EIw'')'' - Nw'' + c\dot{w} + m\ddot{w} = f(x, t) \tag{6.3.64}$$

Under certain circumstances, solutions in series form are possible. Let us once again turn to the normal modes, φ_i, and frequencies, ω_i, determined from the separated form of the undamped system of equations

$$(EIW'')'' - NW'' + mW = 0 \tag{6.3.65}$$

Let us assume a solution of the form of Equation 6.3.39, or

$$w(x, t) = \Sigma \, \xi_i(t)\varphi_i(x) \tag{6.3.66}$$

If we substitute this into Equation 6.3.64 and go through the same routine as given in Equations 6.3.40 through 6.3.44, we obtain

$$\Sigma \, \ddot{\xi}_i \int \varphi_i\varphi_j m \, dx + \Sigma \, \dot{\xi}_i \int \varphi_i\varphi_j c \, dx + \Sigma \, \xi_i\omega_j^2 \int \varphi_i\varphi_j m \, dx = \int f(x, t)\varphi_j \, dx \tag{6.3.67}$$

In general, the normal modes are not orthogonal when weighted with the damping coefficient; that is,

$$\int \varphi_i\varphi_j c \, dx \neq 0 \tag{6.3.68}$$

for any combination of i and j; however, when c is proportional to m, for example,

$$c = \beta m \tag{6.3.69}$$

then

$$\int \varphi_i\varphi_j c \, dx = \beta \int \varphi_i\varphi_j m \, dx = 0, \qquad i \pm j$$
$$= \beta M_j = C_j, \qquad i = j \tag{6.3.70}$$

and we get

$$\ddot{\xi}_j \int \varphi_j^2 m \, dx + \dot{\xi}_j \beta \int \varphi_j^2 m \, dx + \xi_j\omega_j^2 \int \varphi_j^2 m \, dx = \int f(x, t)\varphi_j \, dx \tag{6.3.71}$$

or

$$M_j\ddot{\xi}_j + C_j\dot{\xi}_j + M_j\omega_j^2\xi_j = \Xi_j(t) \tag{6.3.72}$$

This is an uncoupled set of equations for which each equation can be solved by standard techniques. After solving the individual equations, the results are assembled in the series form of Equation 6.3.66 to form the complete solution.

This is the form of damping, called *proportional damping*, usually studied, because it can be treated mathematically in a convenient way. Unlike the elastic and inertia properties of a structure for which the mathematical model is a highly accurate representation of the true physical model, the true damping properties are not accurately modeled by proportional damping. The actual damping mechanisms are beyond the space available and the scope of this text. The damping described here does give great insight into the damping process and, therefore, is useful. We shall return to it when we study approximate methods in a later section.

We may note that when Equation 6.3.72 is rewritten

$$\ddot{\xi}_j + \beta\dot{\xi}_j + \omega_j^2\xi_j = \frac{\Xi_j}{M_j} \tag{6.3.73}$$

it may be compared with the mass-spring-damper system of Figure 6.3.4. The equation of this system is

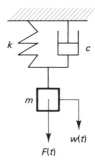

Figure 6.3.4 Damped Harmonic Oscillator.

$$m\ddot{w} + c\dot{w} + kw = F(t) \longrightarrow \ddot{w} + \beta\dot{w} + \omega^2 w = \frac{F(t)}{m} \tag{6.3.74}$$

where

$$\beta = \frac{c}{m}, \qquad \omega^2 = \frac{k}{m} \tag{6.3.75}$$

Thus, all that we know about this system will help us here.

First, we find the homogenous solution by assuming an exponential form. The characteristic equation for the homogenous part of Equation 6.3.73 is

$$r^2 + \beta r + \omega_j^2 = 0 \tag{6.3.76}$$

The roots are

$$r_1 = -\frac{\beta}{2} + \sqrt{\frac{\beta^2}{4} - \omega_j^2}, \qquad r_2 = -\frac{\beta}{2} - \sqrt{\frac{\beta^2}{4} - \omega_j^2} \tag{6.3.77}$$

When

$$\beta^2 \geq 4\omega_j^2 \tag{6.3.78}$$

both roots are real and negative, and the solution to the homogenous part is

$$\xi_j = A_j e^{r_1 t} + B_j e^{r_2 t} \tag{6.3.79}$$

Both terms represent an exponential decay with time. When

$$\beta^2 \leq 4\omega_j^2 \tag{6.3.80}$$

both roots are complex,

$$r_1 = -\frac{\beta}{2} + i\chi, \qquad r_2 = -\frac{\beta}{2} + i\chi, \qquad \chi = \sqrt{\omega_j^2 - \frac{\beta^2}{4}} \tag{6.3.81}$$

and the solution may be written

$$\xi_j(t) = e^{-(\beta/2)t}(A_j e^{i\chi} + B_j e^{-i\chi}) = e^{-(\beta/2)t}(A_j \sin \chi t + B_j \cos \chi t) \tag{6.3.82}$$

This is an oscillatory motion of frequency χ multiplied by an exponential decay with time.

Next we must find the particular part of the solution. For more complete details, see Reference 22 or a similar text. For harmonic forcing, the particular solution is obtained by assuming the following form for the trial solution:

$$(\xi_j)_p = D_j \sin (\Omega t - \psi) \tag{6.3.83}$$

When this is substituted in Equation 6.3.73 and the results collected, we find that

$$D_j = \frac{\Gamma_j}{\sqrt{[1 - (\Omega^2/\omega_j^2)]^2 + (\beta^2 \Omega^2/\omega_j^2)}}, \qquad \tan \psi = \frac{\beta\Omega/\omega_j^2}{1 - (\Omega^2/\omega_j^2)} \tag{6.3.84}$$

At this time, we recognize that the homogeneous solution will die out in time, leaving only the particular solution, which for this reason is called the *steady-state* solution. It is instructive to plot the amplitude of the steady-state solution, or more precisely D_j/Γ_j, versus Ω/ω_j, which is shown in Figure 6.3.5 for several values of β. The effect of the damping is shown clearly. The resonance peaks are now finite.

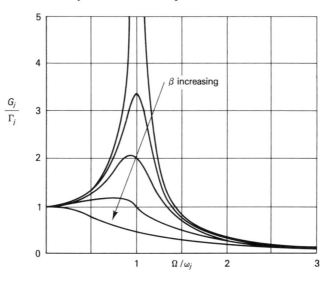

Figure 6.3.5 Resonance with Damping.

In a beam with damping and a narmonic load, the steady-state deflection is given by

$$w(x, t) = \Sigma D_j \sin (\Omega t - \psi) \varphi_j(x) \tag{6.3.85}$$

For other dynamic loadings, appropriate particular solutions must be found, and the transient response is

$$w(x, t) = \Sigma [e^{-(\beta/2)t}(A_j \sin \chi t + B_j \cos \chi t) + (\xi_j)_p] \varphi_j(x) \tag{6.3.86}$$

Appropriate initial conditions must be specified to find the constants A_j and B_j. We shall not give any examples here, but the subject of transient response will come up again when we take up approximate methods.

6.3.5 Dynamic Instability of Beams

In Section 6.3.2, we dropped the axial force from Equation 6.3.2. Now let us return it in the form of a compressive axial internal force $P = -N$. The equation of motion becomes

$$(EIw'')'' + Pw'' + m\ddot{w} = f(x, t) \tag{6.3.87}$$

First, for free motion we have $f(x, t) = 0$ and can separate variables using Equation 6.3.8. We get

$$(EIW'')'' + PW'' - m\omega^2 W = 0, \qquad \ddot{T} + \omega^2 T = 0 \tag{6.3.88}$$

Clearly, the time dependence is mathematically the same. To examine the deflection, let us assume EI is a constant. Then,

$$W'''' + \lambda^2 W'' - \alpha^4 W = 0 \tag{6.3.89}$$

where

$$\lambda^2 = \frac{P}{EI}, \qquad \alpha^4 = \frac{m\omega^2}{EI} \tag{6.3.90}$$

The characteristic equation is

$$r^4 + \lambda^2 r^2 - \alpha^4 = 0 \tag{6.3.91}$$

which has four roots in one real pair and one imaginary pair as follows:

$$\pm\kappa_1 = \pm\sqrt{\frac{-\lambda^2 + \sqrt{\lambda^4 + 4\alpha^4}}{2}}, \qquad \pm i\kappa_2 = \pm\sqrt{\frac{-\lambda^2 - \sqrt{\lambda^4 + 4\alpha^4}}{2}} \tag{6.3.92}$$

Thus, the solution is

$$W(x) = c_3 \sinh \kappa_1 x + c_4 \cosh \kappa_1 x + c_5 \sin \kappa_2 x + c_6 \cos \kappa_2 x \tag{6.3.93}$$

To continue it is best to look at a specific case.

Example 6.3.9

Find the natural modes and frequencies for a simply supported beam with an axial internal force P.

Solution: The boundary conditions are

$$W(0) = 0, \qquad W''(0) = 0, \qquad W(a) = 0, \qquad W''(a) = 0 \qquad \text{(a)}$$

Upon substitution of Equation 6.3.93 into the boundary conditions, we get

$$\begin{bmatrix} 0 & 1 & 0 & 1 \\ 0 & \kappa_1^2 & 0 & -\kappa_2^2 \\ \sinh \kappa_1 a & \cosh \kappa_1 a & \sin \kappa_2 a & \cos \kappa_2 a \\ \kappa_1^2 \sinh \kappa_1 a & \kappa_1^2 \cosh \kappa_1 a & -\kappa_2^2 \sin \kappa_2 a & -\kappa_2^2 \cos \kappa_2 a \end{bmatrix} \begin{bmatrix} c_3 \\ c_4 \\ c_5 \\ c_6 \end{bmatrix} = \begin{bmatrix} 0 \\ 0 \\ 0 \\ 0 \end{bmatrix} \qquad \text{(b)}$$

From this, we can conclude that $c_4 = c_6 = 0$ and therefore

$$\begin{bmatrix} \sinh \kappa_1 a & \sin \kappa_2 a \\ \kappa_1^2 \sinh \kappa_1 & -\kappa_2^2 \sin \kappa_2 a \end{bmatrix} \begin{bmatrix} c_3 \\ c_5 \end{bmatrix} = \begin{bmatrix} 0 \\ 0 \end{bmatrix} \qquad \text{(c)}$$

and from this it may be concluded that

$$c_3 = 0, \qquad c_5 \sin \kappa_2 a = 0 \qquad \text{(d)}$$

and therefore

$$\sin \kappa_2 a = 0 \longrightarrow \kappa_2 a = i\pi \longrightarrow \kappa_2 = \frac{i\pi}{a} \qquad \text{(e)}$$

From Equations 6.3.92 and (e), we can conclude that

$$\alpha^4 = \kappa_2^2(\kappa_2^2 - \lambda^2) \longrightarrow \omega_i^2 = \frac{i^2\pi^2}{ma^2}\left(\frac{i^2\pi^2 EI}{a^2} - P\right) \qquad \text{(f)}$$

Note that as P gets larger the natural frequency gets smaller; in fact, for the first frequency,

$$\omega_1 \longrightarrow 0 \quad \text{as} \quad P \longrightarrow \frac{\pi^2 EI}{a^2} \qquad \text{(g)}$$

Thus, the frequency goes to zero when the internal axial load is equal to the buckling load.

The natural modes are

$$W(x) = c_5 \sin \kappa_2 x \qquad \text{(h)}$$

The conclusion reached in Example 6.3.9 about the buckling load can be generalized. The axial load that causes the first natural frequency to go to zero is the buckling load. This is, in fact, an often used definition of buckling. We can also note that the modes and frequencies determined from Equation 6.3.89 can be used in Equations 6.3.26 and 6.3.30 for satisfying the initial conditions and in Equations 6.3.45 and 6.3.39 for finding the forced motion.

6.4 WORK AND ENERGY METHODS IN BEAM BENDING VIBRATION

The analytical methods we have been studying, for all their elegance, are very difficult to apply in all but the simplest cases. We have chosen some of those simple cases to illustrate the methods. We must find more powerful methods to handle the more complex structures we encounter in the real world of engineering. Fortunately, both the Rayleigh-Ritz and the finite element methods are as useful in structural dynamics as in structural statics and instability.

The virtual work for a dynamic structure is obtained by adding the negative of the inertia force according to D'Alembert's principle to the other forces in the external virtual work term. We shall recognize inertia forces due to both the distributed mass of the beam, m, and also possible concentrated masses, M_k.

$$\Delta\delta W_e = -\int m\ddot{w}\delta w\ dx - \Sigma\ M_k\ddot{w}\delta w(x_k) \tag{6.4.1}$$

When Equation 5.2.51 is modified to account for the inertia terms, we have

$$\delta W = \delta W_e + \Delta\delta W_e - (\delta U + \Delta\delta U)$$

$$= \int \delta w[f(x,\ t) - m\ddot{w}]\ dx - \Sigma\ M_k\ddot{w}\delta w(x_k) + \Sigma\ F_{zi}\delta w_i \tag{6.4.2}$$

$$+ \Sigma\ M_{yj}(\delta w_j)' - \int N\delta w'w'\ dx - \int EI\delta w''w''\ dx = 0$$

Similarly, the potential energy is

$$\Pi = U + \Delta U - (W_e + \Delta W_e)$$

$$= \frac{1}{2}\int EI(w'')^2\ dx + \frac{1}{2}\int N(w')^2\ dx - \int w[f(x,\ t) - m\ddot{w}]\ dx \tag{6.4.3}$$

$$+ \Sigma\ M_k\ddot{w}w(x_k) - \Sigma\ F_{zi}w_i - \Sigma\ M_{yj}w_j'$$

The solution of the integral equation for virtual work or the minimization of the potential energy proceeds much as in the static case. We use the Rayleigh-Ritz method or the finite element method.

6.4.1 Rayleigh-Ritz Method

In the past, much use has been made of the conventional Rayleigh-Ritz method for structural dynamics. We introduce a set of admissible functions and generalized coordinates of the form

$$\{w\} = [\Omega]\{q\} \tag{6.4.4}$$

exactly as in the static case, but now we must recognize that $w(x,\ t)$ and $q(t)$. The function $\Omega(x)$ has exactly the same requirements to be admissible as before. Also, as before (see Section 3.2.4 and Section 5.2.5),

$$[B] = [D][\Omega] \tag{6.4.5}$$

The strain energy terms in the virtual work expression form the $[K]$ stiffness matrix, the axial force term forms the $[K^d]$ differential stiffness matrix, and the applied load terms form the $\{Q\}$ generalized force matrix. These are identical in every respect to the static case, except that the Q_i may be a function of time since the applied loads may be functions of time.

This leaves the inertia terms to account for. As noted previously they are treated in the virtual work and potential energy expressions as equivalent applied loads by virtue of D'Alembert's principle. But they do not contribute to the $\{Q\}$ matrix because they are displacement dependent or, more properly, they are dependent on the second derivative of displacement with respect to time. When the appropriate substitution of Equation 6.4.4 is made into the virtual work integral and the integration duly performed, we are left with a set of algebraic equations of the form

$$[M]\{\ddot{q}\} + ([K] + [K^d])\{q\} = \{Q\} \tag{6.4.6}$$

where the new term $[M]\{\ddot{q}\}$ is derived from the inertia force terms in the virtual work integral. The definition of $[M]$, called the *mass matrix*, is

$$[M] = \int [\Omega]^T[\Omega]m(x)\ dx + \Sigma\ M_k[\Omega(x_k)]^T[\Omega(x_k)] \tag{6.4.7}$$

where

$$M_{ij} = \int (\Omega_i)\ (\Omega_j)m(x)\ dx + \Sigma\ M_k\Omega_i(x_k)\ \Omega_j(x_k) \tag{6.4.8}$$

If there is no axial load, the equations reduce to

$$[M]\{\ddot{q}\} + [K]\{q\} = \{Q\} \tag{6.4.9}$$

These equations may be used to find the natural frequencies and modes. To do this, we consider the set of homogeneous equations

$$[M]\{\ddot{q}\} + [K]\{q\} = 0 \tag{6.4.10}$$

With the knowledge we now have of free vibrations of elastic systems, we can assume directly a simple harmonic response and let

$$\{q\} = \{a\}\ \sin \omega t \tag{6.4.11}$$

When this is substituted into Equation 6.4.10, we get

$$([K] - \omega^2\ [M])\ \{a\} = 0 \tag{6.4.12}$$

which will have a nontrivial solution only if

$$|\ [K] - \omega^2\ [M]\ | = 0 \tag{6.4.13}$$

When expanded, this determinant gives us a polynomial in ω^2. The positive real roots of this polynomial provide approximations to the natural frequencies. Then each frequency is inserted in Equation 6.4.12 in turn and approximations to the modes are found. The following simple example will help in understanding this process.

Example 6.4.1

Find an approximation to the first two natural frequencies and modes of a uniform cantilever beam of length a using the same one- and two-term admissible functions used in Example 3.2.3.

Solution: Let

$$\Omega_1 = \left(\frac{x}{a}\right)^2, \qquad \Omega_2 = \left(\frac{x}{a}\right)^3 \tag{a}$$

For a one-term solution

$$K_{11} = \frac{4EI}{a^3} \tag{b}$$

from Example 3.2.3. The mass term is

$$M_{11} = \int_0^a \Omega_1^2 m\ dx = \int_0^a \left(\frac{x}{a}\right)^4 m\ dx = \frac{ma}{5} \tag{c}$$

and

$$| [K] - \omega^2[M] | = \frac{4EI}{a^3} - \omega^2 \frac{ma}{5} = 0 \tag{d}$$

or

$$\omega_1^2 = \frac{20EI}{a^4 m} \tag{e}$$

and the normal mode is

$$\varphi_1(x) = \left(\frac{x}{a}\right)^2 \tag{f}$$

For a two-term solution, we obtain the stiffness matrix from Example 3.2.3 and the mass matrix by integration. The remaining mass terms are

$$M_{12} = M_{21} = \int_0^a \Omega_1 \Omega_2 m \, dx = \int_0^a \left(\frac{x}{a}\right)^2 \left(\frac{x}{a}\right)^3 m \, dx = \frac{ma}{6}$$

$$M_{22} = \int_0^a \Omega_2^2 m \, dx = \int_0^a \left(\frac{x}{a}\right)^6 m \, dx = \frac{ma}{7} \tag{g}$$

The two-term matrix equation is

$$\left(\frac{EI}{a^3} \begin{bmatrix} 4 & 6 \\ 6 & 12 \end{bmatrix} - \frac{\omega^2 ma}{210} \begin{bmatrix} 42 & 35 \\ 35 & 30 \end{bmatrix} \right) \begin{bmatrix} a_1 \\ a_2 \end{bmatrix} = \begin{bmatrix} 0 \\ 0 \end{bmatrix} \tag{h}$$

Let

$$\gamma = \frac{\omega^2 ma^4}{210 EI} \tag{i}$$

Then the determinant reduces to

$$\left| \begin{bmatrix} 4 - 420\gamma & 6 - 35\gamma \\ 6 - 35\gamma & 12 - 30\gamma \end{bmatrix} \right| = 0 \tag{j}$$

or

$$35\gamma^2 - 204\gamma + 12 = 0 \tag{k}$$

which has the roots

$$\gamma_1 = 0.0594295 \longrightarrow \omega_1^2 = \frac{12.480 EI}{a^4 m}$$

$$\gamma_2 = 5.7691419 \longrightarrow \omega_2^2 = \frac{1211.52 EI}{a^4 m} \tag{l}$$

The first mode is found from

$$(4 - 420\gamma_1)a_1 + (6 - 35\gamma_1)a_2 = 0 \longrightarrow a_2 = 5.347a_1 \tag{m}$$

and, therefore,

$$\varphi_1(x) = \left(\frac{x}{a}\right)^2 + 5.347 \left(\frac{x}{a}\right)^3 \tag{n}$$

The second mode is found from

$$(4 - 420\gamma_2)a_1 + (6 - 35\gamma_2)a_2 = 0 \longrightarrow a_2 = -12.347a_1 \tag{o}$$

and, therefore,

$$\varphi_2(x) = \left(\frac{x}{a}\right)^2 - 12.347 \left(\frac{x}{a}\right)^3 \tag{p}$$

These approximate values of the first and second frequencies and modes may be compared to the exact answers in Example 6.3.2. Comparing frequencies, we have for the exact values

$$\omega_1^2 = \frac{12.360EI}{a^4 m}, \qquad \omega_2^2 = \frac{485.48EI}{a^4 m} \tag{q}$$

Clearly, the first frequency in the one-term solution is not very accurate, but in the two-term solution it appears to be a close approximation, while the second is not at all close. More terms would be needed to improve the answers; nevertheless, the first frequency is quite close considering the crude approximations chosen as admissible functions.

We may note that normal modes of a beam are admissible functions for any different beam with the same constraints; for example, the normal modes of a uniform beam may be used as admissible functions for a tapered beam with the same geometric boundary conditions. In the next example, we use this fact to show how a problem with a concentrated mass may be solved.

Example 6.4.2

A uniform, simply supported beam has a concentrated mass attached as shown in figure (a). Find the first frequency and mode by the Rayleigh-Ritz method.

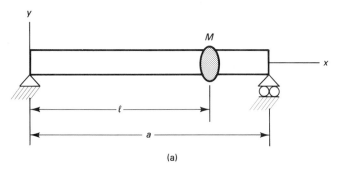

(a)

Solution: Suitable assumed functions are the normal modes for the uniform, simply supported beam without the concentrated mass:

$$\Omega_i = \sin \frac{i\pi x}{a} \tag{a}$$

They have the advantage of being simple, of satisfying natural as well as geometric boundary conditions, and of being orthogonal functions. It follows that

$$K_{ij} = \int_0^a EI\Omega_i''\Omega_j'' \, dx = \frac{i^2 j^2 \pi^4 EI}{a^4} \int_0^a \sin \frac{i\pi x}{a} \sin \frac{j\pi x}{a} \, dx$$

$$= \left(\frac{i\pi}{a}\right)^4 \frac{EIa}{2}, \qquad i = j \tag{b}$$

$$= 0, \qquad\qquad i \neq j$$

Thus, $[K]$ is a diagonal matrix. The mass matrix $[M]$ has components

$$M_{ij} = \int_0^a m \sin\frac{i\pi x}{a} \sin\frac{j\pi x}{a}\, dx + M \sin\frac{i\pi l}{a} \sin\frac{j\pi l}{a}$$

$$= \frac{ma}{2} + M\left(\sin\frac{i\pi l}{a}\right)^2, \qquad i = j \qquad\qquad\text{(c)}$$

$$= M \sin\frac{i\pi l}{a} \sin\frac{j\pi l}{a} \qquad i \neq j$$

In a one-term solution,

$$\omega_1^2 = \frac{K_{ij}}{M_{ij}} = \frac{EI\pi^4}{2a^3}\left[\frac{ma}{2} + M\left(\sin\frac{i\pi l}{a}\right)^2\right] \qquad\qquad\text{(d)}$$

$$\varphi_1(x) = \sin\frac{\pi x}{a}$$

Such a solution might be an adequate approximation for $l = a/2$, since both the approximate normal mode and the selected admissible function are symmetrical about the midpoint of the beam, but it would not be adequate when l is off center. Then at least two terms would be required. In a two- or more term solution, the mass matrix would not be diagonal when the concentrated mass was not at $x = l/2$.

Approximate modes and frequencies found by this method are often used to satisfy initial conditions and forced motion by the normal mode method using the equations we derived in Section 6.3. In other cases, Equations 6.4.9 are solved directly. It should be noted that these equations generally are coupled as noted in Example 6.4.2 and, therefore, are considerably more difficult to solve than the uncoupled Equations 6.3.45. Because of this difficulty, when normal mode methods are not feasible, direct numerical solution of the equations of forced motion by finite difference techniques is often preferred. This type of solution will be taken up later in this chapter.

The Rayleigh-Ritz method has been applied to very simple examples to show how it applies to the eigenvalue problem of vibration. The real value of the method is for solving problems for which the differential equation method is cumbersome or even impossible. Such extensions may be found in the literature. As we have noted, in recent years the special case of Rayleigh-Ritz, the finite element method, is widely used to solve the more complicated problems. We shall turn to that method now.

6.4.2 Finite Element Method for Free Vibration

We can work directly with the equations in finite element form from Section 3.4.1 as adapted for beams in Sections 4.4.2 and 5.2.7. The shape functions and the relation between strain and nodal displacements are exactly the same as used in the static analysis; that is,

$$\{u\} = w = [N]\{r\}, \qquad [B] = [D][N] \qquad\qquad (6.4.14)$$

When these are substituted into the expression for virtual work in Equation 6.4.2, we obtain

$$[M]\{\ddot{r}\} + ([K] + [K^d])\{r\} = \{R\} \qquad\qquad (6.4.15)$$

where the stiffness matrix is the same as that used in static analysis, or

$$[K] = \int [B]^T[G][B] \, dx + \text{spring terms} \tag{6.4.16}$$

and the differential stiffness matrix is the same as that used in static instability analysis, or

$$[K^d] = -P \int [N']^T[N'] \, dx \tag{6.4.17}$$

We now add the mass matrix

$$[M] = \int [N]^T[N]m(x) \, dx + \text{concentrated mass terms} \tag{6.4.18}$$

and the applied load matrix, noting that it is defined by the same integral as in the static case, but it is now a function of time:

$$\{R\} = \int [N]^T f(x, t) \, dx + \text{concentrated force terms} \tag{6.4.19}$$

Corresponding matrices and integrals for a beam element are indicated by using lowercase symbols. For a single beam element, we have the stiffness matrix $[k]$, the differential stiffness matrix $[k^d]$, and the mass matrix $[m]$, which are then assembled for the whole structure. The element shape functions are given in Equation 4.4.23 and repeated here for element i.

$$[n]_i = \frac{1}{L_i^3} [L_i^3 - 3L_i s^2 + 2s^3, \ L_i^3 s - 2L_i^2 s^2 + L_i s^3,$$

$$3L_i s^2 - 2s^3, \ -L_i^2 s^2 + L_i s^3] \tag{6.4.20}$$

From this we obtain the element stiffness matrix in the x-z plane, as shown in Equation 4.4.28.

$$[k]_i = \int EI[n]_i^T[n]_i \, dx = \frac{EI_i}{L_i^3}
\begin{bmatrix}
12 & -6L_i & -12 & -6L_i \\
-6L_i & 4L_i^2 & 6L_i & 2L_i^2 \\
-12 & 6L_i & 12 & 6L_i \\
-6L_i & 2L_i^2 & 6L_i & 4L_i^2
\end{bmatrix} \tag{6.4.21}$$

We can add the differential stiffness as derived in Section 5.2. Letting

$$[k^d] = -P \, [k^D] \tag{6.4.22}$$

from Equation 5.2.81 the element differential stiffness matrix $[k^D]$ is

$$[k^D]_i = \int [n']_i^T[n']_i \, dx = \frac{1}{30L_i}
\begin{bmatrix}
36 & -3L_i & -36 & -3L_i \\
-3L_i & 4L_i^2 & 3L_i & -L_i^2 \\
-36 & 3L_i & 36 & 3L_i \\
-3L_i & -L_i^2 & 3L_i & 4L_i^2
\end{bmatrix} \tag{6.4.23}$$

With these shape functions, the element mass matrix for a uniform distributed mass is

$$[m]_i = \int [n]_i^T[n]_i m \, dx = \frac{m}{420}
\begin{bmatrix}
156 & -22L_i & 54 & 13L_i \\
-22L_i & 4L_i^2 & -13L_i & -3L_i^2 \\
54 & -13L_i & 156 & 22L_i \\
13L_i & -3L_i^2 & 22L_i & 4L_i^2
\end{bmatrix} \tag{6.4.24}$$

We can then assemble the element matrices along with the equivalent nodal load matrix to form the global set of matrix equations shown symbolically in Equation 6.4.15.

For free vibration, we let $\{R\} = \{0\}$ and assume harmonic motion, or

$$\{r\} = \{a\} \sin \omega t \tag{6.4.25}$$

and

$$([K] + [K^d] - \omega^2[M])\{a\} = \{0\} \tag{6.4.26}$$

We shall have a nontrivial solution only if

$$\mid [K] + [K^d] - \omega^2[M] \mid = 0 \tag{6.4.27}$$

Once we have obtained the natural frequencies, ω, we can insert them in turn back in Equations 6.4.26 and find the corresponding modes, $\{a\}$. Note that one of the elements of $\{a\}$ will remain undetermined. Let us now do an example in free vibration.

Example 6.4.3

Find the first two natural frequencies and modes for the beam in Example 5.2.6. First, solve for $P = 0$ and then for $P = 20EI/a^2$.

Solution: First, assemble the mass matrix.

$$[M] = \frac{m}{420}
\begin{bmatrix}
156 & -22\left(\dfrac{a}{2}\right) & 54 & 13\left(\dfrac{a}{2}\right) & 0 & 0 \\[2mm]
-22\left(\dfrac{a}{2}\right) & 4\left(\dfrac{a}{2}\right)^2 & -13\left(\dfrac{a}{2}\right)^2 & -3\left(\dfrac{a}{2}\right)^2 & 0 & 0 \\[2mm]
54 & -13\left(\dfrac{a}{2}\right) & 312 & 0 & 54 & 13\left(\dfrac{a}{2}\right) \\[2mm]
13\left(\dfrac{a}{2}\right) & -3\left(\dfrac{a}{2}\right)^2 & 0 & 8\left(\dfrac{a}{2}\right)^2 & -13\left(\dfrac{a}{2}\right) & -3\left(\dfrac{a}{2}\right)^2 \\[2mm]
0 & 0 & 54 & -13\left(\dfrac{a}{2}\right) & 156 & 22\left(\dfrac{a}{2}\right) \\[2mm]
0 & 0 & 13\left(\dfrac{a}{2}\right) & -3\left(\dfrac{a}{2}\right)^2 & 22\left(\dfrac{a}{2}\right) & 4\left(\dfrac{a}{2}\right)^2
\end{bmatrix} \tag{a}$$

The assembled stiffness and differential stiffness matrices are shown in equation (a) in Example 5.2.6.

When the boundary constraints

$$w_1 = \theta_1 = w_3 = \theta_3 = 0 \tag{b}$$

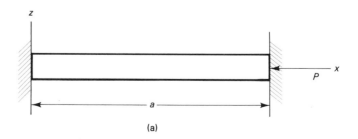

(a)

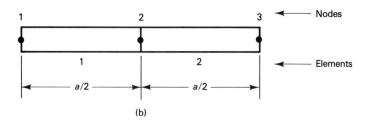

(b)

are applied, the first two rows and columns and the last two rows and columns of the assembled set are partitioned into another set, and the full set of equations reduces to

$$\left(\frac{EI}{(a/2)^3} \begin{bmatrix} 24 & 0 \\ 0 & 8\left(\frac{a}{2}\right)^2 \end{bmatrix} - \frac{P}{15a} \begin{bmatrix} 72 & 0 \\ 0 & 8\left(\frac{a}{2}\right)^2 \end{bmatrix} - \frac{\omega^2 m}{840} \begin{bmatrix} 312 & 0 \\ 0 & 8\left(\frac{a}{2}\right)^2 \end{bmatrix} \right) \begin{bmatrix} q_1 \\ q_2 \end{bmatrix} = \begin{bmatrix} 0 \\ 0 \end{bmatrix} \qquad \text{(c)}$$

To find the natural frequencies, set the determinant of the coefficients to zero. When $P = 0$,

$$\left| \begin{bmatrix} 24 & 0 \\ 0 & 8\left(\frac{a}{2}\right)^2 \end{bmatrix} - \frac{\omega^2 m a^4}{EI \cdot 8 \cdot 840} \begin{bmatrix} 312 & 0 \\ 0 & 8\left(\frac{a}{2}\right)^2 \end{bmatrix} \right| = 0 \qquad \text{(d)}$$

When expanded, we have

$$(24 - 312\gamma)(1 - \gamma)8\left(\frac{a}{2}\right)^2 = 0 \qquad \text{(e)}$$

where

$$\gamma = \frac{\omega^2 m a^4}{EI \cdot 8 \cdot 840} \qquad \text{(f)}$$

For the lowest natural frequency, we get

$$24 - 312\gamma = 0 \longrightarrow \gamma_1 = \frac{1}{13} = 0.0769 \qquad \text{(g)}$$

or

$$\omega_1^2 = \frac{516.92EI}{ma^4} \longrightarrow \omega_1 = 22.74 \sqrt{\frac{EI}{ma^4}} \qquad \text{(h)}$$

The next frequency is

$$1 - \gamma_2 = 0 \longrightarrow \gamma_2 = 1 \qquad \text{(i)}$$

or

$$\omega_2^2 = 6720 \frac{EI}{ma^4} \longrightarrow \omega_2 = 81.98 \sqrt{\frac{EI}{ma^4}} \qquad \text{(j)}$$

The exact answers from the differential equation solution are

$$\omega_1 = 22.37 \sqrt{\frac{EI}{ma^4}}, \qquad \omega_2 = 61.67 \sqrt{\frac{EI}{ma^4}} \qquad \text{(k)}$$

We see that even with only two elements the first frequency is predicted accurately but, as expected, the second is not so good.

To find the first mode, insert γ_1 into equations (e). We get

$$(24 - 312\gamma_1)w_2 + 0 \cdot \theta_2 = 0$$

$$0 \cdot w_2 - (1 - \gamma_1)\theta_2 = 0 \tag{l}$$

These are uncoupled equations, so only w_2 figures in the mode for the first frequency; that is,

$$(24 - 312\gamma_1)w_2 = 0 \tag{m}$$

Therefore,

$$[w(s)] = [n]_1^T\{r_e\}_1 = [n]_1^T \begin{bmatrix} 0 \\ 0 \\ w_2 \\ 0 \end{bmatrix} = w_2 \left(\frac{3as^2}{2} - 2s^3 \right), \qquad \text{element 1}$$

$$\tag{n}$$

$$[w(s)] = [n]_2^T\{r_e\}_2 = [n]_2^T \begin{bmatrix} w_2 \\ 0 \\ 0 \\ 0 \end{bmatrix} = w_2 \left(\frac{a^3}{8} - \frac{3as^2}{2} + 2s^3 \right), \qquad \text{element 2}$$

or, in normal mode format,

$$\varphi_1(s) = \frac{3as^2}{2} - 2s^3, \qquad \text{element 1}$$

$$\tag{o}$$

$$= \frac{a^3}{8} - \frac{3as^2}{2} + 2s^3, \qquad \text{element 2}$$

where s is measured from the left end of each element.

To find the second mode, insert γ_2 into equations (e). We get

$$(24 - 312\gamma_2)w_2 + 0 \cdot \theta_2 = 0$$

$$0 \cdot w_2 - (1 - \gamma_2)\theta_2 = 0 \tag{p}$$

These are uncoupled equations, so that only θ_2 figures in the mode for the second frequency; that is,

$$(1 - \gamma_2)\theta_2 = 0 \tag{q}$$

Therefore,

$$[w(s)] = [n]_1^T\{r_e\}_1 = [n]_1^T \begin{bmatrix} 0 \\ 0 \\ 0 \\ \theta_2 \end{bmatrix} = \theta_2 \left(-\frac{a^2s^2}{4} + \frac{as^3}{2} \right), \qquad \text{element 1}$$

$$\tag{r}$$

$$[w(s)] = [n]_2^T\{r_e\}_2 = [n]_2^T \begin{bmatrix} 0 \\ \theta_2 \\ 0 \\ 0 \end{bmatrix} = \theta_2 \left(\frac{a^3s}{8} - \frac{a^2s^2}{2} + as^3 \right), \qquad \text{element 2}$$

or, in normal mode format,

$$\varphi_2(s) = -\frac{a^2s^2}{4} + \frac{as^3}{2}, \qquad \text{element 1}$$

$$\tag{s}$$

$$= \frac{a^3s}{8} - \frac{a^2s^2}{2} + as^3, \qquad \text{element 2}$$

where s is measured from the left end of each element.

Now, when $P = 20EI/a^2$, the determinant becomes

$$\left| \begin{bmatrix} 12 & 0 \\ 0 & \dfrac{2a^2}{3} \end{bmatrix} - \dfrac{\omega^2 m a^4}{6720EI} \begin{bmatrix} 312 & 0 \\ 0 & 2a^2 \end{bmatrix} \right| = 0 \tag{t}$$

When expanded, we have

$$(12 - 312\gamma)(1 - 3\gamma)\dfrac{2a^2}{3} = 0 \tag{u}$$

where

$$\gamma = \dfrac{\omega^2 m a^4}{6720EI} \tag{v}$$

or

$$12 - 312\gamma = 0 \longrightarrow \gamma_1 = 0.0385 \longrightarrow \omega_1 = 16.08 \sqrt{\dfrac{EI}{ma^4}} \tag{w}$$

$$1 - 3\gamma = 0 \longrightarrow \gamma_2 = \dfrac{1}{3} \longrightarrow \omega_2 = 47.33 \sqrt{\dfrac{EI}{ma^4}} \tag{x}$$

Note the large decrease in the frequencies due to the axial load. The modes may now be determined by the method shown for the first case.

We might note here that the the modes obtained from Equation 6.4.26 are orthogonal. For proof of orthogonality, consider two dissimilar frequencies and modes

$$([K] + [K^d] - \omega_i^2[M])\{a\}_i = \{0\}, \qquad ([K] + [K^d] - \omega_j^2[M])\{a\}_j = \{0\} \tag{6.4.28}$$

If we premultiply the first by $\{a\}_j^T$ and the second by $\{a\}_i^T$, we get

$$\{a\}_j^T ([K] + [K^d] - \omega_i^2[M])\{a\}_i = \{0\}$$
$$\{a\}_i^T ([K] + [K^d] - \omega_j^2[M])\{a\}_j = \{0\} \tag{6.4.29}$$

Since in matrix multiplication

$$[a]^T[b][c] = [c]^T[b][a] \tag{6.4.30}$$

when $[b]$ is symmetric, subtracting the first from the second, we have

$$(\omega_j^2 - \omega_i^2)(\{a\}_j^T[M]\{a\}_i) = \{a\}_j^T ([K] + [K^d])\{a\}_i$$
$$- \{a\}_j^T ([K] + [K^d])\{a\}_i = 0 \tag{6.4.31}$$

or

$$(\{a\}_j^T[M]\{a\}_i) = 0, \qquad \omega_i \neq \omega_j$$
$$= M_j, \qquad \omega_i \neq \omega_j \tag{6.4.32}$$

which is the statement of orthogonality in matrix form.

6.4.3 Eigenvalue Equation Solvers

As was noted for buckling, for problems with many degrees of freedom the extraction of the eigenvalues is not done easily by expanding the determinant. Several methods are available and are discussed in References 23 and 24, among others. The program given in Appendix L, which was mentioned briefly in Section 5.2, is set up to handle both buckling and vibration problems; however, in vibration we usually want to find several eigenvalues and eigenfunctions.

To prepare the equations for the matrix iteration, we let

$$[A] = [K] + [K^d] \tag{6.4.33}$$

and we get

$$([A] - \omega^2[M])\{a\} = \{0\} \tag{6.4.34}$$

Because the method we are about to describe always finds the highest eigenvalue first and and then each succeeding one in descending order, we divide by $-\omega^2$ and get

$$\left([M] - \frac{1}{\omega^2}[A]\right)\{a\} = ([M] - \psi[A])\{a\} = \{0\} \tag{6.4.35}$$

Thus, the highest ψ will be the lowest ω^2. Next we multiply by the inverse of $[A]$, or

$$[A]^{-1}[M]\{a\} = \psi\{a\} \longrightarrow [Y]\{a\} = \psi\{a\} \tag{6.4.36}$$

The iteration proceeds as described in Equations 5.2.84 and 5.2.85 and as shown in Example 5.2.7.

To find the higher eigenvalues, the method uses the property of orthogonality of the eigenfunctions. The trial value for the second eigenfunction is forced to be orthogonal to the first eigenfunction before each iteration. The third eigenfunction is forced to be orthogonal to the first two, and so on. This has been mechanized via a *sweeping matrix* $[S_j]$, which contains the necessary information to ensure orthogonality. The recursion formula to obtain the other eigenvalues is

$$[Y][S_j]\{a\}^i = \{C\}^{i+1} = \psi^{i+1}\{a\}^{i+1} \tag{6.4.37}$$

Each sweeping matrix contains information on the previous eigenfunctions found. The index j refers to the number found.

The program in Appendix L makes use of this method. We do not have space to go into the details of constructing the sweeping matrix here. Check the references.

Example 6.4.4

Consider the free vibration of the tapered cantilever beam in Example 4.4.3 as shown in figure (a). Find the first few natural frequencies and modes using NASTRAN.

Solution: The first six categories of information in Example 4.4.3 carry over unchanged; however, we will not need the coordinates of the cross section since we will not be recovering stress data. The seventh category is not needed, since for free vibration there is no applied load. We do need to invoke the eigenvalue equation solver. A NASTRAN input file has been prepared and is listed in Appendix N. We shall find the frequencies and modes in the range of 1 to 100 Hz.

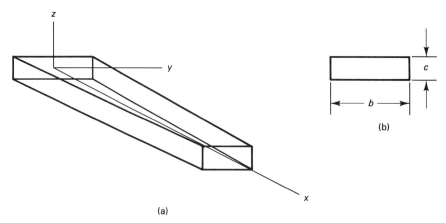

(a)

(b)

The problem has been executed on MSC/NASTRAN and the following frequencies have been found and are listed in units of hertz, or cycles per second.

$$\omega_1 = 6.172 \qquad \omega_2 = 8.111$$

$$\omega_3 = 33.655 \qquad \omega_4 = 38.760$$

$$\omega_5 = 91.122 \qquad \omega_6 = 100.443$$

The problem was remodeled using eight equally spaced elements to see what improvement there would be in the frequencies and also to get better plots of the modes. Remember that the NASTRAN plotter only plots straight lines between nodal displacements; that is, it does not accurately portray the slopes at the nodes. Using eight elements, the frequencies are

$$\omega_1 = 6.160 \qquad \omega_2 = 8.163$$

$$\omega_3 = 33.563 \qquad \omega_4 = 39.066$$

$$\omega_5 = 90.234 \qquad \omega_6 = 100.844$$

Note that the change is not large. Using still more elements would show that the approximation of four or eight elements is very good for just this many frequencies.

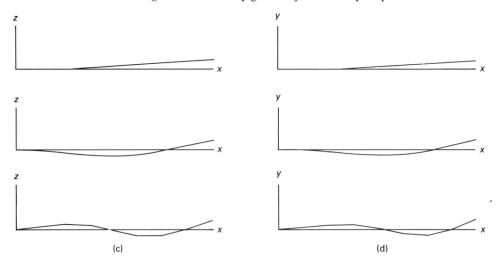

(c)

(d)

The modes are shown in figures (c) and (d). Note that the displacements in the y- and z-directions are uncoupled. The first three modes in the x-z plane are shown in figure (c) and in the x-y plane in figure (d). What we have found are three modes and frequencies in each of the two coordinate directions.

6.4.4 Finite Element Method for Forced Vibration

When $\{R\} \neq \{0\}$, we have forced motion. Equations 6.4.15 can be solved directly, but the equations reduce to a more tractable form if we use normal mode analysis. Following the same pattern as we did with Equations 6.3.39 through 6.3.46, we represent the forced response in terms of a series of normal modes, as follows:

$$\{r\} = [a]\{\xi\} \tag{6.4.38}$$

where the elements of $\{\xi\}$ are as yet unknown function of time and $[a]$ is a rectangular matrix whose columns are the normal modes, or

$$[a] = [\{a\}_1 \ \{a\}_2 \ \{a\}_3 \ldots] \tag{6.4.39}$$

Equations 6.4.15 becomes

$$[M][a]\{\ddot{\xi}\} + [K + K^d][a]\{\xi\} = \{R\} \tag{6.4.40}$$

We premultiply this set of equations with $[a]^T$.

$$[a]^T[M][a]\{\ddot{\xi}\} + [a]^T[K + K^d][a]\{\xi\} = [a]^T\{R\} \tag{6.4.41}$$

First, note that

$$[K + K^d]\{a\}_i = \omega_i^2[M]\{a\}_i \tag{6.4.42}$$

and, therefore,

$$[K + K^d][a] = [\omega^2][M][a] \tag{6.4.43}$$

where

$$[\omega^2] = [\omega_1^2 \ \omega_2^2 \ \omega_3^2 \ldots] \tag{6.4.44}$$

This, together with orthogonality as defined in Equations 6.4.32, permits us to rewrite Equation 6.4.41 to be

$$[M]\{\ddot{\xi}\} + [\omega^2][M]\{\xi\} = \{\Xi\} \tag{6.4.45}$$

where $[M]$ is a diagonal matrix and

$$\{\Xi\} = [a]^T\{R\} \tag{6.4.46}$$

The elements on the diagonal of $[M]$ are M_j as defined in 6.4.32; thus, Equation 6.4.45 is a set of uncoupled second-order ordinary differential equations, each of which is of the form

$$M_j\ddot{\xi}_j + M_j\omega_j^2\xi_j = \Xi_j(t) \tag{6.4.47}$$

This is exactly the same form as Equations 6.3.45 and the methods of solution are exactly the same. Once Equations 6.4.47 have been determined, the solutions for forced harmonic and transient response proceed exactly as in Examples 6.3.7 and 6.3.8.

In the analytical method described in Section 6.3, the work necessary to obtain Equations 6.3.45 can be formidable because of the complexity of the normal modes. In FEM, nothing more than routine matrix multiplication is required.

To add viscous damping to the system, we must go back to the expression for virtual work and add a term:

$$\Delta\delta W_e = -\int c\dot{w}\delta w \, dx - \Sigma \, C_i\dot{w}\delta w(x_i) \qquad (6.4.48)$$

where the C_i's represent concentrated dashpots. When Equation 6.4.14 is introduced and the equations reduced, a term of the form $[C]\{\dot{r}\}$ is added to the global Equations 6.4.15, giving us

$$[M]\{\ddot{r}\} + [C]\{\dot{r}\} + ([K] + [K^d])\{r\} = \{R\} \qquad (6.4.49)$$

where

$$[C] = \int [N]^T[N]c \, dx + \text{concentrated dashpot terms} \qquad (6.4.50)$$

We can subject this to normal mode analysis and get

$$[a]^T[M][a]\{\ddot{\xi}\} + [a]^T[C][a]\{\dot{\xi}\} + [a]^T[K + K^d][a]\{\xi\} = [a]^T\{R\} \qquad (6.4.51)$$

and in the special case when

$$[a]^T[C][a] = [C] \qquad (6.4.52)$$

where $[C]$ is a diagonal matrix, Equation 6.4.51 becomes

$$[M]\{\ddot{\xi}\} + [C]\{\dot{\xi}\} + [\omega^2][M]\{\xi\} = \{\Xi\} \qquad (6.4.53)$$

This is a set of uncoupled equations, each of which is of the form

$$M_j\ddot{\xi}_j + C_j\dot{\xi}_j + M_j\omega_j^2\xi_j = \Xi_j(t) \qquad (6.4.54)$$

which is exactly the same form as Equations 6.3.72. When

$$[C] = \beta \, [M] \qquad (6.4.55)$$

Equation 6.4.54 may be rewritten

$$\ddot{\xi}_j + \beta\dot{\xi}_j + \omega_j^2\xi_j = \frac{\Xi_j}{M_j} \qquad (6.4.56)$$

which is exactly the same as Equation 6.3.73, and the discussion of the solution in Equations 6.3.74 through 6.3.86 fully applies.

While the modal analysis approach is elegant, it often is more efficient to apply a numerical process directly to Equations 6.4.49. This is especially true when the damping matrix does not diagonalize in the modal reduction process. Several approaches to direct numerical integration of these equations have been tried and are outlined in Reference 23 and others. A complete program for direct numerical solution of equations in the form of Equation 6.4.49 is given in Appendix O. Similar methods are also available in the various FEM computer codes such as NASTRAN. We do not have the space to go into this here.

6.5 DYNAMICS OF OTHER STRUCTURAL FORMS

We have spent what may seem like an inordinate amount of time on the dynamics of beams in lateral vibration. What we have done is use the beam to develop many of the classical differential equation and modern numerical approaches to the dynamics of structures. What we have learned may be extended easily to other structural forms.

You have noticed that for all the beams in Sections 6.3. and 6.4 we have assumed that $I_{yz} = 0$ or that the y and z axes are principal axes of the cross section. This has the fortunate result of uncoupling the motion in the y- and z-directions. We could have developed the equations starting with Equations 2.3.35 and adding inertia terms. This would result in coupled equations of motion that would be considerably more difficult to solve. The Rayleigh-Ritz and finite element forms of the coupled equations can also be developed. We shall not pursue this here for want of space.

What we shall do now is adapt these methods to some other popular structural forms. We shall consider the axial and torsional vibration of beams, combined bending and torsion of beams, trusses and frames, plates and shells, and the general solid body.

As with static stable and unstable structures, the classical methods tend to be closely related to the geometrical configuration, thus requiring different approaches for different types of structures, while the numerical methods, particularly FEM, quickly reduce to the same form for all structures. We shall see this develop as we examine several examples.

6.5.1 Axial and Torsional Vibration of Bars

Using D'Alembert's principle, we modify Equation 2.2.14 by adding an inertia term, and we declare $u_0(x, t)$ and $f_x(x, t)$. We get

$$(EAu_0')' = -f_x(x, t) - m\ddot{u}_0 \longrightarrow (EAu_0')' + m\ddot{u}_0 = -f_x(x, t) \qquad (6.5.1)$$

where m is the mass per unit length and is identical to the value used in lateral vibration. For free vibration, we have

$$(EAu_0')' + m\ddot{u}_0 = 0 \qquad (6.5.2)$$

Separation of variables works. Let

$$u_0(x, t) = U_0(x)T(t) \qquad (6.5.3)$$

From this, we get

$$(EAU_0')' - m\omega^2 U_0 = 0, \qquad \ddot{T} + \omega^2 T = 0 \qquad (6.5.4)$$

When EA is constant,

$$U_0'' - \frac{m\omega^2}{EA} U_0 = 0 \longrightarrow U_0'' - \alpha^2 U_0 = 0 \qquad (6.5.5)$$

and the homogeneous solution is

$$U_0(x) = c_3 \sin \alpha x + c_4 \cos \alpha x \qquad (6.5.6)$$

The natural frequencies and modes are found by substituting the homogeneous solution into appropriate boundary conditions. Let us look at an example.

Example 6.5.1

Find the frequencies and modes in axial vibration of a uniform cantilever beam of length a.

Solution: The boundary conditions are

$$U_0(0) = 0, \qquad U_0'(a) = 0 \tag{a}$$

It follows that

$$U_0(0) = c_4, \qquad U_0'(a) = c_3\alpha \cos \alpha a = 0 \tag{b}$$

Therefore,

$$\cos \alpha a = 0 \longrightarrow \alpha a = \frac{(2i-1)^2}{2}\pi = a\sqrt{\frac{m\omega}{EA}} \tag{c}$$

or

$$\omega_i = \frac{(2i-1)^2}{2}\frac{\pi}{a}\sqrt{\frac{EA}{m}} \tag{d}$$

and the mode is

$$U_0(x) = c_3 \sin \alpha x \longrightarrow \varphi_i(x) = \sin \alpha_i x \tag{e}$$

Initial conditions and forced motion can be found by expanding the displacement in a series of normal modes. It can be shown that the modes are orthogonal. We assume

$$u_0(x, t) = \Sigma \, \xi_j(t)\varphi_j(x) \tag{6.5.7}$$

By the same process as in Equations 6.3.39 through 6.3.46, we get the equations for ξ_j to be

$$M_j\ddot{\xi}_j + M_j\omega_j^2\xi_j = \Xi_j \tag{6.5.8}$$

where

$$M_j = \int \varphi_j^2 m \, dx, \qquad \Xi_j = -\int f_x(x, t)\varphi_j \, dx \tag{6.5.9}$$

In fact, once the modes and frequencies are known for any linear elastic structure, the forced motion equation reduces to Equation 6.5.8 with M_j and Ξ_j suitably defined for that structure. We shall see some additional examples.

In fact, the next example is torsional vibration. From Equation 2.2.22, we obtain

$$(GJ\beta')' + I_m\ddot{\beta} = -t_x(x, t) \tag{6.5.10}$$

where I_m is the mass moment of inertia per unit length about the x-axis. For free motion after separation of variables, where

$$\beta(x, t) = \Phi(x)T(t) \tag{6.5.11}$$

we get

$$(GJ\Phi')' + I_m\Phi = 0, \qquad \ddot{T} + \omega^2 T = 0 \tag{6.5.12}$$

When GJ is constant,

$$\Phi'' - \frac{I_m \omega^2}{GJ} \Phi = 0 \longrightarrow \Phi'' - \alpha^2 \Phi = 0 \qquad (6.5.13)$$

and the homogeneous solution is

$$\Phi(x) = c_3 \sin \alpha x + c_4 \cos \alpha x \qquad (6.5.14)$$

The solution for free and forced motion proceeds exactly as described previously.

The finite element form for axial and torsional vibration is also a simple extension of the static case. In both cases, the FEM equations are

$$[M]\{\ddot{r}\} + [K]\{r\} = \{R\} \qquad (6.5.15)$$

where, for axial vibration,

$$[M] = \int [N]^T[N] m \, dx, \quad [K] = \int [B]^T[G][B] \, dx, \quad \{R\} = \int [N]^T f_x(x, t) \, dx \qquad (6.5.16)$$

and, for torsional vibration,

$$[M] = \int [N]^T[N] I_m \, dx, \quad [K] = \int [B]^T[G][B] \, dx, \quad \{R\} = \int [N]^T t_x(x, t) \, dx \qquad (6.5.17)$$

At the element level, the shape function and the stiffness matrix for an axial element are given in Equation 3.4.31 and are repeated here.

$$[n]_i = \begin{bmatrix} 1 - \dfrac{s}{L_i} & \dfrac{s}{L_i} \end{bmatrix}, \qquad [k]_i = \frac{EA}{L_i} \begin{bmatrix} 1 & -1 \\ -1 & 1 \end{bmatrix} \qquad (6.5.18)$$

The mass matrix is

$$[m]_i = \frac{mL_i}{6} \begin{bmatrix} 2 & 1 \\ 1 & 2 \end{bmatrix} \qquad (6.5.19)$$

and the equivalent element nodal loads are

$$\{R\} = \int [n]_i^T f_x(x, t) \, dx \qquad (6.5.20)$$

The element shape functions and the stiffness matrix for the torsional case are given in Equations 4.4.8 and 4.4.10 and are repeated here.

$$[n]_i = \begin{bmatrix} 1 - \dfrac{s}{L_i} & \dfrac{s}{L_i} \end{bmatrix}, \qquad [k]_i = \frac{GJ}{L_i} \begin{bmatrix} 1 & -1 \\ -1 & 1 \end{bmatrix} \qquad (6.5.21)$$

The mass matrix, in this case the mass moment of inertia matrix, is

$$[m]_i = \frac{I_m L_i}{6} \begin{bmatrix} 2 & 1 \\ 1 & 2 \end{bmatrix} \qquad (6.5.22)$$

and the equivalent nodal loads are

$$\{R\} = \int [n]_i^T t_x(x, t) \, dx \qquad (6.5.23)$$

The assembly of the element matrices, the imposition of boundary conditions, and the solution for free and forced motion proceed exactly as in the case of the lateral vibration of the beam. Since no new ground is exposed here, we shall not give an example.

The uncoupled equations for axial, torsional, and bending vibration can be combined into a single set of FEM equations. The static case is presented in Section 4.4.3, the displacement matrix is given in Equation 4.4.29 and the stiffness matrix in Equation

4.4.31. The combined mass matrix can be collected from the individual mass matrices with ease. Although this information is combined into a single matrix for the mass and also the stiffness, the two lateral, the axial, and the torsional motions remain uncoupled. There are circumstances in which coupling can occur. One such case is examined in the next section.

6.5.2 Coupled Bending and Torsion

In beams with simple cross sections, for example, symmetrical cross sections, the shear center coincides with the centroid of the area of the cross section. In beams made of a uniform material with no offset nonstructural mass, the center of mass and the centroid of area of the cross section also coincide. In such beams, the torsional and bending vibrations are uncoupled. When these two centers do not coincide, the motion is coupled. To see this, we first define the *elastic axis* of the beam to be the loci of shear centers and the *mass axis* to be the loci of the centers of mass of the cross section. A beam with offset elastic and mass axes is shown in Figure 6.5.1.

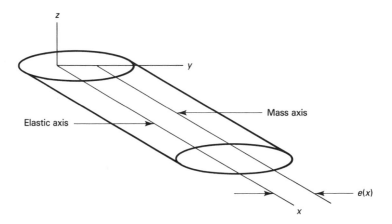

Figure 6.5.1 Beam with Offset Mass Axis.

Let us not include an internal axial force; it can be introduced easily if needed. In this case, the two axes are offset an amount $e(x)$ in the y-direction. The lateral inertia is contributed both by lateral acceleration and rotational acceleration so that the beam bending equation becomes

$$(EIw'')'' = f(x, t) - m\ddot{w} - em\ddot{\beta} \tag{6.5.24}$$

and the torsional equation becomes

$$(GJ\beta')' = -t_x(x, t) + I_m\ddot{\beta} + em\ddot{w} \tag{6.5.25}$$

When rewritten in standard form, the coupled equations are

$$(EIw'')'' + m\ddot{w} + em\ddot{\beta} = f(x, t)$$

$$(GJ\beta')' - I_m\ddot{\beta} - em\ddot{w} = -t_x(x, t) \tag{6.5.26}$$

These are coupled equations, clearly, which can be solved under certain circumstances. Usually, it is more feasible to go directly to the FEM formulation. Neglecting the axial force, concentrated lateral forces, springs, and concentrated masses, the virtual work for such a structure is

$$\delta W = \int \delta w[f(x, t) - m\ddot{w} - em\ddot{\beta}]\, dx + \int \delta\beta[t_x(x, t) - I_m\ddot{\beta} - em\ddot{w}]\, dx$$
$$- \int EI\delta w''w''\, dx - \int GJ\delta\beta'\beta'\, dx = 0 \qquad (6.5.27)$$

If we select the same shape functions for bending and torsion that were used in the uncoupled case, that is, Equations 6.4.20 and 6.5.21, and insert them into this expression for virtual work, we get appropriate definitions for $[m]$, $[k]$, and $\{R_e\}$.

These can then be assembled to obtain the global equations

$$[M]\{\ddot{r}\} + [K]\{r\} = \{R\} \qquad (6.5.28)$$

From this point, the solution proceeds in standard form. These equations are very useful in the analysis of helicopter and turbine blades, high aspect ratio wings, and many other structures.

6.5.3 Trusses and Frames

The equations for the dynamic behavior of trusses and frames are a simple extension of the static case. One needs first to define an appropriate mass matrix in order to add the inertia term to the known stiffness term. For trusses, the element mass matrix is given in Equation 6.5.19. When these are assembled using appropriate coordinate transformations described in Sections 3.4.8 and 4.3, we get the mass matrix for the global structure.

For frames, since the elements in a frame tend to have both lateral and axial loads and displacements, we use the beam element mass matrix given in Equation 6.4.24, combined with the axial element given in Equation 6.5.19. Once again, we must use the appropriate coordinate transformations in assembling the matrices.

For an example of frame vibration analysis, we shall go directly to a problem solved by a commercial FEM code.

Example 6.5.2

The frame that was analyzed in Example 4.4.4 for static deflection and stress and in Example 5.2.10 for buckling is to be checked for natural frequencies and modes. The frame is shown in figure (a).

Solution: The first six items of information required for preparation of the input file for NASTRAN are given in the two examples; however, because we desire several modes we shall use 12 elements instead 6 to model the frame. This will give us more degrees of freedom to more accurately describe the higher modes and will also give us better plots from the NASTRAN plotter. The elements will be of equal length, and therefore the additional nodes will lie halfway between the existing nodes.

This problem has been executed in MSC/NASTRAN; the input file is listed in Appendix N; the first four frequencies are shown below in units of hertz.

$$\omega_1 = 27.17$$
$$\omega_2 = 74.59$$
$$\omega_3 = 108.85$$
$$\omega_4 = 162.84$$

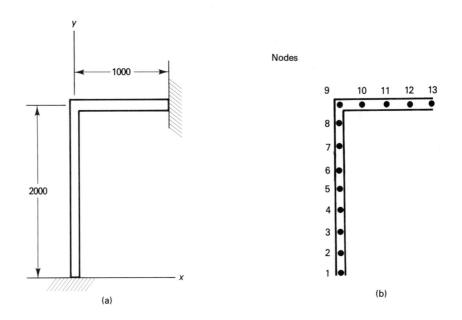

(a)

Nodes

(b)

The modes from the NASTRAN plotter are shown in figure (c). Once again, it is noted that the plots do not represent rotations at the nodes. They only attempt to give an approximation for a quick visual understanding of the shapes.

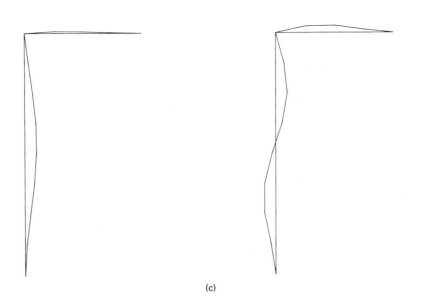

(c)

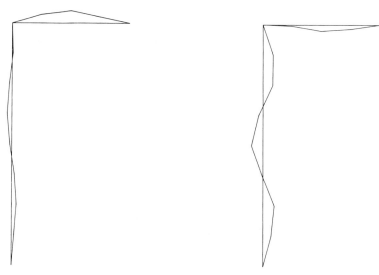

(c) cont'd.

6.5.4 Plates and Shells

The differential equation for the lateral vibration of plates is obtained simply by adding the reversed effective inertia force to Equation 2.4.13, or

$$D\nabla^4 w = p(x, y, t) - \rho h\ddot{w} \longrightarrow D\nabla^4 w + \rho h\ddot{w} = p(x, y, t) \qquad (6.5.29)$$

where now w and p are functions of time as well as the spatial coordinates, and ρ is the mass density and h is the plate thickness. We can separate variables in the free motion case by letting

$$w(x, y, t) = W(x, y)T(t) \qquad (6.5.30)$$

and we obtain

$$D\nabla^4 W - \rho h\omega^2 W = 0, \qquad T + \omega^2 T = 0 \qquad (6.5.31)$$

The motion is harmonic and the natural modes and frequencies can be found from the first of Equations 6.5.31. It can be shown that normal modes are orthogonal as defined by the double integral

$$\int_0^a \int_0^b \varphi_i\varphi_j\rho h\, dx\, dy = 0, \qquad i \neq j$$
$$= M_j, \qquad i = j \qquad (6.5.32)$$

If we let

$$w(x, y, t) = \Sigma\, \xi_i(t)\varphi_i(x, y) \qquad (6.5.33)$$

substitute this into Equation 6.5.25, multiply by φ_j, and integrate over the plate, we get

$$M_j\ddot{\xi}_j + M_j\omega_j^2\xi_j = \Xi_j \qquad (6.5.34)$$

where

$$\Xi_j(t) = \int_0^a \int_0^b p(x, y, t)\varphi_j\,dx\,dy \tag{6.5.35}$$

As we might suspect from our experience with static plate deflection and buckling, only the rectangular plate with simply supported edges is handled easily by analytical methods in Cartesian coordinates.

Example 6.5.3

Consider the free vibration of a rectangular plate with all edges simply supported.

Solution: From Equations 2.4.28 and Example 5.3.2, we can anticipate a solution of the form

$$W(x, y) = A_{rs} \sin\frac{r\pi x}{a} \sin\frac{s\pi y}{b} \tag{a}$$

we substitute this in the equations of motion, we get

$$D\pi^4 \left(\frac{r^4}{a^4} - 2\frac{r^2 s^2}{a^2 b^2} + \frac{s^4}{b^4} \right) - \rho h \omega^2 = 0 \tag{b}$$

or

$$\omega_j^2 = \frac{D\pi^4}{\rho h} \left(\frac{r^2}{a^2} + \frac{s^2}{b^2} \right)^2 \tag{c}$$

These are the natural frequencies and equations (a) are the corresponding modes, or in normalized form

$$\varphi_j = \sin\frac{r\pi x}{a} \sin\frac{s\pi y}{b} \tag{d}$$

The notation must be made clear. For each integer j, which is the index for the modes and frequencies, there is a pair of integers r and s. Note the series of half-wave forms of the modes. The lowest frequency is $r = s = 1$. If $b > a$, the next lowest frequency is $r = 1$ and $s = 2$, which corresponds to one half-wave in the x-direction and two half-waves in the y-direction. The third frequency is $r = 2$ and $s = 1$, for two half-waves in the x-direction and one half-wave in the y-direction. The fourth frequency is $r = 2$ and $s = 2$, and so on. Similar wave patterns are found in the vibration of plates with other shapes and boundary conditions.

Most problems in plate vibration must be solved by approximate methods for which those based on virtual work are very suitable. Historically, the Rayleigh–Ritz method has been used frequently. Today most problems are solved by FEM. Using the shape functions described in Section 4.7, the element mass matrix and equivalent nodal load matrix are defined by

$$[m]_i = \int\int [n]_i^T [n]_i \rho(x, y)\,dx\,dy$$
$$\{R_e\}_i = \int\int [n]_i^T p(x, y, t)\,dx\,dy \tag{6.5.36}$$

When these and the stiffness matrix $[k]_i$ are assembled, the standard FEM in global form is obtained. Solutions then proceed as in any FEM vibration problem.

For an example, we shall consider the rectangular plates that were studied for buckling in Chapter 5.

Example 6.5.4

Find the first four frequencies and modes for the rectangular plate in Example 5.3.3 and the plate with a hole in Example 5.3.4. Use NASTRAN.

Solution: To modify the input file, we need only drop the initial loading information and request vibration frequencies and modes in place of buckling loads and modes. This has been done; MSC/NASTRAN input files have been prepared and are listed in Apendix N, the problems have been executed, and the results are given next.

The first four frequencies are

Plate without hole	Plate with hole
$\omega_1 = 7.825$	$\omega_1 = 7.599$
$\omega_2 = 16.911$	$\omega_2 = 15.344$
$\omega_3 = 24.886$	$\omega_3 = 20.511$
$\omega_4 = 37.449$	$\omega_4 = 34.234$

The corresponding modes are shown in figure (a) for the plate without a hole and in figure (b) for the plate with a hole.

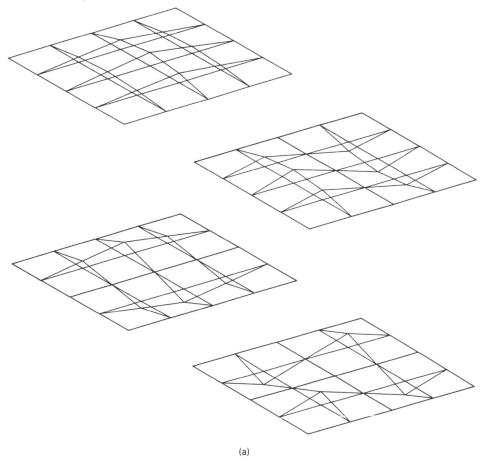

(a)

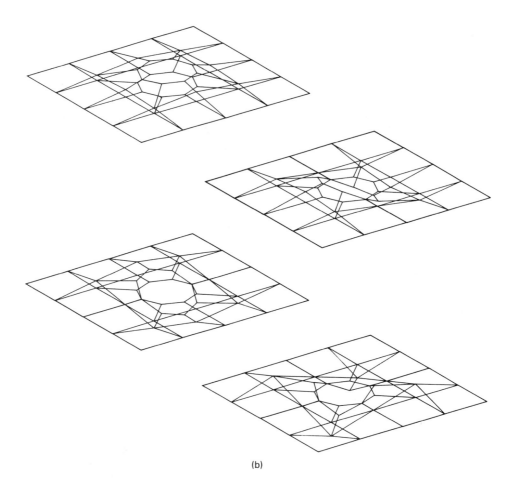

(b)

The elements used in plate analysis that exhibit combined plate and membrane effects may also be used in shell analysis. See Section 4.7.2.

6.5.5 Three-Dimensional Solid Bodies

The use of solid elements such as the tetrahedron and hexahedron in vibration analysis has added a new dimension in structural dynamics. Up to the advent of FEM, little could be done to study the dynamic behavior of solid bodies. Exact analytical methods were virtually useless. The Rayleigh-Ritz method based on global admissible functions had limited utility. Experimental methods were the main resort; however, experimental methods required that the part be built first. Now we see that FEM vibration analysis is a simple extension of static analysis as far as forming the input file is concerned. To show this we shall do an example.

Example 6.5.5

Find the first three natural frequencies and modes of the aluminum lug shown in figure (a). Dimensions are in millimeters.

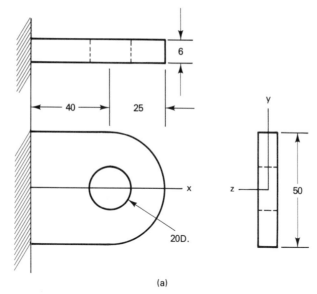

(a)

Solution: Solid eight node hexagonal elements are used to model the lug. For even a simple three-dimensional structure the division into elements, the numbering of the nodes and elements, and the entry of the nodal coordinates and the element connectivity can be a very tedious job. A graphics-based pre-processor that does these tasks automatically was employed. The element pattern is shown in figure (b). There is only one element through the depth of the model. The node and element numbering, and the listing of the nodal coordinates and the element connectivity will not be shown here. This information can be retrieved from the NASTRAN input file listing given in Appendix N, Section N.5.1.

The first three natural frequencies of the system are

$$\omega_1 = 1670.0 \text{ hertz}$$

$$\omega_2 = 4377.7 \text{ hertz} \qquad \text{(a)}$$

$$\omega_3 = 8333.9 \text{ hertz}$$

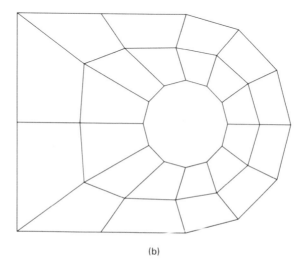

(b)

The corresponding modes are shown in figure (c) below.

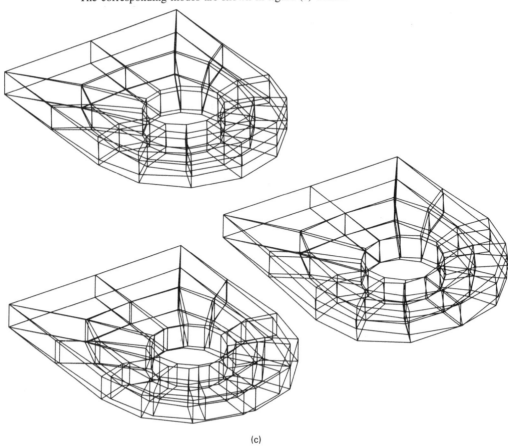

(c)

We shall not try to list the deflections at the nodes for the modes here. The purpose of this example is to illustrate that finding modes and frequencies by finite element analysis using solid elements has become a routine procedure and, except for the more complicated geometry and the larger number of nodes and elements encountered when compared to typical beam and plate problems, it also requires only a simple extension of the input file for the static analysis case.

6.6 CONCLUSIONS AND SUMMARY

The equations for the dynamics of structures can be derived by including the inertia forces as part of the applied load according to D'Alembert's principle. The solution of the resulting equations requires some new methods and techniques, including separation of variables and the use of superposition of eigenfunctions to satisfy initial conditions and the effects of dynamically applied loads. These methods are illustrated primarily by the equations for the lateral vibration of beams. The methods are then extended and generalized to other structural forms.

The differential equation for lateral vibration of a beam is

$$(EIw'')'' - Nw'' + m\ddot{w} = f(x, t) \tag{6.3.2}$$

Appropriate boundary conditions and initial conditions must be supplied to complete the statement of a specific problem. When there are no applied loads, we have the equation for free vibration,

$$(EIw'')'' + m\ddot{w} = 0 \tag{6.3.7}$$

for which we can assume

$$w(x, t) = W(x)T(t) \tag{6.3.8}$$

and use separation of variables. The separated equations are

$$\ddot{T} + \omega^2 T = 0, \qquad (EIW'')'' - m\omega^2 W = 0 \tag{6.3.12}$$

We may note that

$$T(t) = c_1 \sin \omega t + c_2 \cos \omega t \tag{6.3.17}$$

and when EI is constant

$$W(x) = c_3 \sinh \alpha x + c_4 \cosh \alpha x + c_5 \sin \alpha x + c_6 \cos \alpha x \tag{6.3.23}$$

Equation 6.3.23 is inserted in the boundary condition to obtain the natural frequencies and normal modes.

Initial conditions may be satisfied by a superposition of the normal modes as follows. Let

$$w(x, t) = \Sigma (A_i \sin \omega_i t + B_i \cos \omega_i t)\varphi_i(x) \tag{6.3.26}$$

where

$$A_i = \frac{1}{M_i \omega_i} \int w(x, 0)\varphi_i(x)m \, dx$$
$$B_i = \frac{1}{M_i} \int w(x, 0)\varphi_i(x)m \, dx \tag{6.3.30}$$

and

$$M_i = \int [\varphi_i(x)]^2 \, m \, dx \tag{6.3.31}$$

For forced motion, we assume

$$w(x, t) = \Sigma \, \xi_i(t)\varphi_i(x) \tag{6.3.39}$$

and, with the help of the orthogonality relation,

$$\int m\varphi_i\varphi_j \, dx = 0 \tag{6.3.36}$$

obtain

$$M_j\ddot{\xi}_j + M_j\omega_j^2\xi_j = \Xi_j(t) \tag{6.3.45}$$

where

$$M_j = \int \varphi_j^2 m \, dx, \qquad \Xi_j(t) = \int f(x, t) \, \varphi_j \, dx \tag{6.3.46}$$

Particularly interesting results are obtained when we have a harmonic forcing function, or when

$$\Xi_j(t) = Z_j \sin \Omega t \qquad (6.3.50)$$

The solution exhibits the phenomenon of resonance.

When viscous damping is added, the equation of motion becomes

$$(EIw'')'' - Nw'' + c\dot{w} + m\ddot{w} = f(x, t)$$

If we let

$$w(x, t) = \Sigma \, \xi_i(t)\varphi_i(x) \qquad (6.3.66)$$

and consider only damping which is proportional to the mass, or

$$c = \beta m \qquad (6.3.69)$$

then

$$\int \varphi_i\varphi_j c \, dx = \beta \int \varphi_i\varphi_j m \, dx = 0 \qquad i \pm j$$

$$= \beta M_j = C_j, \qquad i = j \qquad (6.3.70)$$

and we get

$$M_j\ddot{\xi}_j + C_j\dot{\xi}_j + M_j\omega_j^2\xi_j = \Xi_j(t) \qquad (6.3.72)$$

The virtual work for the beam in bending vibration is

$$\delta W = \delta W_e + \Delta\delta W_e - (\delta U + \Delta\delta U)$$

$$= \int \delta w[f(x, t) - m\ddot{w}] \, dx - \Sigma \, M_k\ddot{w}\delta w(x_k) + \Sigma \, F_{zi}\delta w_i \qquad (6.4.2)$$

$$+ \Sigma \, M_{yj}(\delta w_j)' - \int N\delta w'w' \, dx - \int EI\delta w''w'' \, dx = 0$$

When we assume admissible functions

$$\{u\} = w = [\Omega]\{q\} \qquad (6.4.4)$$

and apply the Rayleigh-Ritz method, we get

$$[M]\{\ddot{q}\} + ([K] + [K^d])\{q\} = \{Q\} \qquad (6.4.6)$$

Likewise, with the finite element method, we assume

$$\{u\} = w = [N]\{r\} \qquad (6.4.14)$$

and we get

$$[M]\{\ddot{r}\} + ([K] + [K^d])\{r\} = \{Q\} \qquad (6.4.15)$$

where $[K]$ and $[K^d]$ are the same as in the static case, and the mass matrix $[M]$ is assembled from the element mass matrix

$$[m]_i = \int [n]_i^T[n]_i \, m \, dx = \frac{m}{420} \begin{bmatrix} 156 & -22L_i & 54 & 13L_i \\ -22L_i & 4L_i^2 & -13L_i & -3L_i^2 \\ 54 & -13L_i & 156 & 22L_i \\ 13L_i & -3L_i^2 & 22L_i & 4L_i^2 \end{bmatrix} \qquad (6.4.24)$$

The eigenvalues are determined from the determinant

$$\left|[K] + [K^d] - \omega^2[M]\right| = 0 \tag{6.4.27}$$

and the eigenfunctions from the equations

$$([K] + [K^d] - \omega^2[M])\{a\} = \{0\} \tag{6.4.26}$$

The resulting modes are orthogonal:

$$\begin{aligned}
(\{a\}_j^T[M])\{a\}_i) &= 0 && \omega_i \neq \omega_j \\
&= M_j, && \omega_i \neq \omega_j
\end{aligned} \tag{6.4.32}$$

This may be used to reduce the equations for forced motion by modal analysis.

The equation of motion for axial vibration is

$$(EAu_0')' + m\ddot{u}_0 = -f_x(x, t) \tag{6.5.1}$$

and for torsional vibration it is

$$(GJ\beta')' + I_m\ddot{\beta} = -t_x(x, t) \tag{6.5.10}$$

For coupled bending and torsion, the differential equations of motion are

$$\begin{aligned}
(EIw'')'' + m\ddot{w} + em\ddot{\beta} &= f(x, t) \\
(GJ\beta')' - I_m\ddot{\beta} - em\ddot{w} &= -t_x(x, t)
\end{aligned} \tag{6.5.26}$$

and the expression for virtual work is

$$\begin{aligned}
\delta W = {}&\int \delta w(f(x, t) - m\ddot{w} - em\ddot{\beta})\, dx + \int \delta\beta(t_x(x, t) - I_m\ddot{\beta} - em\ddot{w})\, dx \\
&- \int EI\delta w''w''\, dx - \int GJ\delta\beta'\beta'\, dx = 0
\end{aligned} \tag{6.5.27}$$

Finally, the differential equation of motion for flat plates is

$$D\nabla^4 w + \rho h\ddot{w} = p(x, y, t) \tag{6.5.29}$$

In general, the methods explained in detail for the bending vibration of beams are easily adapted to the vibration of other structural forms.

PROBLEMS

1. Find the first three natural frequencies and modes for the fixed-simply supported beam shown by solving the differential equation.

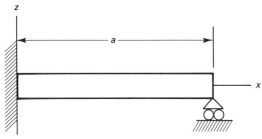

2. Find the expression from which the natural frequencies may be determined for the elastically supported cantilever beam shown.

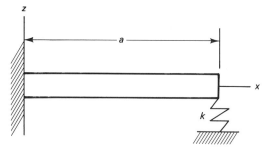

3. Find the first few natural frequencies and modes for a beam which is simply supported on one end and free on the other as shown.

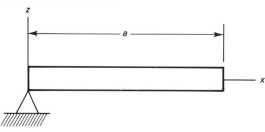

4. Find the motion of a simply supported beam with an applied harmonic loading, as shown where

$$F(x, t) = F_0 \sin \frac{\pi x}{a} \sin \Omega t$$

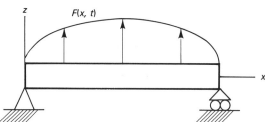

5. Find the motion of a simply supported beam of length a which is at rest at time $t = 0$ and has a uniform distributed load F_0 applied to it in the form of a ramp until time τ and then has the load removed as shown in the following plot of load versus time.

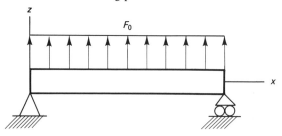

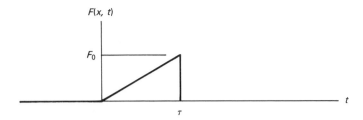

6. Find the first few natural frequencies and modes for a fixed simply supported beam with an applied axial load F where $P < P_{cr}$.

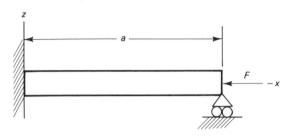

7. Consider the beam shown.

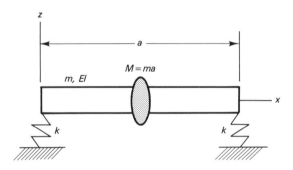

It has a constant stiffness, EI, and distributed mass, m. The concentrated mass at the midpoint has the same value as the total for the beam, i.e., $M = ma$. The Rayleigh-Ritz method is selected and the following assumed functions have been suggested:

$$\Omega_1 = 1 \qquad \Omega_2 = \frac{2x}{a} - 1 \qquad \Omega_{n+2} = \sin\frac{n\pi x}{a}$$

Explain why these functions have been suggested. Find the first three frequencies and modes.

8. Consider the beam shown on the top of page 358. Motion is restricted to the plane.
   (a) Find the finite element equations for the beam with boundary conditions shown. Divide the beam into four elements of equal length. Show how the boundary conditions are applied.
   (b) Find the first two frequencies and modes for the beam, given $a = 1000$ mm, the cross section is a square 10 mm on a side, and the material is aluminum.

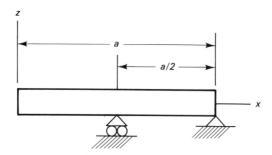

9. A stretched string can be modeled by letting $EI \to 0$ in Equation 6.3.2. Such a string is plucked in the middle so that

$$w(x, 0) = \frac{2x}{a} \qquad 0 \leftarrow x \leftarrow \frac{a}{2}$$

$$= 2\left(1 - \frac{x}{a}\right) \qquad \frac{a}{2} \leftarrow x \leftarrow a$$

$$w(x, 0) = 0$$

as shown by

Find the resulting motion. If the string (for example, a guitar string) is plucked to one side of the midpoint would you expect it to sound differently?

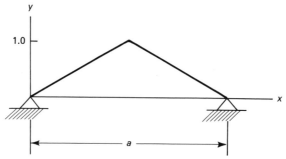

10. A circular shaft fixed at one end has a disk attached to the other as shown. Dimensions are in millimeters and the material is steel.

The disk is twisted 5 degrees about the x-axis and then released. What is the resulting motion?

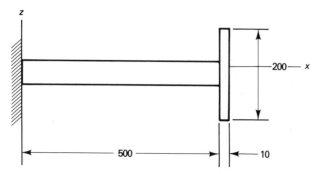

11. Set up the FEM equations for the frequencies and modes of a structure with both truss and beam elements as shown. Member AC is a pin jointed truss member with mass per unit length, m_{AC}, and area, A. Member BCD is a continuous beam with a uniform mass per unit length, m_{bcd}, and a stiffness EI.

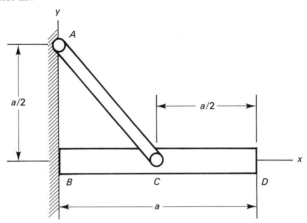

12. A rectangular simply supported flat plate has an applied load

$$p(x, y, t) = p_0 \sin \Omega t$$

Find the resulting motion.

APPENDIXES

APPENDIX A: MATRIX OPERATIONS

A.1 INTRODUCTION

A short review of the basic matrix definitions and operations is given here. While this material is very likely familiar to the reader, this summary will provide a quick reference and ensure that the particular use of notation in this text is understood.

A.2 MATRIX DEFINITIONS

A *rectangular matrix* of *order* $m \times n$ is an array of quantities in m rows and n columns as follows:

$$[a] = \begin{bmatrix} a_{11} & a_{12} & \dots & a_{1n} \\ a_{21} & a_{22} & \dots & a_{2n} \\ \cdot & \cdot & & \cdot \\ \cdot & \cdot & & \cdot \\ \cdot & \cdot & & \cdot \\ a_{m1} & a_{m2} & \dots & a_{mn} \end{bmatrix} \tag{A.1}$$

The element in the mth row and nth column is represented by the notation a_{mn}. A set of square brackets [ ] will denote a rectangular matrix of any order. For convenience, we shall define some special cases and use some special notation as explained in the following.

A *column* matrix has m rows and one column, or an order of $m \times 1$. Curved braces { } are used to designate a column matrix. The element in the mth row is designated a_m. For example, if $a_1 = 1$, $a_2 = 3$, and $a_3 = 2$, the matrix is of order 3×1 and is given by

$$\{a\} = \begin{bmatrix} 1 \\ 3 \\ 2 \end{bmatrix} \tag{A.2}$$

A *row* matrix has 1 row and n columns, for example,

$$[a] = [1 \ 3 \ 2] \tag{A.3}$$

and the element in the nth column is designated a_n.

A *square* matrix has the same number of rows and columns, or $m = n$. For example, a square matrix of order 3 x 3 might look like this

$$[a] = \begin{bmatrix} 1 & 3 & 2 \\ 4 & 8 & 1 \\ 7 & 5 & 2 \end{bmatrix} \tag{A.4}$$

A *symmetrical* matrix is a square matrix that has exactly the same value when rows and columns are interchanged, or

$$a_{mn} = a_{nm} \tag{A.5}$$

For example,

$$[a] = \begin{bmatrix} 1 & 3 & 2 \\ 3 & 8 & 1 \\ 2 & 1 & 2 \end{bmatrix} \tag{A.6}$$

A *diagonal* matrix is a square matrix in which all elements are zero except those on the principal diagonal, or

$$a_{mn} = 0, \quad \text{if } m \neq n \tag{A.7}$$

For example,

$$[a] = \begin{bmatrix} 1 & 0 & 0 \\ 0 & 8 & 0 \\ 0 & 0 & 2 \end{bmatrix} \tag{A.8}$$

The diagonal matrix is always symmetrical. The *identity* matrix is denoted by $[I]$ and is a special case of the diagonal matrix for which

$$a_{mn} = 1, \quad \text{when } m = n \tag{A.9}$$

For example,

$$[I] = \begin{bmatrix} 1 & 0 & 0 \\ 0 & 1 & 0 \\ 0 & 0 & 1 \end{bmatrix} \tag{A.10}$$

It is also frequently called the *unit* matrix. A *single-element* matrix is a *scalar* and may be written with or without brackets.

The determinant $|a|$ formed from the elements of a square matrix is known as the *determinant of* $[a]$.

The *transpose* of a matrix is found by interchanging rows and columns and is denoted with a superscript T. The elements of the transpose of a matrix are found by setting

$$(a_{mn})^T = a_{nm} \tag{A.11}$$

For example,

$$[a] = \begin{bmatrix} 1 & 2 \\ 7 & 9 \\ 6 & 3 \end{bmatrix}, \qquad [a]^T = \begin{bmatrix} 1 & 7 & 6 \\ 2 & 9 & 3 \end{bmatrix} \tag{A.12}$$

The transpose of a row matrix is a column matrix, and vice versa. A symmetric matrix is identical to its transpose.

A.3 MATRIX ALGEBRA

We now define certain rules of matrix mathematics that make it especially useful.

1. *Equality.* Two matrices are equal if they are of the same order and all corresponding elements are equal. That is,

$$[a] = [b], \qquad \text{if } a_{mn} = b_{mn} \tag{A.13}$$

2. *Addition and subtraction.* The matrices must be of the same order. Addition is performed by adding corresponding elements and subtraction by subtracting corresponding elements. Thus,

$$[a] + [b] = [c], \qquad \text{where } c_{mn} = a_{mn} + b_{mn} \tag{A.14}$$

For example,

$$\begin{bmatrix} 2 & 1 & 7 \\ 3 & 6 & 9 \\ 1 & -4 & 0 \end{bmatrix} + \begin{bmatrix} 1 & 6 & 7 \\ 1 & -2 & 0 \\ 0 & 0 & 1 \end{bmatrix} = \begin{bmatrix} 3 & 7 & 14 \\ 4 & 4 & 9 \\ 1 & -4 & 1 \end{bmatrix} \tag{A.15}$$

3. *Multiplication by a scalar.* Any matrix may be multiplied by a scalar by multiplying each element of the matrix by the scalar. Thus,

$$a[b] = [c], \qquad \text{where } c_{mn} = ab_{mn} \tag{A.16}$$

4. *Multiplication.* We denote multiplication of two matrices, say, $[a]$ and $[b]$ by

$$[a][b] = [c] \tag{A.17}$$

provided certain conditions exist. The number of columns in $[a]$ must equal the number of rows in $[b]$. Each element in $[c]$ is obtained by multiplying the elements of the corresponding row in $[a]$ by the elements of the corresponding column in $[b]$ and adding the results according to the rule

$$c_{mk} = \Sigma\, a_{mn} b_{nk} \tag{A.18}$$

For example,

$$\begin{bmatrix} 3 & 2 \\ 1 & 1 \\ 7 & -1 \end{bmatrix} \begin{bmatrix} 1 & 2 & 1 \\ 6 & -3 & 1 \end{bmatrix} = \begin{bmatrix} 3\cdot 1 + 2\cdot 6 & 3\cdot 2 - 2\cdot 3 & 3\cdot 1 + 2\cdot 1 \\ 1\cdot 1 + 1\cdot 6 & 1\cdot 2 - 1\cdot 3 & 1\cdot 1 + 1\cdot 1 \\ 7\cdot 1 - 1\cdot 6 & 7\cdot 2 + 1\cdot 3 & 7\cdot 1 - 1\cdot 1 \end{bmatrix}$$

$$= \begin{bmatrix} 15 & 0 & 5 \\ 7 & -1 & 2 \\ 1 & 17 & 6 \end{bmatrix} \tag{A.19}$$

Note that if the order of $[a]$ is $m \times n$ and the order of $[b]$ is $n \times k$ the order of $[c]$ is $m \times k$.

Matrix multiplication is *associative*, *distributive*, but, in general, not *commutative*. Thus,

$$[a]([b][c]) = ([a][b])[c]$$

$$[a]([b] + [c]) = [a][b] + [a][c] \tag{A.20}$$

$$[a][b] \neq [b][a]$$

5. *Inversion.* Division, as such, is not defined for matrices but is replaced by something called inversion. We denote the inverted matrix with the symbolic form $[a]^{-1}$, and it is defined so that

$$[a]^{-1}[a] = [I] \tag{A.21}$$

The elements of the inverse matrix can be defined algebraically in terms of the original matrix. In practice, a matrix is seldom inverted by the algebraic process. Instead, a numerical process is used. We shall not go into either here.

A.4 PARTITIONED MATRICES

A useful operation with matrices is *partitioning* into *submatrices*. These submatrices may be treated as elements of the parent matrix and manipulated by the rules just reviewed.

$$[a] = \begin{bmatrix} 7 & 5 & 9 & 4 & 3 \\ 3 & 9 & 2 & 7 & 8 \\ 1 & 2 & 8 & 6 & 5 \end{bmatrix} = \begin{bmatrix} [A^{11}] & [A^{12}] \\ [A^{21}] & [A^{22}] \end{bmatrix} \tag{A.22}$$

where

$$[A^{11}] = \begin{bmatrix} 7 & 5 & 9 \\ 3 & 9 & 2 \end{bmatrix}, \qquad [A^{12}] = \begin{bmatrix} 4 & 3 \\ 7 & 8 \end{bmatrix}$$

$$[A^{21}] = [1\ 2\ 8] \qquad [A^{22}] = [6\ 5]$$

A.5 DIFFERENTIATING A MATRIX

To differentiate a matrix, we differentiate each element in the conventional manner. For example, if

$$[a] = \begin{bmatrix} x & x^2 & 3x \\ x^2 & x^4 & 2x \\ 3x & 2x & x^3 \end{bmatrix} \tag{A.23}$$

then

$$\frac{\partial}{\partial x}[a] = \begin{bmatrix} 1 & 2x & 3 \\ 2x & 4x^3 & 2 \\ 3 & 2 & 3x^2 \end{bmatrix} \tag{A.24}$$

In structural theory, we often encounter an expression of the following form,

$$U = \frac{1}{2}\{a\}^T[c]\{a\} \tag{A.25}$$

and we wish to differentiate U with respect to each element of $\{a\}$ or find $\partial U/\partial a_m$ and $\partial^2 U/\partial a_m \partial a_n$. By expanding the above form, that is, by multiplying the matrices in Equation A.24, differentiating each element in turn by each element of $\{a\}$, and reassembling the result, we obtain

$$\frac{\partial U}{\partial a_m} = [c]\{a\}, \qquad \frac{\partial^2 U}{\partial a_m \partial a_n} = [c] \tag{A.26}$$

For example, if $[c]$ is symmetrical and

$$U = \frac{1}{2}[a_1 \; a_2] \begin{bmatrix} c_{11} & c_{12} \\ c_{21} & c_{22} \end{bmatrix} \begin{bmatrix} a_1 \\ a_2 \end{bmatrix} = \frac{1}{2}(c_{11}a_1^2 + 2c_{12}a_1a_2 + c_{22}a_2^2) \tag{A.27}$$

(since $c_{12} = c_{21}$), differentiation yields

$$\frac{\partial U}{\partial a_1} = c_{11}a_1 + c_{12}a_2, \qquad \frac{\partial U}{\partial a_2} = c_{12}a_1 + c_{22}a_2 \tag{A.28}$$

which in matrix form is

$$\begin{bmatrix} \dfrac{\partial U}{\partial a_1} \\ \dfrac{\partial u}{\partial a_2} \end{bmatrix} = \begin{bmatrix} c_{11} & c_{12} \\ c_{12} & c_{22} \end{bmatrix} \begin{bmatrix} a_1 \\ a_2 \end{bmatrix} \tag{A.29}$$

We note further that the second derivatives are

$$\frac{\partial^2 U}{\partial a_1^2} = c_{11}, \qquad \frac{\partial^2 U}{\partial a_1 \partial a_2} = c_{12}, \qquad \frac{\partial^2 U}{\partial a_2^2} = c_{22} \tag{A.30}$$

or, in general,

$$\frac{\partial^2 U}{\partial a_m \partial a_n} = c_{mn} \tag{A.31}$$

A.6 INTEGRATING A MATRIX

Matrix integration is defined to be consistent with matrix differentiation. We integrate term by term. For example, if

$$[a] = \begin{bmatrix} 1 & 2x & 3 \\ 2x & 4x^3 & 2 \\ 3 & 2 & 3x^2 \end{bmatrix} \tag{A.32}$$

then

$$\int [a]\, dx = \begin{bmatrix} x & x^2 & 3x \\ x^2 & x^4 & 2x \\ 3x & 2x & x^3 \end{bmatrix} + [C] \tag{A.33}$$

where $[C]$ are the constants of integration. For definite integrals, each term is evaluated for the limits of integration present.

A.7 SUMMARY OF USEFUL MATRIX RELATIONS

$$[a][I] = [I][a] = [a]$$
$$a([b] + [c]) = a[b] + a[c]$$
$$[a]([b] + [c]) = [a][b] + [a][c]$$
$$[a] + [b] + [c] = [a] + ([b] + [c]) = ([a] + [b]) + [c]$$
$$[a][b][c] = [a]([b][c]) = ([a][b])[c]$$
$$[a] + [b] = [b] + [a]$$
$$[a][b] \neq [b][a]$$
$$([a][b])^T = [b]^T[a]^T$$
$$([a][b])^{-1} = [b]^{-1}[a]^{-1}$$
$$([a]^T)^{-1} = ([a]^{-1})^T$$

APPENDIX B: TRANSFORMATION OF STRESS. PRINCIPAL STRESS

B.1 INTRODUCTION

In normal stress analysis, the stresses as found with respect to coordinate directions chosen for convenience in defining the geometry of the structure or for convenience of analysis may not be the critical values for determining the integrity of the structure. This leads us first to develop equations for the *transformation of stress* at a point; that is, we find the values of stress at a point for any axis orientation in terms of the values of stress for any other axis orientation, or rotation. We then move to a definition of the *principal stresses*. The principal stresses are found to be three values of normal stress on three orthogonal surfaces that have no shear stress and are found at a unique orientation of the axes call the *principal axes*. One principal stress is the maximum normal stress encountered at that point, and since some materials fail according to a maximum normal stress criterion you can readily see its importance.

In standard stress notation, we designate the magnitude and both the direction of the stress and the direction of the normal of the surface on which it acts; hence the double subscript notation. As stated in Section 1.2.1, the state of stress at a point with respect to one orientation of the coordinate system will uniquely determine the state of stress at the same point with respect to any other rotational orientation of the coordinates. Because we must keep track of both the direction of the stress and the direction of the surface on which it acts, the transformation of stress during a coordinate rotation is a bit complicated. Detailed descriptions of the transformations that are summarized next are given in References 1 to 8, among others.

B.2 FORCE COMPONENTS WITH RESPECT TO ROTATED AXES

First we must have a convenient way to quantify the relative position of two orthogonal coordinate systems that have the same origin but that are rotated with respect to one another. One commonly used system is by means of direction cosines. In Figure B.1, we show two sets of axes, primed and unprimed, with the components of a vector F shown in each set. The direction cosines are given in the following table:

	x	y	z
x'	l_1	m_1	n_1
y'	l_2	m_2	n_2
z'	l_3	m_3	n_3

where, for example, $l_1 = \cos xx'$, where xx' signifies the angle between the x and x' axes. Similarly, $m_3 = \cos yz'$, where yz' signifies the angle between the y and z' axes.

You can verify that the relationships between the force components of the vector F in Figure B.1 are given by the matrix transformation

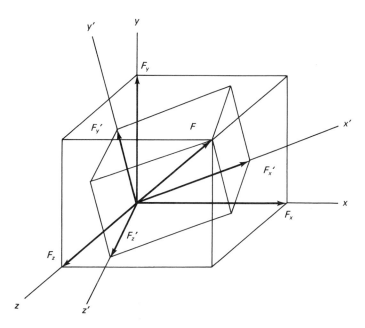

Figure B.1 Rotation of Axes.

$$\{F'\} = [T]\{F\} \tag{B.1}$$

where

$$\begin{bmatrix} F_{x'} \\ F_{y'} \\ F_{z'} \end{bmatrix} = \begin{bmatrix} l_1 & m_1 & n_1 \\ l_2 & m_2 & n_2 \\ l_3 & m_3 & n_3 \end{bmatrix} \begin{bmatrix} F_x \\ F_y \\ F_z \end{bmatrix} \tag{B.2}$$

B.3 TRANSFORMATION OF STRESS

For the transformation of stress, we need to recall the definition of stress as a force per unit area and in the process must specify both the direction of the force and the orientation of the area. Let us look at the equilibrium of a small tetrahedon, as shown in Figure B.2.

In the figure, N_1 is a force per unit area acting on the oblique face, with components X_1, Y_1, Z_1 in the three coordinate directions, x, y, z, respectively. To specify the orientation of the surface, define the direction cosines of the normal to the surface by l_1, m_1, and n_1. Then the components of N_1 may be given by

$$\begin{bmatrix} X_1 \\ Y_1 \\ Z_1 \end{bmatrix} = \begin{bmatrix} l_1\sigma_x + m_1\tau_{xy} + n_1\tau_{zx} \\ l_1\tau_{xy} + m_1\sigma_y + n_1\tau_{yz} \\ l_1\tau_{zx} + m_1\tau_{yz} + n_1\sigma_z \end{bmatrix} \tag{B.3}$$

or, in matrix form,

$$\{N_1\} = [\sigma]\{\ell_1\} \tag{B.4}$$

where

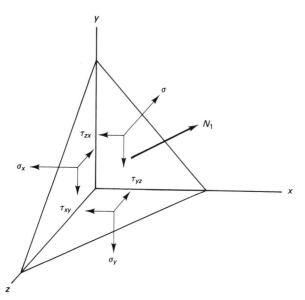

Figure B.2 Element for Geometrical Stress Transformation.

$$[\sigma] = \begin{bmatrix} \sigma_x & \tau_{xy} & \tau_{zx} \\ \tau_{xy} & \sigma_y & \tau_{yz} \\ \tau_{zx} & \tau_{yz} & \sigma_z \end{bmatrix} \tag{B.5}$$

and

$$[\ell_1] = \begin{bmatrix} l_1 \\ m_1 \\ n_1 \end{bmatrix} \tag{B.6}$$

This is a new rectangular form for the stress matrix $[\sigma]$, which contains the same quantities as $\{\sigma\}$, used elsewhere, but for convenience they are placed in this format.

If the normal to the plane on which $\{N_1\}$ acts is taken to be the x'-axis and appropriate directions are chosen for the y'- and z'-axes, two other tetrahedrons can be drawn, one with a normal in the y'-direction and one with a normal in the z'-direction. In this way, two additional matrix expressions may be formed:

$$\{N_2\} = [\sigma]\{\ell_2\}, \qquad \{N_3\} = [\sigma]\{\ell_3\} \tag{B.7}$$

These can be combined to form the x, y, z components of stress on the three faces normal to the x', y', z' axes. First, we define a matrix $[N]$ made up of the three column matrices $\{N_i\}$ and a matrix $[\ell]$ made of the three column matrices $\{\ell_i\}$ as follows:

$$[N] = \begin{bmatrix} X_1 & X_2 & X_3 \\ Y_1 & Y_2 & Y_3 \\ Z_1 & Z_2 & Z_3 \end{bmatrix}, \qquad [\ell] = \begin{bmatrix} l_1 & l_2 & l_3 \\ m_1 & m_2 & m_3 \\ n_1 & n_2 & n_3 \end{bmatrix} \tag{B.8}$$

We can readily see that

$$[\ell] = [T]^T \tag{B.9}$$

and that

$$[N] = [\sigma][T]^T \tag{B.10}$$

Furthermore, we can verify that

$$[\sigma'] = [T][N] = [T][\sigma][T]^T \tag{B.11}$$

where

$$[\sigma'] = \begin{bmatrix} \sigma_{x'} & \tau_{x'y'} & \tau_{z'x'} \\ \tau_{x'y'} & \sigma_{y'} & \tau_{y'z'} \\ \tau_{z'x'} & \tau_{y'z'} & \sigma_{z'} \end{bmatrix} \tag{B.12}$$

Sometimes, this expression is expanded and written in the column matrix form for stress:

$$\{\sigma'\} = [T_\sigma]\{\sigma\} \tag{B.13}$$

where

$$[T_\sigma] = \begin{bmatrix} l_1^2 & m_1^2 & n_1^2 & 2l_1m_1 & 2m_1n_1 & 2n_1l_1 \\ l_2^2 & m_2^2 & n_2^2 & 2l_2m_2 & 2m_2n_2 & 2n_2l_2 \\ l_3^2 & m_3^2 & n_3^2 & 2l_3m_3 & 2m_3n_3 & 2n_3l_3 \\ l_1l_2 & m_1m_2 & n_1n_2 & l_1m_2+l_2m_1 & m_1n_2+m_2n_1 & n_1l_2+n_2l_1 \\ l_2l_3 & m_2m_3 & n_2n_3 & l_2m_3+l_3m_2 & m_2n_3+m_3n_2 & n_2l_3+n_3l_2 \\ l_3l_1 & m_3m_1 & n_3n_1 & l_3m_1+l_1m_3 & m_3n_1+m_1n_3 & n_3l+n_1l_3 \end{bmatrix}$$

B.4 PRINCIPAL STRESSES AND AXES FOR THREE DIMENSIONS

It has been shown, for example, in References 5, 6, and 7, that there are three orthogonal planes on each of which the stress at a point consists of a normal stress and no shearing stresses. These are called the principal stresses, and the rotated coordinate axes that are normal to each of these planes are called the principal axes. In terms of the preceding equations, this may be stated

$$[T_p][\sigma][T_p]^T = [\sigma_p] \tag{B.14}$$

where

$$[\sigma_p] = \begin{bmatrix} \sigma_{p1} & 0 & 0 \\ 0 & \sigma_{p2} & 0 \\ 0 & 0 & \sigma_{p3} \end{bmatrix} \tag{B.15}$$

and $[T_p]$ is the appropriate set of direction cosines for the principal axes. To find the principal stresses and principal axes, we return to Equation B.3. This time we shall declare that the only stress on the oblique surface is a normal stress with a magnitude σ. This makes the normal to the surface one of the principal axes and the direction cosines unknowns along with the value of σ. We can recognize that under these circumstances the three components of stress in the x, y, z directions on the oblique plane have values σl, σm, and σn, or

$$\{N\} = \sigma\{\ell\} = \begin{bmatrix} l \\ m \\ n \end{bmatrix} \tag{B.16}$$

where we have dropped the subscripts. We shall reserve the subcripts on both $\{N_i\}$ and $\{\ell_i\}$ for cases where the amount of rotation is known. In this case, $\{\ell\}$ is an unknown.

Under these circumstances, Equation B.4 becomes

$$[\sigma]\{\ell\} = \sigma\{\ell\} \tag{B.17}$$

or

$$\begin{bmatrix} \sigma_x - \sigma & \tau_{xy} & \tau_{zx} \\ \tau_{xy} & \sigma_y - \sigma & \tau_{yz} \\ \tau_{zx} & \tau_{yz} & \sigma_z - \sigma \end{bmatrix} \begin{bmatrix} 1 \\ m \\ n \end{bmatrix} = \begin{bmatrix} 0 \\ 0 \\ 0 \end{bmatrix} \tag{B.18}$$

This is a classical eigenvalue problem, which is studied in detail is Chapters 5 and 6 of this text. Briefly, this set of equations has as its solution either $\{\ell\} = \{0\}$, or the determinant of the coefficient matrix is zero. It is the latter that provides us with useful information about principal stresses and axes. First, we set the determinant equal to zero:

$$\begin{vmatrix} \sigma_x - \sigma & \tau_{xy} & \tau_{zx} \\ \tau_{xy} & \sigma_y - \sigma & \tau_{yz} \\ \tau_{zx} & \tau_{yz} & \sigma_z - \sigma \end{vmatrix} = 0 \tag{B.19}$$

When expanded, it yields the following cubic equation:

$$\sigma^3 - S_1\sigma^2 + S_2\sigma - S_3 = 0 \tag{B.20}$$

where

$$\begin{aligned} S_1 &= \sigma_x + \sigma_y + \sigma_z \\ S_2 &= \sigma_x\sigma_y + \sigma_y\sigma_z + \sigma_z\sigma_x - \tau_{xy}^2 - \tau_{yz}^2 - \tau_{zx}^2 \\ S_3 &= \sigma_x\sigma_y\sigma_z + 2\tau_{xy}\tau_{yz}\tau_{zx} - \sigma_x\tau_{yz}^2 - \sigma_y\tau_{zx}^2 - \sigma_z\tau_{xy}^2 \end{aligned} \tag{B.21}$$

The three roots of this equation give the three principal stresses. Unfortunately, there is no simple analytical solution for the roots of cubic equations, but simple numerical solutions are possible. A computer program for finding such roots is given in Appendix K.

For each principal stress, there is a set of direction cosines that allows us to orient one of the principal axes. To find each set of direction cosines, we use two of the equations in Equation B.18 (we cannot use all three because the determinant of the coefficients is zero) and the well-known property of direction cosines for orthogonal axes that

$$1^2 + m^2 + n^2 = 1 \tag{B.22}$$

B.5 STRESS TRANSFORMATION AND PRINCIPAL STRESSES FOR TWO DIMENSIONS

The stress transformations and principal stresses for two-dimensional states of stress are dwelt on at great length in every elementary mechanics of materials book and practically every other book on solid mechanics and structures ever written. They may be derived from the equations given previously by setting all components of stress in one coordinate direction, say, the z-direction, equal to zero and simplifying the resulting set of equations. The principal stress eigenvalue problem becomes especially simple, since finding the roots of a quadratic equation results. A quadratic equation does have an exact analytical solution. There is also a very helpful graphical form of the stress transformation called Mohr's circle. We shall not repeat any of this development here since it is so readily available. We shall limit ourselves to repeating the main equations for use as a ready reference.

When the z-components of stress are zero, that is,

$$\sigma_z = \tau_{yz} = \tau_{zx} = 0 \tag{B.23}$$

and we are interested only in what happens in the x-y plane, for example, in plane stress, we have a rotation about one axis, say, the z-axis. Then

$$[T] = \begin{bmatrix} l_1 & m_1 & n_1 \\ l_2 & m_2 & n_2 \\ l_3 & m_3 & n_3 \end{bmatrix} = \begin{bmatrix} \cos\alpha & \sin\alpha & 0 \\ -\sin\alpha & \cos\alpha & 0 \\ 0 & 0 & 1 \end{bmatrix} \tag{B.24}$$

When these values of the direction cosines, along with Equation B.23, are entered into Equation B.14, this yields

$$\sigma_{x'} = \sigma_x \cos^2\alpha + \sigma_y \sin^2\alpha + 2\tau_{xy} \sin\alpha \cos\alpha \tag{B.25}$$

or, using trigonometic identities,

$$\sigma_{x'} = \frac{\sigma_x + \sigma_y}{2} + \frac{\sigma_x - \sigma_y}{2} \cos 2\alpha + \tau_{xy} \sin 2\alpha \tag{B.26}$$

Similarly,

$$\sigma_{y'} = \frac{\sigma_x + \sigma_y}{2} - \frac{\sigma_x - \sigma_y}{2} \cos 2\alpha + \tau_{xy} \sin 2\alpha$$
$$\tau_{x'y'} = -\frac{\sigma_x - \sigma_y}{2} \sin 2\alpha + \tau_{xy} \cos 2\alpha \tag{B.27}$$

To find the principal axes and principal stresses, we set $\tau_{x'y'} = 0$ and obtain

$$\tan 2\alpha_p = \frac{2\tau_{xy}}{\sigma_x - \sigma_y} \tag{B.28}$$

where the subscript p has been added to α to indicate that this is the rotation necessary to define a principal axis. Once α_p is found, it can be substituted in Equations B.27 and B.28 to find the principal stresses.

By combining equations B.27 with B.25 and B.26, it can be shown that the principal stresses are given by the following formula:

$$\sigma_{p1, p2} = \frac{\sigma_x + \sigma_y}{2} \pm \left[\left(\frac{\sigma_x + \sigma_y}{2}\right)^2 + \tau_{xy}^2 \right]^{1/2} \tag{B.29}$$

These are the maximum and minimum normal stresses at a point.

APPENDIX C: TRANSFORMATION OF STRAIN

C.1 INTRODUCTION

Strain at a point, as is the case with stress, varies with the orientation of the coordinate axes. The equations for the *transformation of strain* are given in the following. Arguments for principal strains and principal axes of strain similar to those for stress can be found in various references and will not be given here. Suffice it to say that since in Hooke's law the normal and shear stresses and strains are uncoupled, that is, normal stresses depend only on normal strains, and vice versa, and shearing stresses depend only on shearing strains, and vice versa, the principal axes of strain are identical to the principal axes of stress. Thus the direction cosines for principal axes of stress can be used in the following formulas for the principal axes of strain. A separate development is not needed.

C.2 TRANSFORMATION OF STRAIN

The strain transformations are essentially the transformations of displacement derivatives. The displacements in the primed axes direction can be given in terms of displacements in the unprimed coordinates by the relation

$$\{u'\} = [T]\{u\} \tag{C.1}$$

where $[T]$ is the matrix of direction cosines defined in Appendix B, Equation B.2, repeated here,

$$[T] = \begin{bmatrix} l_1 & m_1 & n_1 \\ l_2 & m_2 & n_2 \\ l_3 & m_3 & n_3 \end{bmatrix} \tag{C.2}$$

and the two displacement matrices are

$$\{u'\} = \begin{bmatrix} u' \\ v' \\ w' \end{bmatrix}, \qquad \{u\} = \begin{bmatrix} u \\ v \\ w \end{bmatrix} \tag{C.3}$$

Using the chain rule for differentiation, we have

$$\frac{\partial u'}{\partial x'} = \frac{\partial u'}{\partial x}\frac{\partial x}{\partial x'} + \frac{\partial u'}{\partial y}\frac{\partial y}{\partial x'}\frac{\partial u'}{\partial z}\frac{\partial z}{\partial x'} \tag{C.4}$$

where

$$\frac{\partial x}{\partial x'} = l_1, \qquad \frac{\partial y}{\partial x'} = m_1, \qquad \frac{\partial z}{\partial x'} = n_1 \tag{C.5}$$

and where, from Equation C.1,

$$\frac{\partial u'}{\partial x} = l_1\frac{\partial u}{\partial x} + m_1\frac{\partial v}{\partial x} + n_1\frac{\partial w}{\partial x} \tag{C.6}$$

Similar relations are found for $\partial u'/\partial y'$ and all other derivatives of displacement used in defining the strains. Then all the strains related to the primed coordinates are defined in terms of the appropriate derivatives of displacement in the primed coordinates, and all the strains related to the unprimed coordinates are defined in terms of the appropriate derivatives of displacement in the unprimed coordinates, and the following equations result:

$$\{\epsilon'\} = [T_\epsilon]\{\epsilon\} \tag{C.7}$$

where

$$[T_\epsilon] = \begin{bmatrix} l_1^2 & m_1^2 & n_1^2 & 2l_1m_1 & 2m_1n_1 & 2n_1l_1 \\ l_2^2 & m_2^2 & n_2^2 & 2l_2m_2 & 2m_2n_2 & 2n_2l_2 \\ l_3^2 & m_3^2 & n_3^2 & 2l_3m_3 & 2m_3n_3 & 2n_3l_3 \\ l_1l_2 & m_1m_2 & n_1n_2 & l_1m_2 + l_2m_1 & m_1n_2 + m_2n_1 & n_1l_2 + n_2l_1 \\ l_2l_3 & m_2m_3 & n_2n_3 & l_2m_3 + l_3m_2 & m_2n_3 + m_3n_2 & n_2l_3 + n_3l_2 \\ l_3l_1 & m_3m_1 & n_3n_1 & l_3m_1 + l_1m_3 & m_3n_1 + m_1n_3 & n_3l_1 + n_1l_3 \end{bmatrix}$$

and

$$\{\epsilon'\} = \begin{bmatrix} \epsilon_x' \\ \epsilon_y' \\ \epsilon_z' \\ \gamma_{x'y'} \\ \gamma_{y'z'} \\ \gamma_{z'x'} \end{bmatrix}, \qquad \{\epsilon\} = \begin{bmatrix} \epsilon_x \\ \epsilon_y \\ \epsilon_z \\ \gamma_{xy} \\ \gamma_{yz} \\ \gamma_{zx} \end{bmatrix} \tag{C.8}$$

C.3 STRAIN TRANSFORMATION FOR TWO DIMENSIONS

When the z-components of strain are zero, that is,

$$\epsilon_z = \tau_{yz} = \tau_{zx} = 0 \tag{C.9}$$

and where rotation is done about the z-axis, we have

$$[T] = \begin{bmatrix} l_1 & m_1 & n_1 \\ l_2 & m_2 & n_2 \\ l_3 & m_3 & n_3 \end{bmatrix} = \begin{bmatrix} \cos\alpha & \sin\alpha & 0 \\ -\sin\alpha & \cos\alpha & 0 \\ 0 & 0 & 1 \end{bmatrix} \tag{C.10}$$

When these values of the direction cosines, along with Equation C.9, are entered into Equation C.7, this yields

$$\epsilon_{x'} = \epsilon_x \cos^2\alpha + \epsilon_y \sin^2\alpha + 2\epsilon_{xy} \sin\alpha \cos\alpha \tag{C.11}$$

or, using trigonometric identities,

$$\epsilon_{x'} = \frac{\epsilon_x + \epsilon_y}{2} + \frac{\epsilon_x - \epsilon_y}{2} \cos 2\alpha + \epsilon_{xy} \sin 2\alpha \tag{C.12}$$

Similarly,

$$\epsilon_{y'} = \frac{\epsilon_x + \epsilon_y}{2} - \frac{\epsilon_x - \epsilon_y}{2} \cos 2\alpha + \epsilon_{xy} \sin 2\alpha$$

$$\epsilon_{x'y'} = -\frac{\epsilon_x - \epsilon_y}{2} \sin 2\alpha + \epsilon_{xy} \cos 2\alpha \tag{C.13}$$

APPENDIX D: MATERIAL PROPERTIES

D.1 INTRODUCTION

Material properties are introduced into the equations of solid mechanics through the stress-strain relations. A very special case of these relations is called Hooke's law and is developed in the main body of the text. Here we expand on the background necessary to understand this relationship and review some extensions of the law to handle a wider class of materials.

D.2 AN EXPERIMENTALLY DETERMINED STRESS-STRAIN CURVE

The particular relations used in this text are suggested by a simple experiment. A long, slender rod made of a homogeneous and isotropic material is stretched by slowly applying a force (so slow there are no recognizable inertia effects) and a force-deflection curve is plotted. This curve is then reduced to a stress-strain curve by dividing the force by the original cross-sectional area of the rod and the deflection by the original length of the rod. (This way of calculating the stress and strain in a slender rod is justified in Section 1.3.2.) This latter curve is found to be the same for any sized rod made from a given material. When the material is one of many commonly used in structures, for example, steel and aluminum and their many alloys, it exhibits the characteristics shown in Figure D.1.

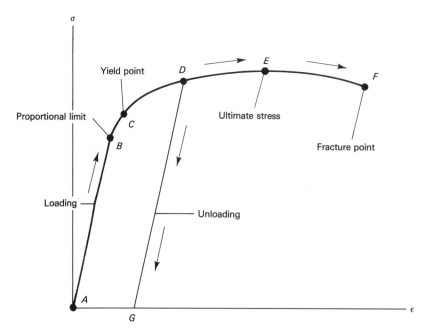

Figure D.1 Typical Stress-Strain Curve.

The curve represented by the heavier line along the path *ABCDEF* is found by plotting data from slowly loading the rod with a positive force. The region *ABC*, that is, from the origin to the *yield point,* is the elastic region. By elastic we mean that the curve is the same for loading as well as unloading. Thus, should we stop loading and start unloading while in that region, the data would follow the same curve. The elastic region is divided into two parts: the linearly elastic part *AB* from the origin to the *proportional limit,* and the nonlinearly elastic part *BC* from the proportional limit to the yield point. In many real materials, the part *BC* is very small, and no distinction need be made between the two points in the analysis of a structure. In some other materials, rubber is an example, the region *AB* is very small and *BC* relatively large.

In the linearly elastic part, stress is proportional to strain, resulting in the familiar Hooke's law relation

$$\sigma_x = E\epsilon_x \tag{D.1}$$

where E is the constant of proportionality called *Young's modulus.* The equation for the nonlinear region depends on the material, and no general formulation is possible. If, during this experiment, the width of the rod is measured, a narrowing or contraction of the cross section is noted. This may be expressed by

$$\epsilon_y = \epsilon_z = -\frac{v}{E}\sigma_x \tag{D.2}$$

where v is called *Poisson's ratio.* These two material constants, whose values are obtained by the experiment just described, are used to develop the *stress-strain equations* or *Hooke's law* for homogeneous and isotropic materials in Section 1.2.3.

The region *CDEF* is the plastic region. Within this region, deformation obtained from loading is permanent. Thus a new curve, for example, *DG,* follows from the data obtained for unloading. Two points of interest in the plastic region are the *ultimate stress* and the *fracture point,* each named for obvious reasons.

All regions are of major importance in solid mechanics; however, this text is limited to the linearly elastic region, with the exception of a couple of very brief forays into the plastic region. Fortunately, most major materials used in structures have a significant linearly elastic range. Furthermore, the operational stress level of many structures, such as those used in aerospace vehicles and automobiles, is kept within this range as a design requirement. Thus, the remainder of this appendix is devoted to a generalization of Hooke's law, which is shown in its most elementary form by Equations D.1 and D.2.

D.3 GENERALIZED HOOKE'S LAW

The most general statement of linear elasticity can be captured by the matrix equation

$$\{\sigma\} = [G]\{\epsilon\} \tag{D.3}$$

or its complementary form

$$\{\epsilon\} = [C]\{\sigma\} \tag{D.4}$$

Since each of the stress and strain matrices are column matrices with six rows, the matrix [G] has 36 elements. For nonhomogeneous materials, some of these elements would be functions of the coordinates. For a given nonhomogeneous and anisotropic material, appropriate constants would have to be found by experiment for each particular case of such a material.

When limited to materials that are homogeneous and anisotropic, each of these elements is a constant. Furthermore, it may be shown that in such a case the matrix [G] or [C] is symmetric. This means that only 21 of the constants are independent. A special case of anisotropic material occurs when material properties vary in three mutually perpendicular directions. Such a material is said to be *orthotropic*. Reflective symmetry arguments lead to the reduction of the number of constants to nine or

$$[G] = \begin{bmatrix} G_{11} & G_{12} & G_{13} & 0 & 0 & 0 \\ G_{21} & G_{22} & G_{23} & 0 & 0 & 0 \\ G_{31} & G_{32} & G_{33} & 0 & 0 & 0 \\ 0 & 0 & 0 & G_{44} & 0 & 0 \\ 0 & 0 & 0 & 0 & G_{55} & 0 \\ 0 & 0 & 0 & 0 & 0 & G_{66} \end{bmatrix} \tag{D.5}$$

where

$$G_{ij} = G_{ji} \tag{D.6}$$

The increased use of layered materials in composite and sandwich construction has increased interest in orthotropic and even more complex anisotropic materials in recent years.

For materials that are homogeneous and isotropic the 36 possible constants reduce to just 2 independent constants, usually E, Young's modulus, and v, Poisson's ratio. Usually, the relations are expressed in terms of three constants, the third being G, the shear modulus, which can be expressed in terms of the other two as

$$G = \frac{E}{2(1 + v)} \tag{D.7}$$

The full value of [G] for a homogeneous and isotropic material is given in Equation 1.2.31 and for [C] in Equation 1.2.30.

D.4 EFFECT OF TEMPERATURE CHANGE

The material properties defined in the preceding are dependent on temperature. Furthermore, the expansion of a material due to temperature change has the effect of an initial strain in a body. Since thermal expansion results only in extensional behavior, Hooke's law for a homogeneous and isotropic body is modified as follows:

$$\{\epsilon\} = [C]\{\sigma\} - \Delta T[\kappa] \tag{D.8}$$

where

$$[\kappa] = \begin{bmatrix} \kappa \\ \kappa \\ \kappa \\ 0 \\ 0 \\ 0 \end{bmatrix} \tag{D.9}$$

where κ is the coefficient of linear thermal expansion and ΔT is the temperature change.

APPENDIX E: EQUATIONS OF ELASTICITY IN CYLINDRICAL AND POLAR COORDINATES

E.1 INTRODUCTION

Many structures are, or contain parts which are, bodies of revolution. It is often more convenient to represent these structures in cylindrical coordinates in three dimensions and polar coordinates in two dimensions than in rectangular cartesian coordinates. References 6, 7, and 8 contain details of the equations and a number of worked examples. Here the equations will be summarized but not derived.

E.2 EQUATIONS IN CYLINDRICAL COORDINATES

The coordinates are r, θ, and z, which are in the radial, angular, and axial directions, respectively. The stress, strain, and displacement matrices are

$$\{\sigma\} = \begin{bmatrix} \sigma_r \\ \sigma_\theta \\ \sigma_z \\ \tau_{r\theta} \\ \tau_{\theta z} \\ \tau_{zr} \end{bmatrix}, \qquad \{\epsilon\} = \begin{bmatrix} \epsilon_r \\ \epsilon_\theta \\ \epsilon_z \\ \gamma_{r\theta} \\ \gamma_{\theta z} \\ \gamma_{zr} \end{bmatrix}, \qquad \{u\} = \begin{bmatrix} u \\ v \\ w \end{bmatrix} \tag{E.1}$$

The equilibrium equations $[E]\{\sigma\} + \{f\} = \{0\}$ are

$$\frac{\partial \sigma_r}{\partial r} + \frac{1}{r}\frac{\partial \tau_{r\theta}}{\partial \theta} + \frac{\sigma_r - \sigma_\theta}{r} + \frac{\partial \tau_{zr}}{\partial z} + f_r = 0$$

$$\frac{1}{r}\frac{\partial \sigma_\theta}{\partial \theta} + \frac{\partial \tau_{r\theta}}{\partial r} + 2\frac{\tau_{r\theta}}{r} + \frac{\partial \tau_{\theta z}}{\partial z} + f_\theta = 0 \tag{E.2}$$

$$\frac{\partial \sigma_z}{\partial z} + \frac{1}{r}\frac{\partial \tau_{\theta z}}{\partial \theta} + \frac{\partial \tau_{zr}}{\partial r} + \frac{\tau_{zr}}{r} + f_z = 0$$

Since cylindrical coordinates are orthogonal, Hooke's law is the same as for rectangular cartesian coordinates with subscripts r, θ, and z replacing x, y, and z, respectively, in Equations 1.2.27 and 1.2.28, or

$$\epsilon_r = \frac{1}{E}[\sigma_r - \upsilon(\sigma_\theta + \sigma_z)], \qquad \epsilon_\theta = \frac{1}{E}[\sigma_\theta - \upsilon(\sigma_z + \sigma_r)]$$

$$\epsilon_z = \frac{1}{E}[\sigma_z - \upsilon(\sigma_r + \sigma_\theta)] \tag{E.3}$$

$$\gamma_{r\theta} = \frac{1}{G}\tau_{r\theta}, \qquad \gamma_{\theta z} = \frac{1}{G}\tau_{\theta z} \qquad \gamma_{zr} = \frac{1}{G}\tau_{zr}$$

The strain-displacement equations are

$$\epsilon_r = \frac{\partial u}{\partial r} \qquad \epsilon_\theta = \frac{u}{r} + \frac{1}{r}\frac{\partial v}{\partial \theta} \qquad \gamma_{r\theta} = \frac{\partial v}{\partial r} + \frac{1}{r}\frac{\partial u}{\partial \theta} - \frac{v}{r}$$

$$\epsilon_z = \frac{\partial w}{\partial z}, \qquad \gamma_{\theta z} = \frac{1}{r}\frac{\partial w}{\partial \theta} + \frac{\partial v}{\partial z}, \qquad \gamma_{zr} = \frac{\partial u}{\partial z} + \frac{\partial w}{\partial r} \tag{E.4}$$

E.3 EQUATIONS IN POLAR COORDINATES

The eight equations of the plane stress case can be expressed in polar coordinates. First we let

$$\sigma_z = \tau_{\theta z} = \tau_{zr} = f_z = 0 \tag{E.5}$$

and we note that in most two-dimensional problems in which polar coordinates are applicable the only body force present is radial in nature; therefore, we let $f_\theta = 0$, as well. The equations of equilibrium then reduce to two:

$$\frac{\partial \sigma_r}{\partial r} + \frac{1}{r}\frac{\partial \tau_{r\theta}}{\partial \theta} + \frac{\sigma_r - \sigma_\theta}{r} + f_r = 0$$

$$\frac{1}{r}\frac{\partial \sigma_\theta}{\partial \theta} + \frac{\partial \tau_{r\theta}}{\partial r} + 2\frac{\tau_{r\theta}}{r} = 0 \tag{E.6}$$

Hooke's law is given by the three coupled equations

$$\epsilon_r = \frac{1}{E}(\sigma_r - \upsilon\sigma_\theta), \qquad \epsilon_\theta = \frac{1}{E}(\sigma_\theta - \upsilon\sigma_r), \qquad \gamma_{r\theta} = \frac{1}{G}\tau_{r\theta} \tag{E.7}$$

Two of the strains are zero:

$$\gamma_{\theta z} = \gamma_{zr} = 0 \tag{E.8}$$

And the three coupled strain-displacement equations are

$$\epsilon_r = \frac{\partial u}{\partial r}, \qquad \epsilon_\theta = \frac{u}{r} + \frac{1}{r}\frac{\partial v}{\partial \theta}, \qquad \gamma_{r\theta} = \frac{\partial v}{\partial r} + \frac{1}{r}\frac{\partial u}{\partial \theta} - \frac{v}{r} \tag{E.9}$$

In addition, we have the uncoupled strain-displacement equation and stress-strain equation

$$\epsilon_z = \frac{\partial w}{\partial z} = -\frac{\upsilon(\sigma_r + \sigma_\theta)}{E} \tag{E.10}$$

The plane strain case is also of interest. We let

$$w = \epsilon_z = \gamma_{\theta z} = \gamma_{zr} = \tau_{\theta z} = \tau_{zr} = f_z = f_\theta = 0 \tag{E.11}$$

The equilibrium equations and the strain-displacement equations are the same as before for plane stress, but the three coupled equations in Hooke's law are given by Equations 1.3.20 with appropriate changes in the subscripts for the stress and strain components, or

$$\sigma_r = (\lambda + 2G)\epsilon_r + \lambda\epsilon_\theta, \qquad \sigma_\theta = \lambda\epsilon_r + (\lambda + 2G)\epsilon_\theta, \qquad \tau_{r\theta} = G\gamma_{r\theta} \tag{E.12}$$

and there is the one uncoupled equation

$$\sigma_z = \lambda(\epsilon_r + \epsilon_\theta) \tag{E.13}$$

E.4 EQUATIONS OF AXISYMMETRY

Of interest is a special case known as *axisymmetry*. In such a case, all quantities are independent of the coordinate θ. Under this circumstance,

$$v = \gamma_{r\theta} = \gamma_{\theta z} = \tau_{r\theta} = \tau_{\theta z} = f_\theta = 0 \tag{E.14}$$

and equilibrium is given by

$$\frac{\partial \sigma_r}{\partial r} + \frac{\sigma_r - \sigma_\theta}{r} + \frac{\partial \tau_{zr}}{\partial z} + f_r = 0$$

$$\frac{1}{r}\frac{\partial \sigma_\theta}{\partial \theta} = 0, \qquad \frac{\partial \sigma_z}{\partial z} + \frac{\partial \tau_{zr}}{\partial r} + \frac{\tau_{zr}}{r} + f_z = 0 \tag{E.15}$$

Hooke's law is

$$\epsilon_r = \frac{1}{E}[\sigma_r - v(\sigma_\theta + \sigma_z)], \qquad \epsilon_\theta = \frac{1}{E}[\sigma_\theta - v(\sigma_z + \sigma_r)]$$

$$\epsilon_z = \frac{1}{E}[\sigma_z - v(\sigma_r + \sigma_\theta)], \qquad \gamma_{zr} = \frac{1}{G}\tau_{zr} \tag{E.16}$$

and the strain-displacement equations are

$$\epsilon_r = \frac{\partial u}{\partial r} \qquad \epsilon_\theta = \frac{u}{r}$$

$$\epsilon_z = \frac{\partial w}{\partial z} \qquad \gamma_{zr} = \frac{\partial u}{\partial z} + \frac{\partial w}{\partial r} \tag{E.17}$$

APPENDIX F: SHEAR FLOW IN MULTICELL BEAMS

F.1 INTRODUCTION

The shear flow analysis for torsional and shear loading can be extended to include multicelled beams. First we shall look at the case for pure torsion and then transverse shear. References 4, 5, and 9 give more complete information on this topic.

F.2 TORSION OF MULTICELLED BEAMS

The following analysis for multicelled beams with a pure torque applied is based on two assumptions: (1) there is a constant shear flow in each cell, and (2) all cells twist the same amount. The following brief summary is given for a three-celled beam but is easily generalized for any number of cells.

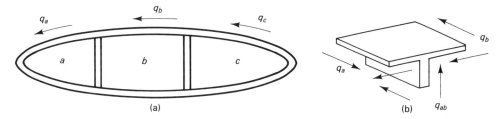

Figure F.1 Torsional Shear Flows in a Multicelled Beam.

If the shear flows in each cell are denoted by q_a, q_b, and q_c, as shown in Figure F.1(a), we can note that at the juncture of the vertical web with the outer shell, shown in Figure F.1(b), from equilibrium

$$q_{ab} = q_a - q_b, \qquad q_{bc} = q_b - q_c \qquad (F.1)$$

These are the values in the vertical webs that are parts of the two adjacent cells.

The applied torque must equal the total torque due to the shear flows, or

$$M_x = 2A_a q_a + 2A_b q_b + 2A_c q_c \qquad (F.2)$$

The condition that all cells twist the same amount is, from Equation 2.3.23,

$$\frac{1}{2A_a G} \oint \frac{q}{h} ds = \frac{1}{2A_b G} \oint \frac{q}{h} ds = \frac{1}{2A_c G} \oint \frac{q}{h} ds \qquad (F.3)$$

Equations F.2 and F.3 are three equations in three unknowns and may be solved for q_a, q_b, and q_c.

F.3 TRANSVERSE SHEAR IN MULTICELLED BEAMS

The shear flow analysis when the loading is a transverse shear force requires defining a starting point for the s-coordinate for each cell. For example, for the three-celled beam in Figure F.2, we have shown a possible point for defining the start of the integration for each cell. At each point, we have the value of the shear flow associated with the constant of integration. These values are unknown until we invoke a twist condition to determine them.

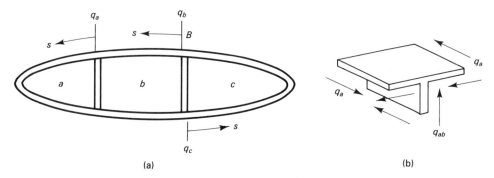

Figure F.2 Transverse Shear in a Multicelled Beam.

The shear flow for each cell is

$$q_i = q_i(0) - \int_0^s \frac{S_y h}{I_{zz}} y \, ds \tag{F.4}$$

However, when we reach a juncture between the outer shell and the vertical web, we must use the equilibrium at the junction depicted in Figure F.2(b).

$$q_{ab} = q_a - q_b, \qquad q_{bc} = q_b - q_c \tag{F.5}$$

where in this case the values of the shear flow are those in the outer shell and web at the junction. To complete the set of equations, we assume that the shear load is through the shear center and there is no twist of the section, or

$$\frac{1}{2A_a G} \oint \frac{q}{h} ds = \frac{1}{2A_b G} \oint \frac{q}{h} ds = \frac{1}{2A_c G} \oint \frac{q}{h} ds = 0 \tag{F.6}$$

The shear center is found from

$$S_y e_z = \int qr \, ds, \qquad S_z e_y = \int qr \, ds \tag{F.7}$$

APPENDIX G: INTERNAL VIRTUAL WORK AND STRAIN ENERGY

G.1 INTRODUCTION

In developing the principle of virtual work, we make use of the fact that the internal virtual work is equivalent to the negative of a change in a quantity called the *strain energy*. In many texts, this equivalence is stated without further evidence and from that moment the reader must accept it. Since most readers are intuitively more comfortable with Newton's laws, we present here a development of that equivalence and of the principle of virtual work using equilibrium as presented by Newton. This should help to increase our confidence in the work and energy principles.

G.2 PRINCIPLE OF VIRTUAL WORK

The virtual work of the surface forces is

$$\delta W_s = \int (X_s\, \delta u + Y_s\, \delta v + Z_s\, \delta w)\, dA \tag{G.1}$$

where integration is carried out over the surface A on which the forces act. The surface forces are given in terms of the components of stress at the surface by Equations 1.2.32 to be

$$X_s = l\sigma_x + m\tau_{xy} + n\tau_{zx}$$
$$Y_s = l\tau_{xy} + m\sigma_y + n\tau_{yz} \tag{G.2}$$
$$Z_s = l\tau_{zx} + m\tau_{yz} + n\sigma_z$$

Thus, the surface forces may be written in terms of the surface stresses, and from a theorem in mathematics called the *divergence theorem,* which relates a surface integral to a volume integral, we have

$$\begin{aligned}
\delta W_s = \int & [(l\sigma_x + m\tau_{xy} + n\tau_{zx})\, \delta u \\
& + (l\tau_{xy} + m\sigma_y + n\tau_{yz})\, \delta v \\
& + (l\tau_{zx} + m\tau_{yz} + n\sigma_z)\, \delta w]\, dA \\
= \int & \left[\frac{\partial}{\partial x} (\sigma_x\, \delta u + \tau_{xy}\, \delta v + \tau_{zx}\, \delta w) \right. \\
& + \frac{\partial}{\partial y} (\tau_{xy}\, \delta u + \sigma_y\, \delta v + \tau_{yz}\, \delta w) \\
& \left. + \frac{\partial}{\partial z} (\tau_{zx}\, \delta u + \tau_{yz}\, \delta v + \sigma_z\, \delta w) \right] dV
\end{aligned} \tag{G.3}$$

where now the integration is carried out over the volume V of the body.

By carrying out the indicated differentiation and regrouping terms, we obtain

$$\delta W_s = \int \left[\left(\frac{\partial \sigma_x}{\partial x} + \frac{\partial \tau_{xy}}{\partial y} + \frac{\partial \tau_{zx}}{\partial z} \right) \delta u + \left(\frac{\partial \tau_{xy}}{\partial x} + \frac{\partial \sigma_y}{\partial y} + \frac{\partial \tau_{yz}}{\partial z} \right) \delta v \right.$$

$$+ \left(\frac{\partial \tau_{zx}}{\partial x} + \frac{\partial \tau_{yz}}{\partial y} + \frac{\partial \sigma_z}{\partial z} \right) \delta w + \left(\sigma_x \frac{\partial \delta u}{\partial x} + \tau_{xy} \frac{\partial \delta v}{\partial x} + \tau_{zx} \frac{\partial \delta w}{\partial x} \right) \qquad \text{(G.4)}$$

$$\left. + \left(\tau_{xy} \frac{\partial \delta u}{\partial y} + \sigma_y \frac{\partial \delta v}{\partial y} + \tau_{yz} \frac{\partial \delta w}{\partial y} \right) + \left(\tau_{zx} \frac{\partial \delta u}{\partial z} + \tau_{yz} \frac{\partial \delta v}{\partial z} + \sigma_z \frac{\partial \delta w}{\partial z} \right) \right] dV$$

To simplify this expression, we note that the quantities in the first three sets of parentheses are related to the body force components by the equilibrium equations given in Equation 1.2.5. Furthermore, if we use the notation

$$\epsilon_x + \delta\epsilon_x = \frac{\partial}{\partial x}(u + \delta u) = \frac{\partial u}{\partial x} + \frac{\partial \delta u}{\partial x} \qquad \text{(G.5)}$$

or

$$\delta\epsilon_x = \frac{\partial \delta u}{\partial x} \qquad \text{(G.6)}$$

and

$$\gamma_{xy} + \delta\gamma_{xy} = \frac{\partial}{\partial x}(v + \delta v) + \frac{\partial}{\partial y}(u + \delta u)$$

$$= \frac{\partial u}{\partial y} + \frac{\partial \delta u}{\partial y} + \frac{\partial v}{\partial x} + \frac{\partial \delta v}{\partial x} \qquad \text{(G.7)}$$

or

$$\delta\gamma_{xy} = \frac{\partial \delta u}{\partial y} + \frac{\partial \delta v}{\partial x} \qquad \text{(G.8)}$$

and so on for the other terms, we can say that

$$\delta W_s = -\int (f_x \, \delta u + f_y \, \delta v + f_z \, \delta w) \, dV$$

$$+ \int (\sigma_x \, \delta\epsilon_x + \sigma_y \, \delta\epsilon_y + \sigma_z \, \delta\epsilon_z \qquad \text{(G.9)}$$

$$+ \tau_{xy} \, \delta\gamma_{xy} + \tau_{yz} \, \delta\gamma_{yz} + \tau_{zx} \, \delta\gamma_{zx}) \, dV$$

The first integral on the right side can be identified as the virtual work of the body forces, δW_b, and the second integral as the virtual work of the internal forces, which we label δU.

Thus, in summary,

$$\delta W_s = -\delta W_b + \delta U \qquad \text{(G.10)}$$

or

$$\delta W = \delta W_s + \delta W_b - \delta U$$

$$= \delta W_e - \delta U = 0 \qquad \text{(G.11)}$$

where

$$\delta W_s = \int (X_s\, \delta u + Y_s\, \delta v + Z_s\, \delta w)\, dA$$

$$\delta W_b = \int (f_x\, \delta u + f_y\, \delta v + f_z\, \delta w)\, dV \tag{G.12}$$

$$\delta U = \int (\sigma_x\, \delta\epsilon_x + \sigma_y\, \delta\epsilon_y + \sigma_z\, \delta\epsilon_z$$

$$+\ \tau_{xy}\, \delta\gamma_{xy} + \tau_{yz}\, \delta\gamma_{yz} + \tau_{zx}\, \delta\gamma_{zx})\, dV$$

The definition of strain energy is

$$U = \frac{1}{2} \int (\sigma_x\epsilon_x + \sigma_y\epsilon_y + \sigma_z\epsilon_z + \tau_{xy}\gamma_{xy} + \tau_{yz}\gamma_{yz} + \tau_{zx}\gamma_{zx})\, dV \tag{G.13}$$

Thus, δU is seen as a change in strain energy during a virtual displacement. The physical significance of strain energy is discussed in detail in Section 3.3.

APPENDIX H: PROGRAM FOR SOLVING SIMULTANEOUS LINEAR ALGEBRAIC EQUATIONS

H.1 INTRODUCTION

There are many methods for solving simultaneous linear algebraic equations. Among the most efficient are those based on Gauss elimination. Of those, a very effective one uses the Gauss-Doolittle factorization. This method is briefly explained in the text in Section 2.5.1. A program listing written in Microsoft BASIC for the Macintosh Computer is given next. This program is easily modified for use on other computers.

H.2 PROGRAM LISTING

```
PRINT "PROGRAM FOR SOLVING SIMULTANEOUS LINEAR ALGEBRAIC EQUATIONS"
PRINT
PRINT "EQUATIONS OF THE FORM [A]{X} = {B} ARE SOLVED WHERE [A] IS SYMMETRICAL"
PRINT
INPUT "THE NUMBER OF EQUATIONS N (N <= 10) IS ",N
REM The DIM statement for arrays must be added to the program if N >10.
PRINT:PRINT "FIRST ENTER THE MATRIX [A]":PRINT
Matrixa:
  FOR I = 1 TO N
  FOR J = I TO N
  PRINT "        A(";I;",";J;") = ";:INPUT" ",A(I,J)
  A(J,I) = A(I,J)
  NEXT J
  NEXT I
  PRINT:PRINT "CHECK FOR ERRORS - IF ANY ENTER 1 AND REENTER ALL TERMS"
  PRINT:PRINT "IF NO ERRORS PRESS RETURN AND CONTINUE"
  INPUT" ",AO
  IF AO = 1 THEN Matrixa
PRINT:PRINT "NEXT ENTER THE {B} MATRIX":PRINT
Matrixb:
  FOR I = 1 TO N
  PRINT "        B(";I;") = ";:INPUT" ",B(I)
  NEXT I
  PRINT:PRINT "CHECK FOR ERRORS - IF ANY ENTER 1 AND REENTER ALL TERMS"
  PRINT:PRINT "IF NO ERRORS PRESS RETURN AND CONTINUE"
  INPUT" ",BO
  IF BO = 1 THEN Matrixb
FOR I = 1 TO N
FOR J = 1 TO N
SUM = A(I,J)
LET K1 = I - 1
IF I = 1 THEN One
FOR K = 1 TO K1
```

```
SUM = SUM − A(K,I)*A(K,J)*A(K,K)
NEXT K
One:
  IF J <> I THEN Two
  A(I,J) = SUM
  GOTO Three
Two:
  A(I,J) = SUM/A(I,I)
Three:
  NEXT J
  NEXT I
FOR I = 1 TO N
SUM = B(I)
K1 = I − 1
IF I = 1 THEN Four
FOR K = 1 TO K1
SUM = SUM − A(K,I)*X(K)
NEXT K
Four:
  X(I) = SUM
  NEXT I
FOR I1 = 1 TO N
I = N − I1 + 1
SUM = X(I)
K2 = I +1
IF I = N THEN Five:
FOR K = K2 TO N
SUM = SUM − A(I,K)*X(K)*A(I,I)
NEXT K
Five:
  X(I) = SUM/A(I,I)
  NEXT I1
PRINT:PRINT "AND THE SOLUTION IS ":PRINT
FOR I = 1 TO N
PRINT "          X(";I;") = ";X(I)
NEXT I
PRINT:PRINT
END
```

APPENDIX I: OTHER WORK AND ENERGY METHODS

I.1 INTRODUCTION

There are two forms of the principle of virtual work. In the one discussed in the main body of Chapter 3, we deal with the virtual work of true forces subjected to virtual displacements. In the other form, we deal with virtual forces subjected to true displacements. The comparable complement to potential energy is called *complementary energy*. We shall briefly outline these principles in the next section. Reference 4 provides an extended discussion of the two sets of principles.

Both work and energy principles have been expounded in the past in several specialized forms. In the interest of providing the reader some insight when it is useful or necessary to consult references that use these specialized forms, we shall mention some of them and provide some perspective on their role in structural analysis in the concluding section of this appendix. Reference 4 also discusses some of these special cases.

I.2 PRINCIPLES OF COMPLEMENTARY VIRTUAL WORK AND ENERGY

Consider a solid body that has deformed to its final equilibrium position under a set of applied surface and body forces. The state of stress in the body is denoted by $\{\sigma\}$. Now let us imagine a new state of stress $\{\sigma + \delta\sigma\}$, where the components of this matrix are $\sigma_x + \delta\sigma_x$, $\sigma_y + \delta\sigma_y$, and so on. The quantities $\delta\sigma_x$, $\delta\sigma_y$, and so on, are called *virtual stresses*. The true stresses satisfy equilibrium, so let us require that the new state of stress does also; thus,

$$[E]\{\sigma + \delta\sigma\} + \{f\} = \{0\} \tag{I.1}$$

from which we can conclude

$$[E]\{\delta\sigma\} = \{0\} \tag{I.2}$$

On the part of the boundary where surface forces are prescribed, the new stresses must be such that the surface forces are not changed (analogous to virtual displacements not violating geometric constraints) as a consequence of the preceding requirement of equilibrium; however, on the part where displacements are prescribed, there may be changes in the stress components. Functions that have this property are called *statically admissible functions*. We also define a set of *virtual surface forces* $\{\delta F_s\}$ and *virtual body forces* $\{\delta f\}$.

Given that we only consider states of stress that are in equilibrium, we seek the conditions that ensure that they also produce strains and displacements which are compatible with the requirements of continuity and constraints. By arguments analogous to those used in establishing the principle of virtual work in Chapter 3, we can show that

$$\delta W^* = \delta W_e^* - \delta U^* = 0$$
$$= \int_{A_1} \{\delta F_s\}^T\{u\}\, dA + \int \{\delta f\}^T\{u\}\, dV - \int \{\delta\sigma\}\{\epsilon\}\, dV = 0 \tag{I.3}$$

where A_1 is that part of the boundary where surface forces are not prescribed. The quantity δW^* is called the *total complementary virtual work*.

$$\delta W_e^* = \int_{A_1} \{\delta F_s\}^T\{u\}\ dA + \int \{\delta f\}^T\{u\}\ dV \tag{I.4}$$

is the *external complementary virtual work*, and

$$\delta U^* = \int \{\delta\sigma\}^T\{\epsilon\}\ dV \tag{I.5}$$

is the *internal complementary virtual work*. Equation I.3 is a symbolic statement of the *principle of complementary virtual work*. In words, this principle is stated as follows:

The strains and displacements of a linear elastic solid body are compatible and consistent with the constraints if and only if the total complementary virtual work is zero for every statically admissible system of virtual stresses.

We may define

$$\Pi^* = U^* - W^*_e \longrightarrow \delta\Pi^* = 0 \tag{I.6}$$

where Π^* is called the *complementary energy*. For a linear elastic solid, the complementary internal energy is

$$U^* = U = \int \{\sigma\}^T\{\epsilon\}\ dV \tag{I.7}$$

and the complementary external energy is

$$W_e^* = \int_{A_1} \{F_s\}^T\{u\}\ dA + \int \{f\}^T\{u\}\ dV \tag{I.8}$$

It may be shown that $\delta\Pi^* = 0$ is equivalent to the following statement:

Of all statically admissible systems of stresses, those that satisfy compatibility make the complementary energy a minimum.

Direct methods for extracting solutions from these principles may be found in the literature, including several of the references. In fact, a finite element method known as the *force method*, in contrast to the one we study in Chapters 3 and 4, which is called the *displacement method*, has been developed. In the early days of FEM, it was nearly as popular, but in a short time the displacement method won out and in most recent books the force method is hardly mentioned, if at all.

I.3 SPECIAL FORMS OF WORK AND ENERGY

In many problems, only surface forces are prescribed, or if displacements are also prescribed, they are given as zero. In such cases

$$\delta W^* = \delta U = 0 \tag{I.9}$$

This is called the *principle of least work*.

Consider a problem in which the applied loads are all concentrated loads, F_i, and, furthermore, we can express the strain energy in terms of the loads. If we define discrete displacements, Δ_i, at the points where the loads are applied and in the direction of the loads, and imagine that the loads, F_i, are caused by prescribed values of Δ_i, the complementary energy is

$$W^* = U(F_i) - \Sigma F_i \Delta_i \tag{I.10}$$

It follows that the minimum of the complementary energy may be found by

$$\frac{\partial W^*}{\partial F_j} = \frac{\partial U}{\partial F_j} - \Delta_j = 0 \tag{I.11}$$

or

$$\frac{\partial U}{\partial F_j} = \Delta_j \tag{I.12}$$

It matters not that the displacements are actually the unknowns and the forces are the knowns. This is known as *Castigliano's second theorem*.

There is a similar formulation based on minimum potential energy. Consider a problem in which the applied loads are all concentrated loads, F_i, and, furthermore, we can express the strain energy in terms of the discrete displacements, Δ_i. The potential energy of the system is

$$W = U(\Delta_i) - \Sigma F_i \Delta_i \tag{I.13}$$

It follows that the minimum of the potential energy may be found by

$$\frac{\partial W}{\partial \Delta_j} = \frac{\partial U}{\partial \Delta_j} - F_j = 0 \tag{I.14}$$

or

$$\frac{\partial U}{\partial \Delta_j} = F_j \tag{I.15}$$

This is known as *Castigliano's first theorem*.

We shall mention two more special cases. Consider a solid body that has at least one applied concentrated load, F, and a corresponding displacement, Δ, that acts at the point and in the direction of the force. Since the virtual displacement in the principle of virtual work is arbitrary, let us select one in which F is the only external force to be subjected to a virtual displacement, $\delta\Delta$; thus,

$$\delta W = F \, \delta\Delta - \int \{\delta\epsilon\}^T \{\sigma\} \, dV = 0 \tag{I.16}$$

We can go further and let the virtual displacement be unity; thus,

$$F = \int \{\delta\epsilon\}^T \{\sigma\} \, dV \tag{I.17}$$

where $\{\delta\epsilon\}^T$ is the virtual strain associated with the virtual displacement just described. This is the general form of a method known as the *dummy unit displacement method*.

In an analogous situation, we once again consider a solid body that has at least one applied concentrated load, F, and a corresponding displacement, Δ, that acts at the point and in the direction of the force. This time we note that in the principle of complementary virtual work the virtual stress field is arbitrary as long as it satisfies equilibrium. Let us select as the virtual stresses those that result from a virtual force δF; then

$$\delta W^* = \delta F \Delta - \int \{\delta\sigma\}^T \{\epsilon\} \, dV = 0 \tag{I.18}$$

And let us select this virtual force to be unity; thus,

$$\Delta = \int \{\delta\sigma\}^T \{\epsilon\} \, dV \tag{I.19}$$

where $\{\delta\sigma\}^T$ is the virtual stress associated with the virtual force just described. This is the general form of a method known as the *dummy unit load method*.

We shall not attempt to elaborate on these two methods, but the reader is alerted to their existence. Examples of the use of both methods in practical analysis may be found in Reference 4 and some of the other references.

APPENDIX J: NASTRAN INPUT FILES FOR STATIC STRESS AND DEFLECTION ANALYSIS

J.1 INTRODUCTION

Here we shall show and discuss input files for a commercial FEM computer code for solving examples in Chapter 4. One of the most popular commercial FEM computer codes is the version of NASTRAN developed and supported by the MacNeal-Schwendler Corporation, which will be referred to as MSC/NASTRAN. NASTRAN was originally a product of NASA, and the version developed under their sponsorship, called COSMIC NASTRAN, is still available; however, the MSC version includes many enhancements and is the one we shall use for illustration. It is representative of other codes that contain all the same basic features in their input files. References 16 and 23 contain additional information.

    Note: In all the files in this appendix, the line numbers are *not* part of the file. They are there to assist in the discussion that follows each listing.

J.2 EXAMPLE 4.3.1: SPACE TRUSS

J.2.1 Input File Listing

```
 1 ID EXAMPLE SPCTRS1
 2 TIME 1
 3 SOL 24
 4 CEND
 5 TITLE=SPACE TRUSS
 6 ECHO=BOTH
 7 DISPLACEMENT=ALL
 8 STRESS=ALL
 9 FORCE=ALL
10 SPCFORCES=ALL
11 SUBCASE 1
12 SUBTITLE=SHEAR LOAD
13 LOAD=67
14 SUBCASE 2
15 SUBTITLE=TORSIONAL LOAD
16 LOAD=69
17 OUTPUT(PLOT)
18 PLOTTER NASTRAN
19 SET 53 INCLUDE ALL
20 FIND SCALE ORIGIN 7
21 PLOT LABEL BOTH
22 PLOT STATIC DEFORMATION 0 SET 53
23 BEGIN BULK
24 GRID,1,,0.0,-500.0,500.0,,123456
```

```
25  GRID,2,,0.0,-500.0,-500.0,,123456
26  GRID,3,,0.0,500.0,-500.0,,123456
27  GRID,4,,0.0,500.0,500.0,,123456
28  GRID,5,,2000.0,-250.0,250.0
29  GRID,6,,2000.0,-250.0,-250.0
30  GRID,7,,2000.0,250.0,-250.0
31  GRID,8,,2000.0,250.0,250.0
32  CROD,1,23,1,5
33  CROD,2,23,2,6
34  CROD,3,23,3,7
35  CROD,4,23,4,8
36  CROD,5,24,2,5
37  CROD,6,24,3,6
38  CROD,7,24,4,7
39  CROD,8,24,1,8
40  CROD,9,24,1,6
41  CROD,10,24,2,7
42  CROD,11,24,3,8
43  CROD,12,24,4,5
44  CROD,13,23,5,6
45  CROD,14,23,6,7
46  CROD,15,23,7,8
47  CROD,16,23,8,5
48  CROD,17,23,1,2
49  CROD,18,23,2,3
50  CROD,19,23,3,4
51  CROD,20,23,4,1
52  PROD,23,12,471.24
53  PROD,24,12,314.16
54  MAT1,12,68950.0,,0.3
55  FORCE,67,5,0,1.0,0.0,0.0,1000.0
56  FORCE,67,6,0,1.0,0.0,0.0,1000.0
57  FORCE,67,7,0,1.0,0.0,0.0,1000.0
58  FORCE,67,8,0,1.0,0.0,0.0,1000.0
59  FORCE,69,5,0,1.0,0.0,-1000.0,0.0
60  FORCE,69,6,0,1.0,0.0,00.0,-1000.0
61  FORCE,69,7,0,1.0,0.0,1000.0,0.0
62  FORCE,69,8,0,1.0,0.0,0.0,1000.0
63  ENDDATA
```

J.2.2 Discussion

The first four lines of this file are known in MSC/NASTRAN as the Executive Control Deck. Line 1 is just a file ID; line 2 puts a time limit in minutes on the CPU time allowed to run the problem to prevent an excessive run time in case of an error in the system; line 3 states the equation solver to be used, which in this case is very like the one in Appendix H; and line 4 signals that this phase of the input file is ending.

Lines 5 through 22 are part of the Case Control Deck. Essentially, these lines specify the content and format of the output files. The title and subtitle in lines 5, 12, and 15 appear at appropriate places in the output file; line 6 asks that the input file be shown as entered and in an alphabetically sorted form; line 7 asks that all the nodal displacements be listed; lines 8 and 9 ask that all the element stresses and stress resultants be listed; and line 10 asks that the constraint, or support, forces be shown. Lines 11 through 16 identify the two loading cases to be considered; the 67 and 69 will appear later in the file to identify two separate loading cases. Lines 17 through 22 instruct the plotter to produce the figures showing the deflections for the two loading cases shown in Example 4.3.1.

Lines 23 through 66 are called the Bulk Data Deck. Here is where the seven categories of input information listed in Example 4.3.1 are entered in the file. Line 23 signals that the data are coming and line 66 tells that it is at an end. Lines 24 through 31 enter the nodal numbers and their coordinates (nodes are called GRIDs in NASTRAN) and the nodal constraint information. The only thing not obvious here is the constraints. Very simply, 123456 means no displacement in x-, y-, and z-directions and no rotations about the x-, y-, and z-axis at those nodes.

Lines 32 through 51 identify the type of element (CROD is the axial line element in NASTRAN), its number, a number identifying where to find its geometric properties, and the associated nodes. Lines 52 and 53 first identify that geometric property information is coming (PROD), then give a number associated with certain elements, give another number that tells where to find material property information, and then give the geometric properties, which in this case is the cross-sectional area of the element and no more. Line 54 contains the material properties of Young's modulus and Poisson's ratio. The shear modulus is calculated. Finally, the forces in the two load cases are given in lines 55 through 63.

It is interesting to note that to do the second part of Example 4.3.1, where one member is removed, we need only delete line 36 and rerun the problem. It would be desirable to change the ID in line 1 and the title in line 5 for identification purposes. This is one of the great advantages of FEM computer codes over previous methods. Modification of the structure adds very little work for the analyst.

J.3 EXAMPLE 4.4.3: TAPERED BEAM

J.3.1 Input File Listing

```
 1 ID EXAMPLE TPRDBM
 2 TIME 1
 3 SOL 24
 4 CEND
 5 TITLE=TAPERED CANTILEVER BEAM.
 6 ECHO=BOTH
 7 DISPLACEMENT=ALL
 8 STRESS=ALL
 9 ELFORCE=ALL
10 SPCFORCES=ALL
11 SUBCASE 1
12 LOAD = 42
13 SUBTITLE=THREE DIMENSIONAL LOADS.
14 OUTPUT(PLOT)
15 PLOTTER NASTRAN
```

16 SET 53 INCLUDE ALL
17 AXES MY,X,Z
18 VIEW 0.0,0.0,0.0
19 FIND SCALE ORIGIN 7
20 PLOT STATIC DEFORMATION 0 SET 53
21 AXES Z,X,Y
22 VIEW 0.0,0.0,0.0
23 FIND SCALE ORIGIN 7
24 PLOT LABEL BOTH
25 PLOT STATIC DEFORMATION 0 SET 53
26 BEGIN BULK
27 PARAM,AUTOSPC,YES
28 GRID,1,,0.0,0.0,0.0,,123456
29 GRID,2,,1000.0,0.0,0.0,,0.0
30 GRID,3,,2000.0,0.0,0.0,,0.0
31 GRID,4,,3000.0,0.0,0.0,,0.0
32 GRID,5,,4000.0,0.0,0.0,,0.0
33 CBAR,1,11,1,2,1.0,1.0,0.0
34 CBAR,2,12,2,3,1.0,1.0,0.0
35 CBAR,3,13,3,4,1.0,1.0,0.0
36 CBAR,4,14,4,5,1.0,1.0,0.0
37 PBAR,11,33,14062.5,2.317E07,1.406E07,5.19E07,,,+11
38 +11,70.3125,50.0,−70.3125,50.0,−70.3125,−50.0,70.3125,−50.0
39 PBAR,12,33,12187.5,1.509E07,1.219E07,3.04E07,,,+12
40 +12,60.9375,50.0,−60.9375,50.0,−60.9375,−50.0,60.9375,−50.0
41 PBAR,13,33,10312.5,0.914E07,1.031E07,1.54E07,,,+13
42 +13,51.5625,50.0,−51.5625,50.0,−51.5625,−50.0,51.5625,−50.0
43 PBAR,14,33,0.84375,0.501E07,0.844E07,1.31E07,,,+1
44 +14,42.1875,50.0,−42.1875,50.0,−42.1875,−50.0,42.1875,−50.0
45 MAT1,33,68950.0,,0.3
46 FORCE,42,1,0,1.0,0.0,0.0,250.0,0.0
47 FORCE,42,2,0,1.0,0.0,0.0,500.0,0.0
48 FORCE,42,3,0,1.0,0.0,0.0,500.0,5000.0
49 FORCE,42,4,0,1.0,0.0,0.0,500.0,0.0
50 FORCE,42,5,0,1.0,0.0,0.0,250.0,−3000.0
51 MOMENT,42,1,0,1.0,0.0,0.0,0.0,41650.0
52 MOMENT,42,5,0,1.0,200000.0,0.0,−41650.0
53 ENDDATA

J.3.2 Discussion

The discussion of lines 1 to 4, the Executive Control Deck, and lines 5 to 25 the Case
Control Deck, are covered by the comments in Section J.2.2. Lines 26 to 53, the Bulk
Data Deck, have a few differences that need explanation. Line 27 activates a function that
sweeps out singularities that might inadvertently creep into the problem. In this problem
there are none, and the line is not needed. It is often included to help catch mistakes. In
lines 28 to 32 the nodes are entered as before. In lines 33 to 36 we identify the elements
and the nodes that go with each element. The last three entries in each line are used to
orient the cross section axes so that the proper moments of inertia can be assigned in the
property cards.

The property cards in lines 37 to 44 include area, moments of inertia about the y- and z-axes, the torsional constant, and the points on the cross section at which the stresses are to be reported. The material card is given in line 45, and the force and moment data in lines 46 to 52.

J.4 EXAMPLE 4.4.4: TWO–DIMENSIONAL FRAME

J.4.1 Input File Listing

```
 1 ID EXAMPLE FRAME2D
 2 TIME 1
 3 SOL 24
 4 CEND
 5 TITLE=FRAME ANALYSIS
 6 ECHO=BOTH
 7 DISPLACEMENT=ALL
 8 STRESS=ALL
 9 FORCE=ALL
10 SPCFORCES=ALL
11 SUBCASE 1
12 LOAD=67
13 SUBTITLE=TWO CONCENTRATED LOADS
14 OUTPUT(PLOT)
15 PLOTTER NASTRAN
16 SET 53 INCLUDE ALL
17 AXES Z,X,Y
18 VIEW 0.,0.,0.
19 FIND
20 PLOT LABEL BOTH
21 PLOT STATIC DEFORMATION 0 SET 53
22 BEGIN BULK
23 PARAM,AUTOSPC,YES
24 GRDSET,,,,,,,345
25 GRID,1,,0.0,0.0,0.0,,123456
26 GRID,2,,0.0,500.0,0.0
27 GRID,3,,0.0,1000.0,0.0
28 GRID,4,,0.0,1500.0,0.0
29 GRID,5,,0.0,2000.0,0.0
30 GRID,6,,500.0,2000.0,0.0
31 GRID,7,,1000.0,2000.0,0.0,,123456
32 CBAR,1,23,1,2,1.0,1.0,0.0
33 CBAR,2,23,2,3,1.0,1.0,0.0
34 CBAR,3,23,3,4,1.0,1.0,0.0
35 CBAR,4,23,4,5,1.0,1.0,0.0
36 CBAR,5,23,5,6,1.0,1.0,0.0
37 CBAR,6,23,6,7,1.0,1.0,0.0
38 PBAR,23,12,576.0,27648.0,27648.0,,,,+P1023
39 +P1023,12.0,12.0,-12.0,12.0,-12.0,-12.0,12.0,-12.0
40 MAT1,12,68950.0,,0.3
```

```
41  FORCE,67,3,0,1.0,6000.0,0.0,0.0
42  FORCE,67,5,0,1.0,0.0,−6000.0,0.0
43  ENDDATA
```

J.4.2 Discussion

There is little difference between this file and the one for the tapered beam. This shows that the amount of work in setting up a FEM problem is about the same for problems that differ a great deal.

J.5 EXAMPLE 4.4.5: SPACE FRAME

J.5.1 Input File Listing

```
 1  ID EXAMPLE SPCFRM1
 2  TIME 1
 3  SOL 24
 4  CEND
 5  TITLE=SPACE FRAME
 6  ECHO=BOTH
 7  DISPLACEMENT=ALL
 8  STRESS=ALL
 9  FORCE=ALL
10  SPCFORCES=ALL
11  SUBCASE 1
12  SUBTITLE=SHEARLOAD
13  LOAD=67
14  SUBCASE 2
15  SUBTITLE=TORSIONALLOAD
16  LOAD=69
17  OUTPUT(PLOT)
18  PLOTTER NASTRAN
19  SET 53 INCLUDE ALL
20  FIND SCALE ORIGIN 7
21  PLOT LABEL BOTH
22  PLOT STATIC DEFORMATION 0 SET 53
23  BEGIN BULK
24  PARAM,AUTOSPC,YES
25  GRID,1,,0.0,−500.0,500.0,,123456
26  GRID,2,,0.0,−500.0,−500.0,,123456
27  GRID,3,,0.0,500.0,−500.0,,123456
28  GRID,4,,0.0,500.0,500.0,,123456
29  GRID,5,,2000.0,−250.0,250.0
30  GRID,6,,2000.0,−250.0,−250.0
31  GRID,7,,2000.0,250.0,−250.0
32  GRID,8,,2000.0,250.0,250.0
33  CBAR,1,23,1,5,6
34  CBAR,2,23,2,6,5
35  CBAR,3,23,3,7,8
```

```
36  CBAR,4,23,4,8,7
37  CBAR,5,24,2,5,6
38  CBAR,6,24,3,6,5
39  CBAR,7,24,4,7,8
40  CBAR,8,24,1,8,7
41  CBAR,9,25,1,6,5
42  CBAR,10,25,2,7,8
43  CBAR,11,25,3,8,7
44  CBAR,12,25,4,5,6
45  CBAR,13,26,5,6,2
46  CBAR,14,26,6,7,3
47  CBAR,15,26,7,8,4
48  CBAR,16,26,8,5,6
49  CBAR,17,23,1,2,5
50  CBAR,18,23,2,3,4
51  CBAR,19,23,3,4,8
52  CBAR,20,23,4,1,5
53  PBAR,23,12,471.24,53014.4,53014.4,106028.8,,,+A23
54  +A23,0.0,15.0,-15.0,0.0,0.0,-15.0,15.0,0.0
55  PBAR,24,12,314.16,15708.0,15708.0,31416.0,,,+A24
56  +A24,0.0,10.0,-10.0,0.0,0.0,-10.0,10.0,0.0
57  PBAR,25,12,314.16,15708.0,15708.0,31416.0,,,+A25
58  +A25,0.0,10.0,-10.0,0.0,0.0,-10.0,10.0,0.0
59  PBAR,26,12,471.24,53014.4,53014.4,106028.8,,,+A26
60  +A26,0.0,15.0,-15.0,0.0,0.0,-15.0,15.0,0.0
61  MAT1,12,68950.0,,0.3
62  FORCE,67,5,0,1.0,0.0,0.0,1000.0
63  FORCE,67,6,0,1.0,0.0,0.0,1000.0
64  FORCE,67,7,0,1.0,0.0,0.0,1000.0
65  FORCE,67,8,0,1.0,0.0,0.0,1000.0
66  FORCE,69,5,0,1.0,0.0,-1000.0,0.0
67  FORCE,69,6,0,1.0,0.0,00.0,-1000.0
68  FORCE,69,7,0,1.0,0.0,1000.0,0.0
69  FORCE,69,8,0,1.0,0.0,0.0,1000.0
70  ENDDATA
```

J.5.2 Discussion

This file differs very little from that for the space truss in Section J.2.1. The CROD elements have been replaced by the CBAR elements in lines 33 to 52. This requires the addition of terms in those lines to account for the orientation of the cross-sectional moments of inertia. It also requires replacement of the PROD lines with PBAR lines and the addition of information needed in those lines. The rest of the input file is essentially the same.

J.6 EXAMPLE 4.6.1: PLANE STRESS

J.6.1 Input File Listing

```
 1 ID EXAMPLE PLSTRESS
 2 TIME 1
 3 SOL 24
 4 CEND
 5 TITLE=PLANE STRESS
 6 ECHO=BOTH
 7 DISPLACEMENT=ALL
 8 STRESS=ALL
 9 ELFORCE=ALL
10 SPCFORCES=ALL
11 SUBCASE 1
12 SUBTITLE= EDGE LOADING
13 LOAD= 67
14 OUTPUT(PLOT)
15 PLOTTER NASTRAN
16 SET 53 INCLUDE ALL
17 AXES Z,X,Y
18 VIEW 0.,0.,0.
19 FIND
20 PLOT LABEL BOTH
21 PLOT STATIC DEFORMATION 0 SET 53
22 BEGIN BULK
23 PARAM,AUTOSPC,YES
24 GRDSET,,,,,,,3456
25 GRID,1,,0.0,150.0,0.0,,13456
26 GRID,2,,0.0,325.0,0.0,,13456
27 GRID,3,,0.0,500.0,0.0,,13456
28 GRID,4,,57.4,138.6,0.0
29 GRID,5,,250.0,500.0,0.0
30 GRID,6,,106.1,106.1,0.0
31 GRID,7,,303.0,303.0,0.0
32 GRID,8,500.0,500.0,0.0
33 GRID,9,,138.6,57.4,0.0
34 GRID,10,,500.0,250.0,0.0
35 GRID,11,,150.0,0.0,0.0,,23456
36 GRID,12,,325.0,0.0,0.0,,23456
37 GRID,13,,500.0,0.0,0.0,,23456
38 GRID,14,,191.0,325.0
39 GRID,15,,325.0,191.2
40 CQUAD4,1,29,1,4,14,2
41 CQUAD4,2,29,2,14,5,3
42 CQUAD4,3,29,4,6,7,14
43 CQUAD4,4,29,14,7,8,5
44 CQUAD4,5,29,6,9,15,7
45 CQUAD4,6,29,7,15,10,8
46 CQUAD4,7,29,9,11,12,15
47 CQUAD4,8,29,15,12,13,10
```

```
48 PSHELL,29,13,10.
49 MAT1,13,68950.0,,0.3
50 FORCE,67,8,0,1.0,125000.0,0.0,0.0
51 FORCE,67,10,0,1.0,250000.0,0.0,0.0
52 FORCE,67,13,01.0,125000.0,0.0,0.0
53 ENDDATA
```

J.6.2 Discussion

This is our first listing of a plane stress element. Actually, the element is the plate element with deflections and rotations out of the plane suppressed with the GRDSET line. The CQUAD4 lines show the four nodes that go with this quadrilateral element. Otherwise, there is little new in the listing.

J.7 EXAMPLE 4.6.2: SOLID ELEMENT

J.7.1 Input File Listing

```
 1 ID EXAMPLE BRACKET
 2 TIME 1
 3 SOL 24
 4 CEND
 5 TITLE=BRACKET
 6 ECHO=BOTH
 7 DISPLACEMENT=ALL
 8 STRESS=ALL
 9 FORCE=ALL
10 SPCFORCES=ALL
11 SUBCASE 1
12 LOAD = 67
13 SUBTITLE=SOLID ELEMENTS
14 OUTPUT(PLOT)
15 PLOTTER NASTRAN
16 SET 53 INCLUDE ALL
17 FIND SCALE ORIGIN 7
18 PLOT LABEL BOTH
19 PLOT STATIC DEFORMATION 0 SET 53
20 BEGIN BULK
21 PARAM,AUTOSPC,YES
22 GRDSET,,,,,,,456
23 GRID,1,,   0.0,0.0,0.0,,123456
24 GRID,2,,10.0,0.0,0.0
25 GRID,3,,20.0,0.0,0.0
26 GRID,4,,30.0,0.0,0.0
27 GRID,5,,40.0,0.0,0.0
28 GRID,6,,   0.0,0.0,10.0,,123456
29 GRID,7,,10.0,0.0,10.0
30 GRID,8,,20.0,0.0,10.0
31 GRID,9,,30.0,0.0,10.0
```

```
32  GRID,10,,40.0,0.0,10.0
33  GRID,11,,   0.0,0.0,20.0,,123456
34  GRID,12,,10.0,0.0,20.0
35  GRID,13,,   0.0,0.0,30.0,,123456
36  GRID,14,,10.0,0.0,30.0
37  GRID,15,,   0.0,0.0,40.0,,123456
38  GRID,16,,10.0,0.0,40.0
39  GRID,21,,   0.0,10.0,0.0,,123456
40  GRID,22,,10.0,10.0,0.0
41  GRID,23,,20.0,10.0,0.0
42  GRID,24,,30.0,10.0,0.0
43  GRID,25,,40.0,10.0,0.0
44  GRID,26,,   0.0,10.0,10.0,,123456
45  GRID,27,,10.0,10.0,10.0
46  GRID,28,,20.0,10.0,10.0
47  GRID,29,,30.0,10.0,10.0
48  GRID,30,,40.0,10.0,10.0
49  GRID,31,,   0.0,10.0,20.0,,123456
50  GRID,32,,10.0,10.0,20.0
51  GRID,33,,   0.0,10.0,30.0,,123456
52  GRID,34,,10.0,10.0,30.0
53  GRID,35,,   0.0,10.0,40.0,,123456
54  GRID,36,,10.0,10.0,40.0
55  GRID,37,,20.0,10.0,20.0
56  GRID,38,,30.0,10.0,20.0
57  GRID,39,,20.0,10.0,30.0
58  GRID,41,,   0.0,20.0,0.0,,123456
59  GRID,42,,10.0,20.0,0.0
60  GRID,43,,20.0,20.0,0.0
61  GRID,44,,30.0,20.0,0.0
62  GRID,45,,40.0,20.0,0.0
63  GRID,46,,   0.0,20.0,10.0,,123456
64  GRID,47,,10.0,20.0,10.0
65  GRID,48,,20.0,20.0,10.0
66  GRID,49,,30.0,20.0,10.0
67  GRID,50,,40.0,20.0,10.0
68  GRID,51,,   0.0,20.0,20.0,,123456
69  GRID,52,,10.0,20.0,20.0
70  GRID,53,,   0.0,20.0,30.0,,123456
71  GRID,54,,10.0,20.0,30.0
72  GRID,55,,   0.0,20.0,40.0,,123456
73  GRID,56,,10.0,20.0,40.0
74  GRID,57,,20.0,20.0,20.0
75  GRID,58,,30.0,20.0,20.0
76  GRID,59,,20.0,20.0,30.0
77  CHEXA,1,23,1,2,22,21,6,7,+A
78  +A,27,26
79  CHEXA,2,23,2,3,23,22,7,8,+B
80  +B,28,27
81  CHEXA,3,23,3,4,24,23,8,9,+C
82  +C,29,28
```

```
 83  CHEXA,4,23,4,5,25,24,9,10,+D
 84  +D,30,29
 85  CHEXA,5,23,6,7,27,26,11,12,+E
 86  +E,32,31
 87  CHEXA,6,23,11,12,32,31,13,14,+F
 88  +F,34,33
 89  CHEXA,7,23,13,14,34,33,15,16,+G
 90  +G,36,35
 91  CHEXA,8,23,21,22,42,41,26,27,+AA
 92  +AA,47,46
 93  CHEXA,9,23,22,23,43,42,27,28,+BB
 94  +BB,48,47
 95  CHEXA,10,23,23,24,44,43,28,29,+CC
 96  +CC,49,48
 97  CHEXA,11,23,24,25,45,44,29,30,+DD
 98  +DD,50,49
 99  CHEXA,12,23,26,27,47,46,31,32,+EE
100  +EE,52,51
101  CHEXA,13,23,31,32,52,51,33,34,+FF
102  +FF,54,53
103  CHEXA,14,23,33,34,54,53,35,36,+GG
104  +GG,56,55
105  CHEXA,15,23,22,23,43,42,32,37,+HH
106  +HH,57,52
107  CHEXA,16,23,23,24,44,43,28,29,+II
108  +II,59,58
109  CPENTA,17,23,29,30,38,49,50,58
110  CHEXA,18,23,32,37,57,52,34,39,+JJ
111  +JJ,59,54
112  CPENTA,19,23,37,38,39,57,58,59
113  CPENTA,20,23,34,39,36,54,59,56
114  PSOLID,23,12
115  MAT1,12,68950.0,,0.3
116  FORCE,67,7,0,1.0,0.0,0.0,-50.0
117  FORCE,67,8,0,1.0,0.0,0.0,-100.0
118  FORCE,67,9,0,1.0,0.0,0.0,-100.0
119  FORCE,67,10,0,1.0,0.0,0.0,-50.0
120  FORCE,67,27,0,1.0,0.0,0.0,-50.0
121  FORCE,67,28,0,1.0,0.0,0.0,-100.0
122  FORCE,67,29,0,1.0,0.0,0.0,-100.0
123  FORCE,67,30,0,1.0,0.0,0.0,-50.0
124  ENDDATA
```

J.7.2 Discussion

The main price you pay with a solid element is the need to use a large number of elements. Specifying the nodes that go with each eight-noded element used here is a bit tedious. Problems like this begin to illustrate the need for automatic mesh generation using graphical input.

Note the use of two different elements. One is a hexahedral element identified in the lines starting with CHEXA, and the other is wedge shaped and identified in the lines beginning with CPENTA.

J.8 EXAMPLE 4.7.1. PLATE BENDING

J.8.1 Input File Listing

```
 1 ID EXAMPLE PLTBEND2
 2 TIME 1
 3 SOL 24
 4 CEND
 5 TITLE=PLATE BENDING TWO
 6 ECHO=BOTH
 7 DISPLACEMENT=ALL
 8 STRESS=ALL
 9 ELFORCE=ALL
10 SPCFORCES=ALL
11 SUBCASE 1
12 SUBTITLE= RING LOADING
13 LOAD=22
14 OUTPUT(PLOT)
15 PLOTTERNASTRAN
16 SET 47 INCLUDE ALL
17 AXES MY,X,Z
18 FIND SCALE ORIGIN 7
19 PLOT STATIC DEFORMATION 0 SET 47
20 AXES Z,X,Y
21 VIEW 0.,0.,0.
22 FIND SCALE ORIGIN 7
23 PLOT LABEL BOTH
24 BEGIN BULK
25 PARAM,AUTOSPC,YES
26 GRDSET,,,,,,,6
27 GRID,1,,0.0,150.0,0.0,,156
28 GRID,2,,0.0,325.0,0.0,,156
29 GRID,3,,0.0,500.0,0.0,,156
30 GRID,4,,57.4,138.6,0.0
31 GRID,5,,250.0,500.0,0.0
32 GRID,6,,106.1,106.1,0.0
33 GRID,7,,303.0,303.0,0.0
34 GRID,8,,500.0,500.0,0.0,,3456
35 GRID,9,,138.6,57.4,0.0
36 GRID,10,,500.0,250.0,0.0,,3456
37 GRID,11,,150.0,0.0,0.0,,246
38 GRID,12,,325.0,0.0,0.0,,246
39 GRID,13,,500.0,0.0,0.0,,23456
40 GRID,14,,191.0,325.0,0.0
41 GRID,15,,325.0,191.2,0.0
42 GRID,16,,0.0,237.5,0.0,,156
```

```
43  GRID,17,,123.3,231.8,0.0
44  GRID,18,,204.55,204.55,0.0
45  GRID,19,,231.8,123.3,0.0
46  GRID,20,,237.5,0.0,0.0,,246
47  CQUAD4,1,88,1,4,17,16
48  CQUAD4,2,88,2,14,5,3
49  CQUAD4,3,88,4,6,18,17
50  CQUAD4,4,88,14,7,8,5
51  CQUAD4,5,88,6,9,19,18
52  CQUAD4,6,88,7,15,10,8
53  CQUAD4,7,88,9,11,20,19
54  CQUAD4,8,88,15,12,13,10
55  CQUAD4,9,88,16,17,14,2
56  CQUAD4,10,88,17,18,7,14
57  CQUAD4,11,88,18,19,15,7
58  CQUAD4,12,88,19,20,12,15
59  PSHELL,88,76,10.0,76
60  MAT1,76,68950.0,,0.3
61  FORCE,22,1,0,1.0,0.0,0.0,-3.125
62  FORCE,22,4,0,1.0,0.0,0.0,-6.25
63  FORCE,22,6,0,1.0,0.0,0.0,-6.25
64  FORCE,22,9,0,1.0,0.0,0.0,-6.25
65  FORCE,22,11,0,1.0,0.0,0.0,-3.1255
66  ENDDATA
```

J.8.2 Discussion

The plate problem differs from the plane stress problem in Example 4.6.1 only in the specification of the constraints and the loading. The PSHELL card does require the separate specification of material properties for bending and for membrane action.

J.9 EXAMPLE 4.8.1. REINFORCED PLATE

J.9.1 The Input File Listing

```
1   ID EXAMPLE PLTBEND2
2   TIME 1
3   SOL 24
4   CEND
5   TITLE=PLATE BENDING. REINFORCED HOLE.
6   ECHO=BOTH
7   DISPLACEMENT=ALL
8   STRESS=ALL
9   ELFORCE=ALL
10  SPCFORCES=ALL
11  SUBCASE 1
12  SUBTITLE= RING LOADING
13  LOAD=22
14  OUTPUT(PLOT)
```

```
15  PLOTTER NASTRAN
16  SET 47 INCLUDE ALL
17  AXES MY,X,Z
18  FIND SCALE ORIGIN 7
19  PLOT STATIC DEFORMATION 0 SET 47
20  AXES Z,X,Y
21  VIEW 0.,0.,0.
22  FIND SCALE ORIGIN 7
23  PLOT LABEL BOTH
24  BEGIN BULK
25  PARAM,AUTOSPC,YES
26  GRDSET,,,,,,,6
27  GRID,1,,0.0,150.0,0.0,,156
28  GRID,2,,0.0,325.0,0.0,,156
29  GRID,3,,0.0,500.0,0.0,,156
30  GRID,4,,57.4,138.6,0.0
31  GRID,5,,250.0,500.0,0.0
32  GRID,6,,106.1,106.1,0.0
33  GRID,7,,303.0,303.0,0.0
34  GRID,8,,500.0,500.0,0.0,,3456
35  GRID,9,,138.6,57.4,0.0
36  GRID,10,,500.0,250.0,0.0,,3456
37  GRID,11,,150.0,0.0,0.0,,246
38  GRID,12,,325.0,0.0,0.0,,246
39  GRID,13,,500.0,0.0,0.0,,23456
40  GRID,14,,191.0,325.0,0.0
41  GRID,15,,325.0,191.2,0.0
42  GRID,16,,0.0,237.5,0.0,,156
43  GRID,17,,123.3,231.8,0.0
44  GRID,18,,204.55,204.55,0.0
45  GRID,19,,231.8,123.3,0.0
46  GRID,20,,237.5,0.0,0.0,,246
47  CQUAD4,1,88,1,4,17,16
48  CQUAD4,2,88,2,14,5,3
49  CQUAD4,3,88,4,6,18,17
50  CQUAD4,4,88,14,7,8,5
51  CQUAD4,5,88,6,9,19,18
52  CQUAD4,6,88,7,15,10,8
53  CQUAD4,7,88,9,11,20,19
54  CQUAD4,8,88,15,12,13,10
55  CQUAD4,9,88,16,17,14,2
56  CQUAD4,10,88,17,18,7,14
57  CQUAD4,11,88,18,19,15,7
58  CQUAD4,12,88,19,20,12,15
59  CBAR,13,99,1,4,1.0,1.0,0.0,,+A
60  +A,,,0.0,0.0,-15.0,0.0,0.0,-15.0
61  CBAR,14,99,4,6,1.0,1.0,0.0,,+B
62  +B,,,0.0,0.0,-15.0,0.0,0.0,-15.0
63  CBAR,15,99,6,9,1.0,1.0,0.0,,+C
64  +C,,,0.0,0.0,-15.0,0.0,0.0,-15.0
65  CBAR,16,99,9,11,1.0,1.0,0.0,,+D
```

```
66  +D,,,0.0,0.0,-15.0,0.0,0.0,-15.0
67  PSHELL,88,76,10.0,76
68  PBAR,99,76,200.0,6666.7,1666.7,,,,+P
69  +P,-5.0,-10.0,5.0,-10.0
70  MAT1,76,68950.0,,0.3
71  FORCE,22,1,0,1.0,0.0,0.0,-3.125
72  FORCE,22,4,0,1.0,0.0,0.0,-6.25
73  FORCE,22,6,0,1.0,0.0,0.0,-6.25
74  FORCE,22,9,0,1.0,0.0,0.0,-6.25
75  FORCE,22,11,0,1.0,0.0,0.0,-3.1255
76  ENDDATA
```

J.9.2 Discussion

The input file differs from the one in Section J.8.1 only in the addition of the CBAR elements in lines 59 to 66 and the PBAR information in lines 68 to 69.

APPENDIX K: PROGRAM FOR FINDING THE ROOTS OF AN EQUATION

K.1 INTRODUCTION

Frequently, in solving a differential equation for its eigenvalues we must find the roots of a transcendental equation. For example, the solution for the buckling load in Example 5.2.2 requires us to find the first nontrivial positive real root, $X_1 = (\lambda a)_1$, of the equation

$$f(X) = X \sin X + 2(\cos X - 1) = 0 \qquad (K.1)$$

and the solution for the natural frequencies in Example 6.3.3 requires us to find several of the positive real roots, $X_i = (\alpha a)_i$, of the equation

$$f(X) = \cos X + \frac{1}{\cosh X} \qquad (K.2)$$

When eigenvalues are found from equations based on the Rayleigh-Ritz and FEM methods, by setting up the algebraic equations and expanding the determinant of the coefficients, we get polynomial equations for which we must find the positive real roots. These are of the form

$$f(X) = a_1 X^n + a_2 X^{n-1} + a_3 X^{n-2} + \cdots + a_n = 0 \qquad (K.3)$$

where n is the order of the determinant. We usually use a method for extracting roots, which is explained in Appendix L, that works directly on the eigenvalue matrix when working with the Rayleigh-Ritz and FEM methods, but an occasion may arise for finding the roots from the polynomial expansion.

A computer program, given later, has been written for finding such roots. The method explained here works equally well on transcendental and polynomial equations. It is written in Microsoft BASIC for the Macintosh computer. The program may be easily adapted for other implementations of BASIC.

K.2 THE HALF-INTERVAL SEARCH

One way of finding the roots is to plot the function $f(X)$, starting at $X = 0$ and continuing for $X = \Delta X$, then $X = 2\Delta X$, and so on. When the plot shows $f(X)$ crossing the axis, and hence changing sign, we know that $f(X) = 0$ and the root is somewhere between the values of X just before and just after the change in sign. We can continue until we find the next crossing and then have narrowed the region for the second root, and so on. To improve accuracy, we can go to a finer scale on the plot and use smaller increments than ΔX as we near the crossing of the axis. The following program, which uses the *half-interval search method,* is an automated way to do this.

The program starts at some lower limit, which the operator suggests. In buckling and vibration analysis, we are interested only in positive real roots and then usually only in the lowest root for buckling and the lower several roots for vibration. Thus, we often choose zero for the lower limit. The operator is then requested by the program to give the step size and the number of steps to establish a range of values in which roots are to be sought. If N is the number of steps, I is the step size, A is the lower limit, and B the upper limit of the range, then

$$B = A + NI \tag{K.4}$$

Given this information, the program proceeds to calculate the value of $f(X)$, first at A, then at $A + I$, then at $A + 2I$, and so on. At each step it checks the sign of $f(X)$ against the previous value, and when it encounters a change in sign it knows the plot of the function has crossed the axis and a root is somewhere between the two intervals. The program then divides this last interval in half; that is, it backs up a half-interval and calculates $f(X)$ and checks the sign again. Depending on what it finds, it either backs up a half of a half-interval or it advances a half of a half-interval and checks the sign again. It continues halving the previous interval until told to stop; thus, it homes in on the value of X that makes $f(X) = 0$.

Actually, we need a measure to determine when we are near enough to zero. This is done at each step that spans a change in sign by comparing the ratio of the difference in the last two values of X to the sum of the same two values of X with some number stated in the program. When this ratio is smaller than the given number, the program prints the value of X used for the last interval as the root and continues on to look for the next root. As this program is written, $D(1)$ is the next to last value of X found, $D(3)$ is the last value of X found, and 0.000005 is the accuracy sought, or

$$\text{ABS}(D(1) - D(3))/\text{ABS}(D(1) + \text{ABS}(D(3)) < 0.000005 \tag{K.5}$$

The program can be changed to reflect a different standard of accuracy.

K.3 THE PROGRAM ROOTS

The program as written here has Equation K.2 written in. Because this particular version of BASIC does not have hyperbolic functions built in, it was necessary to construct the hyperbolic cosine from exponentials. The expressions for hyperbolic sine and cosine are

$$\text{SINH}(X) = (\text{EXP}(X) - \text{EXP}(-X))/2$$
$$\text{COSH}(X) = (\text{EXP}(X) + \text{EXP}(-X))/2 \tag{K.6}$$

To solve for the roots for a different equation, that equation must replace the one shown. Just change the third line in the listing below by

$$\text{DEF FN F(X)} = \text{your equation} \tag{K.7}$$

and run the program. You will be requested to supply input, which you will type in from the keyboard.

The program listing is as follows:

```
REM We shall find the positive real roots of a characteristic
REM equation by the half-interval search method.
```

```
DEF FN F(X) = COS(X) + 1/((EXP(X) + EXP(−X))/2)
PRINT "ENTER DATA AS REQUESTED BELOW"
PRINT:INPUT "THE LOWER LIMIT OF X IS ",A
PRINT:INPUT "THE NUMBER OF STEPS IS ",N
PRINT:INPUT "THE STEP SIZE IS ",I
B = A + N*I
PRINT:PRINT "THE UPPER LIMIT OF X IS ",B
PRINT
FOR J = 1 TO N
XONE = A + (J − 1)*I
XTWO = A + J*I
IF FN F(XONE) = 0 THEN Root1
SIGNONE = SGN(FN F(XONE))
SIGNTWO = SGN(FN F(XTWO))
IF SIGNONE*SIGNTWO >= 0 THEN Index
D(2 + SIGNONE) = XONE
D(2 − SIGNONE) = XTWO
Find:
   HALF = (D(1) +D(3))/2
   SIGNHALF = SGN(FN F(HALF))
   IF SIGNHALF = 0 THEN Root2
   D(2 + SIGNHALF) = HALF
   IF ABS(D(1) − D(3))/ABS(D(1) + D(3)) < .000005 THEN Root2
   GOTO Find
Root1:
   PRINT "ROOT = ",XONE
   LPRINT "ROOT = ",XONE
   GOTO Index
Root2:
   PRINT "ROOT = ",HALF
   LPRINT "ROOT = ",HALF
Index:
   NEXT J
PRINT:PRINT "DONE"
END
```

APPENDIX L: PROGRAM FOR FINDING EIGENVALUES AND EIGENFUNCTIONS

L.1 INTRODUCTION

A computer code for solving the matrix form of the equations for eigenvalues and eigenfunctions has been prepared to handle simple buckling problems, as discussed in Section 5.2.7, and vibration problems, as discussed in Chapter 6. For static instability analysis of simply loaded structures, the matrix equation to be solved is Equation 5.2.59 for the Rayleigh-Ritz method and Equation 5.2.76 for the finite element method. These are identical in form, so let us use

$$([K] - P[K^D])\{r\} = \{0\} \tag{L.1}$$

For free vibration analysis, the Rayleigh-Ritz method reduces all differential and integral equation mathematical models for small deflections of linear elastic structures to Equation 6.4.12, or the finite element form

$$([K] - \omega^2[M])\{r\} = \{0\} \tag{L.2}$$

These can be combined in a single equation, Equation 6.4.26:

$$([K] - P[K^D] - \omega^2[M])\{r\} = \{0\} \tag{L.3}$$

In the preceding, $[K]$ is exactly the same stiffness matrix as in the static case, $[M]$ is the mass matrix, ω is the natural frequency, $[K^D]$ is the differential stiffness matrix, which includes the mechanism for instability, and $\{r\}$ is the set of nodal displacements. The matrices $[K]$, $[K^D]$, and $[M]$ are symmetrical.

All three are classical matrix eigenvalue equations. A single computer program is given next that uses a form of solution called the *inverse power method* (see, for example, Reference 23) to solve all the preceding forms of the equations. The symbols used in the program depart slightly from standard finite element notation used in the text because of the symbols available in the BASIC program. This program also prompts the user to enter the necessary data, so no programming skill is needed to use it.

```
PRINT "EIGENVALUES AND FUNCTIONS FOR VIBRATION AND BUCKLING"
PRINT:PRINT "THE SET OF EQUATIONS ([K] − P[KD] −(F^b2)[M]){Θ} = {0} IS SOLVED"
PRINT
PRINT "WHERE      [K] IS THE STIFFNESS MATRIX"
PRINT "           [KD] IS THE DIFFERENTIAL STIFFNESS MATRIX"
PRINT "           [M] IS THE MASS MATRIX"
PRINT "           {Θ} IS THE DEFLECTION (MODE) MATRIX"
PRINT "           P IS THE AXIAL COMPRESSIVE LOAD"
PRINT "           F^2 IS THE NATURAL FREQUENCY SQUARED"
PRINT:PRINT "NOTE THAT [K], [KD], AND [M] ARE ALL SYMMETRIC MATRICES"
PRINT:PRINT "THE NUMBER OF EQUATIONS N (N<=10>) IS ";:INPUT" ",N
PRINT:PRINT "FIRST ENTER THE STIFFNESS MATRIX [K]":PRINT
BU = 1
```

```
Stiffness:
  FOR I = 1 TO N
  FOR J = I TO N
  PRINT "              K(";I;",";J;") = ";:INPUT" ",K(I,J)
  K(J,I) = K(I,J)
  NEXT J
  NEXT I
  PRINT:PRINT"CHECK ENTRIES - IF ANY ERROR PRESS 0 AND REENTER ALL TERMS"
  PRINT:PRINT"IF NO ERROR PRESS 1 AND CONTINUE"
  INPUT" ", KO
  IF KO = 0 THEN Stiffness
PRINT:PRINT "IF THERE IS A DIFFERENTIAL STIFFNESS MATRIX LET AD = 0"
PRINT:PRINT "IF THERE IS NO DIFFERENTIAL STIFFNESS LET AD = 1"
INPUT"AD = ",AD
IF AD = 1 THEN Mass
PRINT:PRINT "NOW ENTER THE DIFFERENTIAL STIFFNESS MATRIX [KD]":PRINT
Differential:
  FOR I = 1 TO N
  FOR J = I TO N
  PRINT "              KD(";I;",";J;") = ";:INPUT" ",KD(I,J)
  KD(J,I) = KD(I,J)
  NEXT J
  NEXT I
  PRINT:PRINT"CHECK ENTRIES - IF ANY ERROR PRESS 0 AND REENTER ALL TERMS"
  PRINT:PRINT"IF NO ERROR PRESS 1 AND CONTINUE"
  INPUT" ", KDO
  IF KDO = 0 THEN Differential
PRINT:PRINT "IF THIS IS A STATIC BUCKLING PROBLEM LET BU = 0"
PRINT:PRINT "IF THIS IS A VIBRATION PROBLEM LET BU = 1"
PRINT:INPUT"BU = ",BU
IF BU = 0 THEN Replace
Mass:
  PRINT:PRINT "NEXT ENTER THE MASS MATRIX [M]":PRINT
Massagain:
  FOR I = 1 TO N
  FOR J = I TO N
  PRINT "              M(";I;",";J;") = ";:INPUT" ",M(I,J)
  M(J,I) = M(I,J)
  NEXT J
  NEXT I
  PRINT:PRINT"CHECK ENTRIES - IF ANY ERROR PRESS 0 AND REENTER ALLTERMS"
  PRINT:PRINT"IF NO ERROR PRESS 1 AND CONTINUE"
  INPUT" ", MO
  IF MO = 0 THEN Massagain
IF AD = 1 THEN Solve
PRINT:INPUT"        P = ",P
FOR I = 1 TO N
FOR J = I TO N
K(I,J) = K(I,J) - P*KD(I,J)
K(J,I) = K(I,J)
NEXT J
```

```
NEXT I
GOTO Solve
Replace:
  FOR I = 1 TO N
  FOR J = I TO N
  LET M(I,J) = KD(I,J)
  M(J,I) = M(I,J)
  NEXT J
  NEXT I
Solve:
  REM E1 is the iteration error
  E1 = .000001
  FOR I = 1 TO N
  K1(I,I) = 1
  NEXT I
  FOR J = 1 TO N
  FOR I = J TO N
  IF K(I,J) <> 0 THEN Invert
  NEXT I
  PRINT:PRINT "SINGULAR MATRIX"
  GOTO Nearend
Invert:
  FOR K = 1 TO N
  S = K(J,K)
  K(J,K) = K(I,K)
  K(I,K) = S
  S = K1(J,K)
  K1(J,K) = K1(I,K)
  K1(I,K) = S
  NEXT K
  T = 1 /K(J,J)
  FOR K = 1 TO N
  K(J,K) = T*K(J,K)
  K1(J,K) = T*K1(J,K)
  NEXT K
  FOR L = 1 TO N
  IF L = J THEN Passon
  T = −K(L,J)
  FOR K = 1 TO N
  K(L,K) = K(L,K) + T*K(J,K)
  K1(L,K) = K1(L,K) + T*K1(J,K)
  NEXT K
Passon:
  NEXT L
  NEXT J
  PRINT:PRINT "STIFFNESS MATRIX HAS BEEN INVERTED"
FOR I = 1 TO N
FOR J = 1 TO N
S1 = 0
FOR K = 1 TO N
S1 = S1 + K1(I,K)*M(K,J)
```

```
                    D(I,J) = S1
                    NEXT K
                    NEXT J
                    NEXT I
                    R1 = 0:M2 = N
                    FOR M1 = 1 TO M2
                    PRINT:PRINT "MODE ";M1
                    I1 = 0
                    FOR J = 1 TO N
                    V(M1,J) = 1
                    NEXT J
                    Iteration:
                      FOR I = 1 TO N
                      S1 = 0
                      FOR J = 1 TO N
                      S1 = D(I,J)*V(M1,J) + S1
                      NEXT J
                      V1(M1,I) = S1
                      NEXT I
                      FOR I = 1 TO N
                      V(M1,I) =V1(M1,I)
                      NEXT I
                      I1 = I1 + 1:S2 = V(M1,1)
                      F(M1) = S2
                      FOR I = 1 TO N
                      V(M1,I) = V(M1,I)/S2
                      NEXT I
                      IF I1 < 20 THEN Nostop
                      IF I1 < 100 THEN Checkconverge
                      PRINT:PRINT "NOT CONVERGING":GOTO Nearend
                    Checkconverge:
                      IF ABS(1 −R1/F(M1)) <= E1 THEN Converge
                    Nostop:
                      R1 = F(M1):GOTO Iteration
                    Converge:
                      C1 = 0
                      FOR I = 1 TO N
                      IF ABS(V(M1,I)) <= C1 THEN Tryagain
                      C1 = ABS(V(M1,I))
                    Tryagain:
                      NEXT I
                      FOR I = 1 TO N
                      V(M1,I) = V(M1,I)/C1
                      NEXT I
                      PRINT:PRINT "NUMBER OF ITERATIONS";I1
                      PRINT:PRINT "THE EIGENVALUE IS ";1/F(M1)
                      PRINT:PRINT "THE EIGENVECTOR IS ":PRINT
                      FOR I = 1 TO N
                      PRINT "          Θ(";I;") = ";V(M1,I)
                      NEXT I
                      PRINT:PRINT
```

```
IF M1 = N THEN Nearend
IF BU = 0 THEN Nearend
PRINT "PRESS 1 TO CONTINUE...":INPUT" ",CO
FOR I = 1 TO N
S(I,I) = 1
FOR J = I+1 TO N
S(I,J) = 0
S(J,I) = S(I,J)
NEXT J
NEXT I
FOR I = 1 TO N
S1 = 0
FOR J = 1 TO N
S1 = M(I,J)*V(M1,J) + S1
NEXT J
K1(M1,I) = S1
NEXT I
H1 = M1+1
H2 = M1 −1
FOR J = H1 TO N
T1 = 0:B = 0
IF M1 = 1 THEN Checkone
FOR I = 1 TO H2
T1 = T1 + K1(M1,I)*X(I,J)
B = B + K1(M1,I)*X(I,M1)
NEXT I
Checkone:
  B1 = B + K1(M1,M1)
  S(M1,J) = −(T1 + K1(M1,J))/B1
  NEXT J
  S(M1,M1) = 0
  IF M1 < > 1 THEN Checktwo
  FOR I = 1 TO N
  FOR J = 1 TO N
  L(I,J) = S(I,J)
  NEXT J
  NEXT I
  GOTO Checkthree
Checktwo:
  FOR I = 1 TO N
  FOR J = 1 TO N
  S1 = 0
  FOR K = 1 TO N
  S1 = L(I,K)*S(K,J) + S1
  NEXT K
  T(I,J) = S1
  NEXT J
  NEXT I
  FOR I = 1 TO N
  FOR J = 1 TO N
  L(I,J) = T(I,J)
```

```
        NEXT J
        NEXT I
Checkthree:
        C3 = 0
        FOR I = 1 TO M1
        U(L) = 0
        FOR J = H1 TO N
        U(L) = U(L) + L(L,J)
        NEXT J
        C3 = C3 + K1(M1,L)*U(L)
        NEXT L
        C2 = 0
        FOR I = H1 TO N
        C2 = C2 −K1(M1,I)
        NEXT I
        D1 = C3 − C2
        PRINT:PRINT "CHECK1 = ";C3,"CHECK2 = ";C2
        PRINT:PRINT "THE DIFFERENCE IS ";D1
        PRINT:PRINT "IF THE ITERATION IS CONVERGING DIFFERENCE IS NEAR ZERO"
        FOR I1 = 1 TO N
        FOR J1 = 1 TO N
        S1 = 0
        FOR K1 = 1 TO N
        S1 = D(I1,K1)*S(K1,J1) + S1
        NEXT K1
        P(I1,J1) = S1
        NEXT J1
        NEXT I1
        FOR I1 = 1 TO N
        FOR J1 = 1 TO N
        E(I1,J1) = S(I1,J1)
        X(I1,J1) = L(I1,J1)
        D(I1,J1) = P(I1,J1)
        NEXT J1
        NEXT I1
        NEXT M1
Nearend:
        END
```

APPENDIX M: NASTRAN INPUT FILES FOR BUCKLING ANALYSIS

M.1 INTRODUCTION

The NASTRAN input files for buckling are very similar to those for static stress analysis since the information about nodes, elements, and restraints is identical for a given structure. The loading, the methods of solution, and the output requests do differ.

M.2 EXAMPLE 5.2.9: TAPERED BEAM

M.2.1 Input File Listing

```
 1 ID EXAMPLE TPRDBMBK
 2 TIME 1
 3 SOL 5
 4 ..BM 2 RFA RF5D33
 5 CEND
 6 TITLE=TAPERED CANTILEVER COLUMN.
 7 ECHO=BOTH
 8 SUBCASE 1
 9 SUBTITLE=STRESS ANALYSIS
10 DISPLACEMENT=ALL
11 FORCE=ALL
12 LOAD = 42
13 SUBCASE 2
14 SUBTITLE=BUCKLING ANALYSIS
15 METHOD=25
16 DISPLACEMENT=ALL
17 FORCE=ALL
18 OUTPUT(PLOT)
19 PLOTTER NASTRAN
20 SET 53 INCLUDE ALL
21 AXES MY,X,Z
22 VIEW 0.0,0.0,0.0
23 FIND
24 PLOT LABEL BOTH
25 PLOT MODAL DEFORMATION 0, SET 53
26 BEGIN BULK
27 PARAM,AUTOSPC,YES
28 GRID,1,,0.0,0.0,0.0,,123456
29 GRID,2,,1000.0,0.0,0.0
30 GRID,3,,2000.0,0.0,0.0
31 GRID,4,,3000.0,0.0,0.0
32 GRID,5,,4000.0,0.0,0.0
33 CBAR,1,11,1,2,1.0,1.0,0.0
```

```
34 CBAR,2,12,2,3,1.0,1.0,0.0
35 CBAR,3,13,3,4,1.0,1.0,0.0
36 CBAR,4,14,4,5,1.0,1.0,0.0
37 PBAR,11,33,14062.5,2.317E07,1.406E07,5.19E07
38 PBAR,12,33,12187.5,1.509E07,1.219E07,3.04E07
39 PBAR,13,33,10312.5,0.914E07,1.031E07,1.54E07
40 PBAR,14,33,0.84375,0.501E07,0.844E07,1.31E07
41 MAT1,33,68950.0,,0.3
42 EIGB,25,INV,0.0,1.0,1,,,,+ADD
43 +ADD,MAX
44 FORCE,42,3,0,1.0,-2.0,0.0,0.0
45 FORCE,42,5,0,1.0,-1.0,0.0,0.0
46 ENDDATA
```

M.2.2 Discussion

Compare this with the input file for the static stress analysis of the same structure in bending given in Appendix J, Section J.3.1. We shall assume the reader is familiar with Appendix J and, therefore, comment only on differences found here. Line 3 identifies the equation solver used for buckling. At this writing, line 4 is necessary to make the solver work and is explained in the NASTRAN manuals. Subcase 1 in lines 9 through 12 sets up the stable static analysis and output using the reference loads found in lines 44 and 45. This solution is then used in subcase 2 for the buckling analysis set up in lines 13 through 17. Since there is more than one method within the eigenvalue solver, line 15 permits a particular method to be called in line 42. Lines 18 through 25 instruct the plotter to present the mode shapes.

Lines 26 through 36 are the same in both files. Lines 37 through 40 are simplified from comparable lines in Appendix J because we are not looking for stresses on the cross section. In line 42, the inverse power method is invoked, and information is given on where to look for the lowest buckling load and on how to normalize the buckling mode.

Note that once an input file is created for static stress analysis a few changes adapt it for buckling as well.

M.3 EXAMPLE 5.2.10: TWO-DIMENSIONAL FRAME

M.3.1 Input File Listing

```
 1 ID EXAMPLE FRMBUCK
 2 TIME 2
 3 SOL 5
 4 ..BM 2 RFA RF5D33
 5 CEND
 6 TITLE=FRAME BUCKLING
 7 ECHO=BOTH
 8 SUBCASE 1
 9 SUBTITLE=INITIAL STRESS ANALYSIS.
10 DISPLACEMENT=ALL
```

```
11 FORCE=ALL
12 LOAD = 67
13 SUBCASE 2
14 SUBTITLE=BUCKLING ANALYSIS.
15 METHOD=25
16 DISPLACEMENT=ALL
17 FORCE=ALL
18 SUBTITLE=LINEAR STRESS ANALYSIS
19 OUTPUT(PLOT)
20 PLOTTER NASTRAN
21 SET 53 INCLUDE ALL
22 AXES Z,X,Y
23 VIEW 0.,0.,0.
24 FIND
25 PLOT LABEL BOTH
26 PLOT MODAL DEFORMATION 0 SET 53
27 BEGIN BULK
28 PARAM,AUTOSPC,YES
29 GRID,1,, 0.0,0.0,0.0,,123456
30 GRID,2,,0.0,500.0,0.0,,345
31 GRID,3,,0.0,1000.0,0.0,,345
32 GRID,4,,0.0,1500.0,0.0,,345
33 GRID,5,,0.0,2000.0,0.0,,345
34 GRID,6,,500.0,2000.0,0.0,,345
35 GRID,7,,1000.0,2000.0,0.0,,123456
36 CBAR,1,23,1,2,1.0,1.0,0.0
37 CBAR,2,23,2,3,1.0,1.0,0.0
38 CBAR,3,23,3,4,1.0,1.0,0.0
39 CBAR,4,23,4,5,1.0,1.0,0.0
40 CBAR,5,23,5,6,1.0,1.0,0.0
41 CBAR,6,23,6,7,1.0,1.0,0.0
42 PBAR,23,12,576.0,27648.0,27648.0
43 MAT1,12,68950.0,,0.3
44 EIGB,25,INV,0.0,1.0,1,,,,+ADD
45 +ADD,MAX
46 FORCE,67,3,0,1.0,1.0,0.0,0.0
47 FORCE,67,5,0,1.0,0.0,-1.0,0.0
48 ENDDATA
```

M.3.2 Discussion

The comments made previously for the tapered column generally apply to this case also. The second case presented in this example uses the identical input file with line 46 removed.

M.4 EXAMPLE 5.3.3: RECTANGULAR PLATE

M.4.1 Input File Listing

```
 1 ID EXAMPLE PLTBK
 2 TIME 2
 3 SOL 5
 4 ..BM 2 RFA RF5D33
 5 CEND
 6 TITLE=PLATE BUCKLING
 7 ECHO=BOTH
 8 SUBCASE 1
 9 SUBTITLE=INITIAL STRESS ANALYSIS.
10 DISPLACEMENT=ALL
11 FORCE=ALL
12 LOAD = 67
13 SUBCASE 2
14 SUBTITLE=BUCKLING ANALYSIS.
15 METHOD=25
16 DISPLACEMENT=ALL
17 FORCE=ALL
18 OUTPUT(PLOT)
19 PLOTTER NASTRAN
20 SET 53 INCLUDE ALL
21 FIND SCALE ORIGIN 7
22 PLOT MODAL DEFORMATION 0 SET 53
23 AXES Z,X,Y
24 VIEW 0.0,0.0,0.0
25 FIND SCALE ORIGIN 7
26 PLOT SHRINK
27 PLOT LABEL BOTH
28 BEGIN BULK
29 PARAM,AUTOSPC,YES
30 GRDSET,,,,,,,6
31 GRID,1,,-500.0,-500.0,0.0,,3456
32 GRID,2,,-250.0,-500.0,0.0,,3456
33 GRID,3,,0.0,-500.0,0.0,,3456
34 GRID,4,,250.0,-500.0,0.0,,3456
35 GRID,5,,500.0,-500.0,0.0,,3456
36 GRID,6,,-500.0,-250.0,0.0,,346
37 GRID,7,,-250.0,-250.0,0.0
38 GRID,8,,0.0,-250.0,0.0
39 GRID,9,,250.0,-250.0,0.0
40 GRID,10,,500.0,-250.0,0.0,,346
41 GRID,11,,-500.0,0.0,0.0,,346
42 GRID,12,,-250.0,0.0,0.0
43 GRID,13,,0.0,0.0,0.0,,126
44 GRID,14,,250.0,0.0,0.0
45 GRID,15,,500.0,0.0,0.0,,2346
46 GRID,16,,-500.0,250.0,0.0,,346
47 GRID,17,,-250.0,250.0,0.0
```

```
48 GRID,18,,0.0,250.0,0.0
49 GRID,19,,250.0,250.0,0.0
50 GRID,20,,500.0,250.0,0.0,,346
51 GRID,21,,-500.0,500.0,0.0,,3456
52 GRID,22,,-250.0,500.0,0.0,,3456
53 GRID,23,,0.0,500.0,0.0,,3456
54 GRID,24,,250.0,500.0,0.0,,3456
55 GRID,25,,500.0,500.0,0.0,,3456
56 CQUAD4,1,23,1,2,7,6
57 CQUAD4,2,23,2,3,8,7
58 CQUAD4,3,23,3,4,9,8
59 CQUAD4,4,23,4,5,10,9
60 CQUAD4,5,23,6,7,12,11
61 CQUAD4,6,23,7,8,13,12
62 CQUAD4,7,23,8,9,14,13
63 CQUAD4,8,23,9,10,15,14
64 CQUAD4,9,23,11,12,17,16
65 CQUAD4,10,23,12,13,18,17
66 CQUAD4,11,23,13,14,19,18
67 CQUAD4,12,23,14,15,20,19
68 CQUAD4,13,23,16,17,22,21
69 CQUAD4,14,23,17,18,23,22
70 CQUAD4,15,23,18,19,24,23
71 CQUAD4,16,23,19,20,25,24
72 PSHELL,23,12,1.0,12
73 MAT1,12,68950.0,,0.3
74 EIGB,25,INV,0.0,1.0,1,,,,+ADD
75 +ADD,MAX
76 FORCE,67,1,0,1.0,0.125,0.0,0.0
77 FORCE,67,6,0,1.0,0.25,0.0,0.0
78 FORCE,67,11,0,1.0,0.25,0.0,0.0
79 FORCE,67,16,0,1.0,0.25,0.0,0.0
80 FORCE,67,21,0,1.0,0.125,0.0,0.0
81 FORCE,67,5,0,1.0,-0.125,0.0,0.0
82 FORCE,67,10,0,1.0,-0.25,0.0,0.0
83 FORCE,67,15,0,1.0,-0.25,0.0,0.0
84 FORCE,67,20,0,1.0,-0.25,0.0,0.0
85 FORCE,67,25,0,1.0,-0.125,0.0,0.0
86 ENDDATA
```

M.4.2 Discussion

The invocation of buckling analysis is the same for the plate as for the beam and frame. Again only a few changes from the input file for static analysis are needed.

M.5 EXAMPLE 5.3.4: RECTANGULAR PLATE WITH HOLE

M.5.1 Input File Listing

```
 1 ID EXAMPLE PLTBKWHL
 2 TIME 2
 3 SOL 5
 4 ..BM 2 RFA RF5D33
 5 CEND
 6 TITLE=PLATE BUCKLING WITH HOLE
 7 ECHO=BOTH
 8 SUBCASE 1
 9 SUBTITLE=INITIAL STRESS ANALYSIS.
10 DISPLACEMENT=ALL
11 FORCE=ALL
12 LOAD = 67
13 SUBCASE 2
14 SUBTITLE=BUCKLING ANALYSIS.
15 METHOD=25
16 DISPLACEMENT=ALL
17 FORCE=ALL
18 OUTPUT(PLOT)
19 PLOTTER NASTRAN
20 SET 53 INCLUDE ALL
21 FIND SCALE ORIGIN 7
22 PLOT MODAL DEFORMATION 0 SET 53
23 AXES Z,X,Y
24 VIEW 0.0,0.0,0.0
25 FIND SCALE ORIGIN 7
26 PLOT SHRINK
27 PLOT LABEL BOTH
28 BEGIN BULK
29 PARAM,AUTOSPC,YES
30 GRDSET,,,,,,,6
31 GRID,1,,-500.0,-500.0,0.0,,3456
32 GRID,2,,-250.0,-500.0,0.0,,3456
33 GRID,3,,0.0,-500.0,0.0,,3456
34 GRID,4,,250.0,-500.0,0.0,,3456
35 GRID,5,,500.0,-500.0,0.0,,3456
36 GRID,6,,-500.0,-250.0,0.0,,346
37 GRID,7,,-250.0,-250.0,0.0
38 GRID,8,,0.0,-250.0,0.0
39 GRID,9,,250.0,-250.0,0.0
40 GRID,10,,500.0,-250.0,0.0,,346
41 GRID,11,,-500.0,0.0,0.0,,346
42 GRID,12,,-250.0,0.0,0.0
43 GRID,13,,-150.0,0.0,0.0,,126
44 GRID,14,,250.0,0.0,0.0
45 GRID,15,,500.0,0.0,0.0,,2346
46 GRID,16,,-500.0,250.0,0.0,,346
47 GRID,17,,-250.0,250.0,0.0
```

```
48  GRID,18,,0.0,250.0,0.0
49  GRID,19,,250.0,250.0,0.0
50  GRID,20,,500.0,250.0,0.0,,346
51  GRID,21,,-500.0,500.0,0.0,,3456
52  GRID,22,,-250.0,500.0,0.0,,3456
53  GRID,23,,0.0,500.0,0.0,,3456
54  GRID,24,,250.0,500.0,0.0,,3456
55  GRID,25,,500.0,500.0,0.0,,3456
56  GRID,26,,-106.066,-106.066,0.0
57  GRID,27,,0.0,-150.0,0.0
58  GRID,28,,106.066,-106.066,0.0
59  GRID,29,,150.0,0.0,0.0
60  GRID,30,,106.066,106.066,0.0
61  GRID,31,,0.0,150.0,0.0
62  GRID,32,,-106.066,106.066,0.0
63  CQUAD4,1,23,1,2,7,6
64  CQUAD4,2,23,2,3,8,7
65  CQUAD4,3,23,3,4,9,8
66  CQUAD4,4,23,4,5,10,9
67  CQUAD4,5,23,6,7,12,11
68  CQUAD4,6,23,8,9,28,27
69  CQUAD4,7,23,28,9,14,29
70  CQUAD4,8,23,9,10,15,14
71  CQUAD4,9,23,11,12,17,16
72  CQUAD4,10,23,29,14,19,30
73  CQUAD4,11,23,31,30,19,18
74  CQUAD4,12,23,14,15,20,19
75  CQUAD4,13,23,16,17,22,21
76  CQUAD4,14,23,17,18,23,22
77  CQUAD4,15,23,18,19,24,23
78  CQUAD4,16,23,19,20,25,24
79  CQUAD4,17,23,32,31,18,17
80  CQUAD4,18,23,12,13,32,17
81  CQUAD4,19,23,7,26,13,12
83  CQUAD4,20,23,7,8,27,26
84  PSHELL,23,12,1.0,12
85  MAT1,12,68950.0,,0.3
86  EIGB,25,INV,0.0,1.0,1,,,,+ADD
87  +ADD,MAX
88  FORCE,67,1,0,1.0,0.125,0.0,0.0
89  FORCE,67,6,0,1.0,0.25,0.0,0.0
90  FORCE,67,11,0,1.0,0.25,0.0,0.0
91  FORCE,67,16,0,1.0,0.25,0.0,0.0
92  FORCE,67,21,0,1.0,0.125,0.0,0.0
93  FORCE,67,5,0,1.0,-0.125,0.0,0.0
94  FORCE,67,10,0,1.0,-0.25,0.0,0.0
95  FORCE,67,15,0,1.0,-0.25,0.0,0.0
96  FORCE,67,20,0,1.0,-0.25,0.0,0.0
97  FORCE,67,25,0,1.0,-0.125,0.0,0.0
98  ENDDATA
```

M.5.2 Discussion

The only things that change from the plate without a hole are some of the nodes and elements. Everything before the bulk data deck is the same except for changes in titles, and everything after the element connectivity is unchanged.

APPENDIX N: NASTRAN INPUT FILES FOR VIBRATION ANALYSIS

N.1 INTRODUCTION

The NASTRAN input files for vibration are very similar to those for static stress analysis since the information about nodes, elements, and restraints is identical for a given structure. The loading, the methods of solution, and the output requests do differ.

N.2 EXAMPLE 6.4.4: TAPERED BEAM

N.2.1 Input File Listing

```
 1 ID EXAMPLE TPRDBMVB
 2 TIME 1
 3 SOL 3
 4 CEND
 5 TITLE=TAPERED CANTILEVER BEAM
 6 ECHO=BOTH
 7 SUBTITLE=NORMAL MODES AND FREQUENCIES
 8 DISPLACEMENT=ALL
 9 METHOD=75
10 OUTPUT(PLOT)
11 PLOTTER NASTRAN
12 SET 87 INCLUDE ALL
13 AXES MY,X,Z
14 VIEW 0.0,0.0,0.0
15 FIND
16 PLOT LABEL BOTH
17 PLOT MODAL DEFORMATION 0, 1 THRU 6
18 BEGIN BULK
19 PARAM,AUTOSPC,YES
20 GRID,1,,0.0,0.0,0.0,,123456
21 GRID,2,,1000.0,0.0,0.0
22 GRID,3,,2000.0,0.0,0.0
23 GRID,4,,3000.0,0.0,0.0
24 GRID,5,,4000.0,0.0,0.0
25 CBAR,1,11,1,2,1.0,1.0,0.0
26 CBAR,2,12,2,3,1.0,1.0,0.0
27 CBAR,3,13,3,4,1.0,1.0,0.0
28 CBAR,4,14,4,5,1.0,1.0,0.0
29 PBAR,11,33,14062.5,2.317E07,1.406E07,5.19E07
30 PBAR,12,33,12187.5,1.509E07,1.219E07,3.04E07
31 PBAR,13,33,10312.5,0.914E07,1.031E07,1.54E07
32 PBAR,14,33,0.84375,0.501E07,0.844E07,1.31E07
33 MAT1,33,68950.0,,0.3
```

```
34 EIGR,75,INV,0.0,110.0,6,,,,+ADD
35 +ADD,MAX
36 PARAM,COUPMASS,1
37 ENDDATA
```

N.2.2 Discussion

Compare this with the input file for the static stress analysis of the same structure in bending given in Appendix J, Section J.3.1. We shall assume the reader is familiar with Appendix J and, therefore, comment only on differences found here. Line 3 identifies the equation solver used for vibration. Since there is more than one method within the eigenvalue solver, line 9 permits a particular method to be called in line 34. In line 34, the inverse power method is invoked and information is given on where to look for the first six natural frequencies. Lines 10 through 17 instruct the plotter to present the first six mode shapes.

Note that once an input file is created for static stress analysis a few changes adapt it for vibration as well, since the lines entering the nodes, elements, and restraints are the same.

N.3 EXAMPLE 6.5.2: TWO-DIMENSIONAL FRAME

N.3.1 Input File Listing

```
1 ID EXAMPLE FRMVIB
2 TIME 1
3 SOL 3
4 CEND
5 TITLE=FRAME VIBRATION
6 SUBTITLE=NORMAL MODES AND FREQUENCIES
7 DISPLACEMENT=ALL
8 METHOD=36
9 OUTPUT(PLOT)
10 PLOTTER NASTRAN
11 SET 22 INCLUDE ALL
12 AXES Z,X,Y
13 VIEW 0.,0.,0.
14 FIND
15 PLOT LABEL BOTH
16 PLOT MODAL DEFORMATION 0, 1 THRU 4
17 BEGIN BULK
18 PARAM,AUTOSPC,YES
19 GRID,1,,0.0,0.0,0.0,,123456
20 GRID,2,,0.0,250.0,0.0,,345
21 GRID,3,,0.0,500.0,0.0,,345
22 GRID,4,,0.0,750.0,0.0,,345
23 GRID,5,,0.0,1000.0,0.0,,345
24 GRID,6,,0.0,1250.0,0.0,,345
25 GRID,7,,0.0,1500.0,0.0,,345
```

```
26 GRID,8,,0.0,1750.0,0.0,,345
27 GRID,9,,0.0,2000.0,0.0,,345
28 GRID,10,,250.0,2000.0,0.0,,345
29 GRID,11,,500.0,2000.0,0.0,,345
30 GRID,12,,750.0,2000.0,0.0,,345
31 GRID,13,,1000.0,2000.0,0.0,,123456
32 CBAR,1,23,1,2,1.0,1.0,0.0
33 CBAR,2,23,2,3,1.0,1.0,0.0
34 CBAR,3,23,3,4,1.0,1.0,0.0
35 CBAR,4,23,4,5,1.0,1.0,0.0
36 CBAR,5,23,5,6,1.0,1.0,0.0
37 CBAR,6,23,6,7,1.0,1.0,0.0
38 CBAR,7,23,7,8,1.0,1.0,0.0
39 CBAR,8,23,8,9,1.0,1.0,0.0
40 CBAR,9,23,9,10,1.0,1.0,0.0
41 CBAR,10,23,10,11,1.0,1.0,0.0
42 CBAR,11,23,11,12,1.0,1.0,0.0
43 CBAR,12,23,12,13,1.0,1.0,0.0
44 PBAR,23,12,576.0,27648.0,27648.0
45 MAT1,12,206800 .0,,0.3
46 EIGB,36,INV,0.0,180.0,4,,,,+ADD
47 +ADD,MAX
48 PARAM,COUPMASS,1
49 ENDDATA
```

N.3.2 Discussion

The comments made previously for the tapered column generally apply to this case also.

N.4 EXAMPLE 6.5.4: RECTANGULAR PLATE WITH AND WITHOUT A HOLE

N.4.1 Input File Listing

First, we give the listing for the plate without a hole.

```
 1 ID EXAMPLE PLTVIB
 2 TIME 2
 3 SOL 3
 4 CEND
 5 TITLE=PLATE VIBRATION WITHOUT HOLE
 6 ECHO=BOTH
 7 SUBTITLE=NORMAL MODES AND FREQUENCIES
 8 DISPLACEMENT=ALL
 9 METHOD=2
10 OUTPUT(PLOT)
11 PLOTTER NASTRAN
12 SET 1 INCLUDE ALL
```

```
13  FIND SCALE ORIGIN 7
14  PLOT MODAL DEFORMATION 0, 1 THRU 4
15  AXES Z,X,Y
16  VIEW 0.0,0.0,0.0
17  FIND SCALE ORIGIN 7
18  PLOT SHRINK
19  PLOT LABEL BOTH
20  PLOT MODAL DEFORMATION 0 SET 53
21  BEGIN BULK
22  PARAM,AUTOSPC,YES
23  GRDSET,,,,,,,6
24  GRID,1,,−500.0,−500.0,0.0,,3456
25  GRID,2,,−250.0,−500.0,0.0,,3456
26  GRID,3,,0.0,−500.0,0.0,,3456
27  GRID,4,,250.0,−500.0,0.0,,3456
28  GRID,5,,500.0,−500.0,0.0,,3456
29  GRID,6,,−500.0,−250.0,0.0,,346
30  GRID,7,,−250.0,−250.0,0.0
31  GRID,8,,0.0,−250.0,0.0
32  GRID,9,,250.0,−250.0,0.0
33  GRID,10,,500.0,−250.0,0.0,,346
34  GRID,11,,−500.0,0.0,0.0,,346
35  GRID,12,,−250.0,0.0,0.0
36  GRID,13,,0.0,0.0,0.0,,126
37  GRID,14,,250.0,0.0,0.0
38  GRID,15,,500.0,0.0,0.0,,2346
39  GRID,16,,−500.0,250.0,0.0,,346
40  GRID,17,,−250.0,250.0,0.0
41  GRID,18,,0.0,250.0,0.0
42  GRID,19,,250.0,250.0,0.0
43  GRID,20,,500.0,250.0,0.0,,346
44  GRID,21,,−500.0,500.0,0.0,,3456
45  GRID,22,,−250.0,500.0,0.0,,3456
46  GRID,23,,0.0,500.0,0.0,,3456
47  GRID,24,,250.0,500.0,0.0,,3456
48  GRID,25,,500.0,500.0,0.0,,3456
49  CQUAD4,1,23,1,2,7,6
50  CQUAD4,2,23,2,3,8,7
51  CQUAD4,3,23,3,4,9,8
52  CQUAD4,4,23,4,5,10,9
53  CQUAD4,5,23,6,7,12,11
54  CQUAD4,6,23,7,8,13,12
55  CQUAD4,7,23,8,9,14,13
56  CQUAD4,8,23,9,10,15,14
57  CQUAD4,9,23,11,12,17,16
58  CQUAD4,10,23,12,13,18,17
59  CQUAD4,11,23,13,14,19,18
60  CQUAD4,12,23,14,15,20,19
61  CQUAD4,13,23,16,17,22,21
62  CQUAD4,14,23,17,18,23,22
63  CQUAD4,15,23,18,19,24,23
```

```
64 CQUAD4,16,23,19,20,25,24
65 PSHELL,23,12,1.0,12
66 MAT1,12,68950.0,,0.3
67 EIGB,26,INV,0.0,40.0,4,,,,+ADD
68 +ADD,MAX
69 PARAM,COUPMASS,1
70 ENDDATA
```

Next, we give the listing for the plate with a hole.

```
 1 ID EXAMPLE PLTVIBHL
 2 TIME 2
 3 SOL 3
 4 CEND
 5 TITLE=PLATE VIBRATION WITHOUT HOLE
 6 ECHO=BOTH
 7 SUBTITLE=NORMAL MODES AND FREQUENCIES
 8 DISPLACEMENT=ALL
 9 METHOD=2
10 OUTPUT(PLOT)
11 PLOTTER NASTRAN
12 SET 1 INCLUDE ALL
13 FIND SCALE ORIGIN 7
14 PLOT MODAL DEFORMATION 0, 1 THRU 4
15 AXES Z,X,Y
16 VIEW 0.0,0.0,0.0
17 FIND SCALE ORIGIN 7
18 PLOT SHRINK
19 PLOT LABEL BOTH
20 PLOT MODAL DEFORMATION 0 SET 53
21 BEGIN BULK
22 PARAM,AUTOSPC,YES
23 GRDSET,,,,,,,6
24 GRID,1,,-500.0,-500.0,0.0,,3456
25 GRID,2,,-250.0,-500.0,0.0,,3456
26 GRID,3,,0.0,-500.0,0.0,,3456
27 GRID,4,,250.0,-500.0,0.0,,3456
28 GRID,5,,500.0,-500.0,0.0,,3456
29 GRID,6,,-500.0,-250.0,0.0,,346
30 GRID,7,,-250.0,-250.0,0.0
31 GRID,8,,0.0,-250.0,0.0
32 GRID,9,,250.0,-250.0,0.0
33 GRID,10,,500.0,-250.0,0.0,,346
34 GRID,11,,-500.0,0.0,0.0,,346
35 GRID,12,,-250.0,0.0,0.0
36 GRID,13,,-150.0,0.0,0.0,,126
37 GRID,14,,250.0,0.0,0.0
38 GRID,15,,500.0,0.0,0.0,,2346
39 GRID,16,,-500.0,250.0,0.0,,346
40 GRID,17,,-250.0,250.0,0.0
41 GRID,18,,0.0,250.0,0.0
```

```
42  GRID,19,,250.0,250.0,0.0
43  GRID,20,,500.0,250.0,0.0,,346
44  GRID,21,,-500.0,500.0,0.0,,3456
45  GRID,22,,-250.0,500.0,0.0,,3456
46  GRID,23,,0.0,500.0,0.0,,3456
47  GRID,24,,250.0,500.0,0.0,,3456
48  GRID,25,,500.0,500.0,0.0,,3456
49  GRID,26,,-106.066,-106.066,0.0
50  GRID,27,,0.0,-150.0,0.0
51  GRID,28,,106.066,-106.066,0.0
52  GRID,29,,150.0,0.0,0.0
53  GRID,30,,106.066,106.066,0.0
54  GRID,31,,0.0,150.0,0.0
55  GRID,32,,-106.066,106.066,0.0
56  CQUAD4,1,23,1,2,7,6
57  CQUAD4,2,23,2,3,8,7
58  CQUAD4,3,23,3,4,9,8
59  CQUAD4,4,23,4,5,10,9
60  CQUAD4,5,23,6,7,12,11
61  CQUAD4,6,23,8,9,28,27
62  CQUAD4,7,23,28,9,14,29
63  CQUAD4,8,23,9,10,15,14
64  CQUAD4,9,23,11,12,17,16
65  CQUAD4,10,23,29,14,19,30
66  CQUAD4,11,23,31,30,19,18
67  CQUAD4,12,23,14,15,20,19
68  CQUAD4,13,23,16,17,22,21
69  CQUAD4,14,23,17,18,23,22
70  CQUAD4,15,23,18,19,24,23
71  CQUAD4,16,23,19,20,25,24
72  CQUAD4,17,23,32,31,18,17
73  CQUAD4,18,23,12,13,32,17
74  CQUAD4,19,23,7,26,13,12
75  CQUAD4,20,23,7,8,27,26
76  PSHELL,23,12,1.0,12
77  MAT1,12,68950.0,,0.3
78  EIGB,2,INV,0.0,40.0,4,,,,+ADD
79  +ADD,MAX
80  PARAM,COUPMASS,1
81  ENDDATA
```

N.4.2 Discussion

The invocation of vibration analysis is the same for the plate as for the beam and frame. Again, only a few changes from the input file for static analysis are needed.

The only things that change from the plate without a hole are some of the nodes and elements. Everything before the bulk data deck is the same except for changes in titles, and everything after the element connectivity is unchanged.

N.5 EXAMPLE 6.5.5: LUG WITH HOLE

N.5.1 Input File Listing

```
 1  ID EXAMPLE LUG
 2  TIME 2
 3  SOL 3
 4  CEND
 5  TITLE=VIBRATING LUG
 6  SUBTITLE=NORMAL MODES AND FREQUENCIES
 7  ECHO=BOTH
 8  DISPLACEMENT=ALL
 9  METHOD 56
10  SPC=1
11  OUTPUT(PLOT)
12  PLOTTER NASTRAN
13  SET 4 INCLUDE ALL
14  FIND SCALE ORIGIN 7
15  PLOT MODAL DEFORMATION 0, 1 THRU 4
16  BEGIN BULK
17  PARAM,AUTOSPC,YES
18  GRID,1,,65.000,0.0000,3.0000
19  GRID,2,,61.652,12.498,3.0000
20  GRID,3,,52.498,21.652,3.0000
21  GRID,4,,40.000,25.000,3.0000
22  GRID,5,,20.000,25.000,3.0000
23  GRID,6,,0.0000,25.000,3.0000
24  GRID,7,,0.0000,0.0000,3.0000
25  GRID,8,,57.500,0.0000,3.0000
26  GRID,9,,54.734,9.3677,3.0000
27  GRID,10,47.347,15.706,3.0000
28  GRID11,37.746,17.406,3.0000
29  GRID,12,,25.635,17.163,3.0000
30  GRID,13,,15.985,13.573,3.0000
31  GRID,14,,15.000,0.0000,3.0000
32  GRID,15,50.000,0.0000,3.0000
33  GRID,16,,48.661,4.9990,3.0000
34  GRID,17,,44.999,8.6608,3.0000
35  GRID,18,,40.000,10.000,3.0000
36  GRID,19,,35.001,8.6608,3.0000
37  GRID,20,,31.339,4.9990,3.0000
38  GRID,21,,30.000,0.0000,3.0000
39  GRID,22,,65.000,0.0000,-3.0000
40  GRID,23,,61.652,12.498,-3.0000
41  GRID,24,52.498,21.652,-3.0000
42  GRID,25,,40.000,25.000,-3.0000
43  GRID,26,20.000,25.000,-3.0000
44  GRID,27,,0.0000,25.000,-3.0000
45  GRID,28,,0.0000,0.0000,-3.0000
46  GRID,29,,57.500,0.0000 -3.0000
47  GRID,30,,54.734,9.3677,-3.0000
```

```
48  GRID,31,,47.347,15.706,-3.0000
49  GRID,32,,37.746,17.406,-3.0000
50  GRID,33,,25.635,17.163,-3.0000
51  GRID,34,,15.985,13.573,-3.0000
52  GRID,35,,15.000,0.0000,-3.0000
53  GRID,36,,50.000,0.0000,-3.0000
54  GRID,37,,48.661,4.9990,-3.0000
55  GRID,38,,44.999,8.6608,-3.0000
56  GRID,39,,40.000,10.000,-3.0000
57  GRID,40,,35.001,8.6608,-3.0000
58  GRID,41,,31.339,4.9990,-3.0000
59  GRID,42,,30.000,0.0000,-3.0000
60  GRID,43,,61.652,-12.498,3.0000
61  GRID,44,,52.498,-21.652,3.0000
62  GRID,45,,40.000,-25.000,3.0000
63  GRID,46,,20.000,-25.000,3.0000
64  GRID,47,,0.0000,-25.000,3.0000
65  GRID,48,,54.734,-9.3677,3.0000
66  GRID,49,,47.347,-15.706,3.0000
67  GRID,50,,37.746,-17.406,3.0000
68  GRID,51,,25.635,-17.163,3.0000
69  GRID,52,,15.985,-13.573,3.0000
70  GRID,53,,48.661,-4.9990,3.0000
71  GRID,54,,44.999,-8.6608,3.0000
72  GRID,55,,40.000,-10.000,3.0000
73  GRID,56,,35.001,-8.6608,3.0000
74  GRID,57,,31.339,-4.9990,3.0000
75  GRID,58,,61.652,-12.498,-3.0000
76  GRID,59,,52.498,-21.652,-3.0000
77  GRID,60,,40.000,-25.000,-3.0000
78  GRID,61,,20.000,-25.000,-3.0000
79  GRID,62,,0.0000,-25.000,-3.0000
80  GRID,63,,54.734,-9.3677,-3.0000
81  GRID,64,,47.347,-15.706,-3.0000
82  GRID,65,,37.746,-17.406,-3.0000
83  GRID,66,,25.635,-17.163,-3.0000
84  GRID,67,,15.985,-13.573,-3.0000
85  GRID,68,,48.661,-4.9990,-3.0000
86  GRID,69,,44.999,-8.6608,-3.0000
87  GRID,70,,40.000,-10.000,-3.0000
88  GRID,71,,35.001,-8.6608,-3.0000
89  GRID,72,,31.339,-4.9990,-3.0000
90  CHEXA,1,1,1,2,9,8,22,23,+EA1
91  +EA1,30,29
92  CHEXA,2,1,2,3,10,9,23,24,+EA2
93  +EA2,31,30
94  CHEXA,3,1,3,4,11,10,24,25,+EA3
95  +EA3,32,31
96  CHEXA,4,1,4,5,12,11,25,26,+EA4
97  +EA4,33,32
98  CHEXA,5,1,5,6,13,12,26,27,+EA5
```

```
 99  +EA5,34,33
100  CHEXA,6,1,6,7,14,13,27,28,+EA6
101  +EA6,35,34
102  CHEXA,7,1,8,9,16,15,29,30,+EA7
103  +EA7,37,36
104  CHEXA,8,1,9,10,17,16,30,31,+EA8
105  +EA8,38,37
106  CHEXA,9,1,10,11,18,17,31,32,+EA9
107  +EA9,39,38
108  CHEXA,10,1,11,12,19,18,32,33,+EA10
109  +EA10,40,39
110  CHEXA,11,1,12,13,20,19,33,34,+EA11
111  +EA11,41,40
112  CHEXA,12,1,13,14,21,20,34,35,+EA12
113  +EA12,42,41
114  CHEXA,13,1,43,1,8,48,58,22,+EA13
115  +EA13,29,63
116  CHEXA,14,1,44,43,48,49,59,58,+EA14
117  +EA14,63,64
118  CHEXA,15,1,45,44,49,50,60,59,+EA15
119  +EA15,64,65
120  CHEXA,16,1,46,45,50,51,61,60,+EA16
121  +EA16,65,66
122  CHEXA,17,1,47,46,51,52,62,61,+EA17
123  +EA17,66,67
124  CHEXA,18,1,7,47,52,14,28,62,+EA18
125  +EA18,67,35
126  CHEXA,19,1,48,8,15,53,63,29,+EA19
127  +EA19,36,68
128  CHEXA,20,1,49,48,53,54,64,63,+EA20
129  +EA20,68,69
130  CHEXA,21,1,50,49,54,55,65,64,+EA21
131  +EA21,69,70
132  CHEXA,22,1,51,50,55,56,66,65,+EA22
133  +EA22,70,71
134  CHEXA,23,1,52,51,56,57,67,66,+EA23
135  +EA23,71,72
136  CHEXA,24,1,14,52,57,21,35,67,+EA24
137  +EA24,72,42
138  PSOLID,1,25
139  MAT1,25,68950.0,,0.3,2.77E-9
140  EIGR,56,INV,1500.0,4000.0,3,,,1.0E-5,+ADD
141  +ADD,MAX
142  SPC,1,6,123456
143  SPC,1,7,123456
144  SPC,1,27,123456
145  SPC,1,28,123456
146  SPC,1,47,123456
148  SPC,1,62,123456
149  PARAM,COUPMAS,1
150  ENDDATA
```

N.5.2 Discussion

The input file for vibration analysis of a structure modeled with solid elements requires only a simple modification from that for static analysis. Comments made for the other structures apply here.

APPENDIX O: PROGRAM FOR TRANSIENT RESPONSE

O.1 INTRODUCTION

When inertia and viscous damping terms and dynamic applied loads are included, the finite element method reduces all partial differential and integral equation mathematical models for small deflections of linear elastic structures to a set of linear ordinary differential equations of the form of Equation 6.4.49

$$[M]\{\ddot{r}\} + [C]\{\dot{r}\} + ([K] + [K^d])\{r\} = \{R\} \tag{O.1}$$

where $[M]$ is the matrix that contains the mass properties, $[C]$ contains the damping coefficients, $[K]$ is the stiffness matrix as in the static case, $[K^d]$ is the differential stiffness matrix, $\{r\}$ is the displacement matrix, and $\{R\}$ is the applied load matrix. The matrices $[M]$, $[C]$, and $[K]$ are symmetrical.

A computer program is given next that uses a form of solution called the Newmark method, which is a form of finite differences in time. A FORTRAN version of this program is given in Reference 23. This program requires that the applied force be entered as a program statement. As listed here, it is programmed for a step input at time $t = 0$ at each node.

The symbols used in the program depart slightly from the standard notation used in the text and Equation O.1 because of the symbols available in the BASIC program.

```
PRINT"SOLVES A SET OF SECOND ORDER DIFFERENTIAL EQUATIONS"
PRINT:PRINT"OF THE FORM [M]{r"} + [C]{r'} + [K]{r} = R(t)   WHERE r(t)"
PRINT:PRINT"GIVEN INITIAL VALUES FOR r(0) = 0 and r'(0) = 0"
PRINT:INPUT "THE NUMBER OF EQUATIONS N (N <= 10) IS ",N
REM The DIM statement for arrays must be added to the program if N > 10.
PRINT:PRINT "FIRST ENTER THE MASS MATRIX [M]":PRINT
Mass:
  FOR I = 1 TO N
  FOR J = I TO N
  PRINT"      M(";I;",";J;") = ";:INPUT" ",M(I,J)
  M(J,I) = M(I,J)
  NEXT J
  NEXT i
  PRINT:PRINT "CHECK ERRORS - IF ANY ENTER 1 AND REENTER ALL TERMS"
  PRINT:PRINT "IF NO ERRORS PRESS RETURN AND CONTINUE"
  INPUT" ",MO
  IF MO = 1 THEN Mass
PRINT:PRINT "NEXT ENTER THE DAMPING MATRIX [C]":PRINT
Damping:
  FOR I = 1 TO N
  FOR J = I TO N
  PRINT"      C(";I;",";J;") = ";:INPUT" ",C(I,J)
  C(J,I) = C(I,J)
  NEXT J
  NEXT I
```

```
      PRINT:PRINT "CHECK FOR ERRORS - IF ANY ENTER 1 AND REENTER ALL TERMS"
      PRINT:PRINT "IF NO ERRORS PRESS RETURN AND CONTINUE"
      INPUT" ",CO
      IF CO = 1 THEN Damping
   PRINT:PRINT "NOW ENTER THE STIFFNESS MATRIX [K]":PRINT
   Stiffness:
      FOR I = 1 TO N
      FOR J = I TO N
      PRINT"       K(";I;",";J;") = ";:INPUT" ",K(I,J)
      K(J,I) = K(I,J)
      NEXT J
      NEXT I
      PRINT:PRINT "CHECK FOR ERRORS - IF ANY ENTER 1 AND REENTER ALL TERMS"
      PRINT:PRINT "IF NO ERRORS PRESS RETURN AND CONTINUE"
      INPUT" ",KO
      IF KO = 1 THEN Stiffness
   PRINT:INPUT"TIME INTERVAL STEP SIZE",DELTAT
   PRINT:INPUT"NUMBER OF STEPS ",NSTEPS
   PRINT:INPUT"INITIAL TIME (USUALLY ZERO) ",T
   REM Set displacement, velocity, and acceleration to zero
   FOR I = 1 TO N
   DIS(I) = 0
   VEL(I) = 0
   ACC(I) = 0
   NEXT I
   PRINT:PRINT"THIS PROGRAM SET FOR A STEP INPUT OF FORCE AT t = 0"
   PRINT"WHICH REMAINS CONSTANT FOR t > 0":PRINT
   Force:
      FOR I = 1 TO N
      PRINT"       R(";I;") = ";:INPUT" ",R(I)
      NEXT I
      PRINT:PRINT "CHECK FOR ERRORS - IF ANY ENTER 1 AND REENTER ALL TERMS"
      PRINT:PRINT "IF NO ERRORS PRESS RETURN AND CONTINUE"
      INPUT" ",RO
      IF RO = 1 THEN Force:
   PRINT:PRINT"THE FORCE INPUT MUST BE REPROGRAMMED TO HANDLE"
   PRINT"OTHER FORCING FUNCTIONS"
   Effective:
      FOR I = 1 TO N
      FOR J = I TO N
      KE(I,J) = K(I,J) + (4/DELTAT^2)*M(I,J) + (2/DELTAT)*C(I,J)
      KE(J,I) = KE(I,J)
      NEXT J
      NEXT I
   Decompose:
      FOR I = 1 TO N
      FOR J = 1 TO N
        SUM = KE(I,J)
        LET K1 = I - 1
        IF I = 1 THEN One
        FOR K = 1 TO K1
```

```
        SUM = SUM − KE(K,I)*KE(K,J)*KE(K,K)
        NEXT K
One:
        IF J <> I THEN Two
        KE(I,J) = SUM
        GOTO Three
Two:
        KE(I,J) = SUM/KE(I,I)
Three:
        NEXT J
        NEXT I
FOR KT = 1 TO NSTEPS
T = T + DELTAT
Loading:
        FOR I = 1 TO N
        F1 = 0:F2 = 0
        FOR J = 1 TO N
        F1 = F1 + M(I,J)*((4/DELTAT^2)*DIS(J) + (4/DELTAT)*VEL(J) + ACC(J))
        F2 = F2 + C(I,J)*((2/DELTAT)*DIS(J) + VEL(J))
        NEXT J
        RE(I) = R(I) + F1 + F2
        NEXT I
REM Program for solving simultaneous algebraic equations
FOR I = 1 TO N
SUM = RE(I)
K1 = I − 1
IF I = 1 THEN Four
FOR K = 1 TO K1
SUM = SUM − KE(K,I)*X(K)
NEXT K
Four:
        X(I) = SUM
        NEXT I
FOR I1 = 1 TO N
I = N − I1 + 1
SUM = X(I)
K2 = I + 1
IF I = N THEN Five:
FOR K = K2 TO N
SUM = SUM − KE(I,K)*X(K)*KE(I,I)
NEXT K
Five:
        X(I) = SUM/KE(I,I)
        NEXT I1
PRINT:PRINT "AND THE SOLUTION IS "
PRINT:PRINT"        STEP NUMBER IS ";KT
PRINT"        TIME IS ";T:PRINT
FOR I = 1 TO N
PRINT "        X(";I;") = ";X(I)
NEXT I
PRINT:PRINT"PRESS RETURN TO CONTINUE":INPUT" ",CN
```

```
FOR I = 1 TO N
ACCN(I) = (4/DELTAT^2)*(X(I) − DIS(I)) − (4/DELTAT)*VEL(I) − ACC(I)
VELN(I) = VEL(I) + (DELTAT/2)*ACC(I) + (DELTAT/2)*ACCN(I)
NEXT I
FOR I = 1 TO N
ACC(I) = ACCN(I)
VEL(I) = VELN(I)
DIS(I) = X(I)
NEXT I
NEXT KT
PRINT:PRINT"DONE":PRINT:PRINT
END
```

REFERENCES

1. Gere, J. M., and Timoshenko, S. P., *Mechanics of Materials,* Brooks/Cole, Monterey, Calif., 1984.

2. Beer, F. P., and Johnston, E. R., *Mechanics of Materials,* McGraw-Hill, New York, 1981.

3. Crandall, S. H., Dahl, N. C., and Lardner, T. J., *An Introduction to the Mechanics of Solids,* McGraw-Hill, New York, 1972.

4. Oden, J. T., and Ripperger, E. A., *Mechanics of Elastic Structures,* 2nd. ed., McGraw-Hill, New York, 1981.

5. Allen, D. H., and Haisler, W. E., *Introduction to Aerospace Structural Analysis,* Wiley, New York, 1985.

6. Timoshenko, S. P., and Goodier, J. N., *Theory of Elasticity,* 2nd ed., McGraw-Hill, New York, 1951.

7. Wang, C.-T., *Applied Elasticity,* McGraw-Hill, New York, 1953.

8. Saada, A. S., *Elasticity: Theory and Applications,* Pergamon Press, Elmsford, N.Y., 1974.

9. Rivello, R. M., *Theory and Analysis of Flight Structures,* McGraw-Hill, New York, 1969.

10. Timoshenko, S. P., and Woinowsky-Krieger, S., *Theory of Plates and Shells,* 2nd. ed., McGraw-Hill, New York, 1959.

11. Szilard, R., *Theory and Analysis of Plates,* Prentice-Hall, Englewood Cliffs, N. J., 1974.

12. Roark, R. J., *Formulas for Stress and Strain,* 3rd ed., McGraw-Hill, New York, 1954.

13. Zienkiewicz, O. C., *The Finite Element Method,* McGraw-Hill, New York, 1977.

14. Cook, R. D., *Concepts and Applications of Finite Element Analysis,* 2nd ed., Wiley, New York, 1981.

15. Weaver, W., Jr., and Johnston, P. R., *Finite Elements for Structural Analysis,* Prentice-Hall, Englewood Cliffs, N. J., 1984.

16. Anderson, W. J., *Finite Elements in Mechanical and Structural Design: Linear Static Analysis,* Automated Analysis Corporation, Ann Arbor, Mich., 1984.

17. Grandin, H., Jr., *Fundamentals of the Finite Element Method,* Macmillan, New York, 1986.

18. Yang, T. Y., *Finite Element Structural Analysis,* Prentice-Hall, Englewood Cliffs, N. J., 1986.

19. Brush, D. O., and Almroth, B. O., *Buckling of Bars, Plates, and Shells,* McGraw-Hill, New York, 1975.

20. Timoshenko, S. P., and Gere, J. M., *Theory of Elastic Stability,* 2nd ed., McGraw-Hill, New York, 1961.

21. Ziegler, H., *Principles of Structural Stability,* Blaisdell, Waltham, Mass., 1968.

22. Meirovitch, L., *Elements of Vibration Analysis,* 2nd ed., McGraw-Hill, New York, 1986.

23. Anderson, W. J., *Finite Elements in Mechanical and Structural Design: Dynamic and Nonlinear Analysis,* Automated Analysis Corporation, Ann Arbor, Mich., 1985.

24. Weaver, W., Jr., and Johnston, P. R., *Structural Dynamics by Finite Elements,* Prentice-Hall, Englewood Cliffs, N. J., 1987.

INDEX

A

Algebraic equations, program for solving
    simultaneous linear, 386–87
Applied loads, 3, 14–15
Axial and torsional vibration of bars, 341–44
Axial deformation, equation for, 67, 114
Axial loads
    finite element method applied to, 148–49, 180–88
    slender bars with, 65–68
    space trusses and, 180–88
Axisymmetry, equations of, 380

B

Beam-column(s)
    definition of, 241
    equation, 245
    solutions for, 251–55
    virtual work for, 255–58
Beams, 61
    boundary conditions and beam applied loads,
        81–89
    buckling and MSC/NASTRAN input files and
        tapered, 416–17
    MSC/NASTRAN input files and tapered, 394–96
    shear flow in multicell, 381–82
    slender bars in bending and shear, 69–81
    vibration analysis and MSC/NASTRAN input files
        and tapered, 424–25
    See also Finite element method, beams and;
        Slender bars; Thin-walled beams
Bending
    coupled bending and torsion, 344–45
    deformation, equation for, 78, 115
    finite element method and simple beams in, 190–97
    MSC/NASTRAN input files and plate, 403–4
    slender bars in, 69–81
    thin plates in, 108–14
Bending vibration of beams by work and energy
        methods, 326
    eigenvalue equation solvers, 337–39
    finite element method for forced vibration, 339–40
    finite element method for free vibration, 331–36
    Rayleigh-Ritz method, 327–31
Bending vibration of slender bars by differential
        equation methods, 303
    beams with damping, 322–25
    derivation of the differential equations, 304–5
    dynamic instability of beams, 325–26
    forced vibration of beams in bending, 315–22
    free vibrations of beams in bending, 305–15
Body forces, 3

F

G

H

I

J

K

L

M

Matrix *(cont.)*
  operations, 360–65
  orthogonal, 169
  partitioned, 363
  summary of useful, relations, 365
  sweeping, 337
Membrane analogy, 50–53
Membrane loads, 276
Mid-edge nodes, 222
Moment
  equilibrium, 6
  twisting, 110
MSC/NASTRAN input files, 182
  plane stress, 399–400
  plate bending, 403–4
  reinforced plate, 404–6
  solid element, 400–403
  space frame, 397–98
  space truss, 392–94
  tapered beam, 394–96
  two-dimensional frame, 396–97
MSC/NASTRAN input files for buckling analysis
  rectangular plate, 419–20
  rectangular plate with hole, 421–23
  tapered beam, 416–17
  two-dimensional frame, 417–18
MSC/NASTRAN input files for vibration analysis
  lug with hole, 430–33
  rectangular plate, 426–29
  tapered beam, 424–25
  two-dimensional frame, 425–26

N

NASTRAN. *See* MSC/NASTRAN input files
Natural coordinates, 213
Natural frequency, 308
Negative surface, 4
Nodal displacement, 147
Nodal loads, equivalent, 147, 159–62
Nodes, 147, 148
  mid-edge, 222
Normal coordinates, 315
Normal mode, 313
Normal strains, 10

O

Orthogonality, 313–14
Orthogonal matrix, 169
Orthotropic, 376

P

Partitioning, 157
Pin-jointed trusses, 89, 167–71
Plane problems in elasticity, 21–36
Plane sections remain plane, 71

Plane strain, 32–36, 56
Plane stress
  elasticity and, 21–32, 55–56
  elements, 216–22
  finite element method and, 208–13
  MSC/NASTRAN input files and, 399–400
  rectangular element and, 209–11
Plates
  bending and MSC/NASTRAN input files, 403–4
  bending and thin, 108–14
  curved, 62
  definition of, 62
  equation for deflection of thin, 111, 116
  finite element method and, 227–32
  MSC/NASTRAN input files and reinforced, 404–6
  vibration and, 347–50
  See also Buckling of plates
Poisson's ratio, 12, 375
Polar coordinates, equations in, 379
Positive surface, 4
Power method, 267
Prandtl torsion function, 47
Principal strain, 11
Principle of least work, 389
Proportional damping, 323–25

R

Rayleigh-Ritz method, 130–39
  buckling of slender bars and, 258–61
  work and energy methods in beam bending
    vibration and, 327–31
Rectangular element, plane stress applications and,
    209–11
Reinforced plate and MSC/NASTRAN input files,
    404–6
Residual stresses, 15
Resonance, 318–19
  peak, 318
Reversed effective forces, 302
Rigid body displacement, 7
Rods, 61
Rotational spring, 85

S

St. Venant's principle, 37–38
Semiinverse method, 42
Separation of variables, 305
Shafts, 61
Shear
  center, 90
  finite element method and simple beams in, 190–97
  flow, 90
  flow in multicell beams, 381–82
  modulus, 13
  slender bars in bending and, 69–81
Shearing strain, 10–11

U

Unstable structures, 241
  See also under Buckling

V

Variables, separation of, 305
Vibration analysis, MSC/NASTRAN input files for, 424–33
Vibration of structures. *See* Structural dynamics
Virtual displacement, 122
Virtual work, principle of
  for beam-columns, 255–58
  definition of, 122
  for a deformable body, 124–27
  energy and principles of complementary, 388–89
  energy methods in beam bending vibration and, 326–40
  equilibrium and, 123
  for a particle, 122–24
  in plate buckling, 283–94
  Rayleigh-Ritz method, 130–39
  for slender bars, 127–30
  strain energy and internal, 383–85
  symbolic representation of, 121–22
Viscous damping, 302–3, 322–23

W

Warping function, 42
Wedge element, 225
Work. *See* Virtual work, principle of

Y

Young's modulus, 12, 375